MODULAR MATHEMATICS
Module A: Pure Maths 1

By the same authors

CORE MATHS FOR A-LEVEL
MATHEMATICS – THE CORE COURSE FOR A-LEVEL
FURTHER PURE MATHEMATICS – with C. Rourke
MATHEMATICS – MECHANICS AND PROBABILITY
FURTHER MECHANICS AND PROBABILITY
APPLIED MATHEMATICS I
APPLIED MATHEMATICS II
PURE MATHEMATICS I
PURE MATHEMATICS II

MODULAR MATHEMATICS
Module A: Pure Maths 1

L. Bostock, B.Sc.

S. Chandler, B.Sc.

Stanley Thornes (Publishers) Ltd

First published in 1990 by Stanley Thornes (Publishers) Ltd,
Old Station Drive, Leckhampton, CHELTENHAM GL53 0DN

Reprinted 1992 (twice)

British Library Cataloguing in Publication Data

Bostock, L. (Linda)
   Modular mathematics.
   Module A, Pure maths
   1. Mathematics
   I. Title   II. Chandler, S. (Suzanne)
   510

   ISBN 0-7487-0068-4

Typeset by Tech-Set, Gateshead, Tyne & Wear, in 10/12pt Times
Printed and bound in Great Britain at The Bath Press, Avon

CONTENTS

PREFACE

This is the first book in a series of modular mathematics courses for use by students wishing to gain academic qualifications beyond GCSE. An AS subject in mathematics requires two modules and an A-level needs four modules.

This book can be used for the pure mathematics content of AS subjects combining pure mathematics with mechanics or with statistics. The next book in this series, *Module B: Pure Mathematics 2*, will complete the work necessary for AS Pure Mathematics as well as the pure mathematics content of A-level Mathematics.

Modules in Mechanics, Statistics and Further Pure Mathematics and Mechanics will be produced to cover a variety of options in AS-level mathematics.

Now that GCSE courses have been introduced it can no longer be assumed that all students start with the algebraic skills and geometric knowledge that used to be expected. Many students now moving on to sixth-form colleges come from a variety of backgrounds, including those who wish to embark on an AS or A-level course from intermediate level GCSE.

In much the same way as multiplication tables are the tools needed to build a mathematics course from 11 to 16, skills in algebraic techniques are the tools necessary for building a body of further mathematical knowledge. This book starts with work designed to help those students acquire a facility in using algebra. The exercises all start with straightforward questions. The more sophisticated exam type questions are given in consolidation sections which appear at regular intervals throughout the book. These are intended for use later, to give practice when confidence and sophistication have been developed. The consolidation sections also include a summary of the work in preceeding chapters and a set of multiple choice questions which are very useful for self-testing even if they do not form part of the examination to be taken.

There are many computer programs that aid in the understanding of mathematics. In particular, a good graph drawing package is

invaluable for investigating graphical aspects of functions. In a few places we have indicated a program that we think is relevant. This is either *Super Graph* or a program from *132 Short Programs for the Mathematics Classroom. Super Graph* by David Tall is a flexible graph drawing package and is available from Abco Design Ltd, Unit 11, Stirling Industrial Centre, Stirling, Boreham Wood, Herts. WD6 2BT. Tel: 081 953 9292.
132 Short Programs for the Mathematics Classroom is published in book form by Stanley Thornes (Publishers) Ltd.

We are grateful to the following Examination Boards for permission to reproduce questions from their past examination papers (part questions are indicated by the suffix p):

University of London (U of L)
Joint Matriculation Board (JMB)
University of Cambridge Local Examinations Syndicate (C)
The Associated Examining Board (AEB)
Welsh Joint Education Committee (WJEC)
Cambridge, Oxford and Southern Universities
   Examinations Council (O/C, SU & C)

1990

L. Bostock
S. Chandler

X

NOTES ON USE OF THE BOOK

Notation

$=$	is equal to	$\in$	is a member of
$\equiv$	is identical to	$:$	is such that
$\approx$	is approximately equal to*	$\mathbb{N}$	the natural numbers
$>$	is greater than	$\mathbb{Z}$	the integers
$\geqslant$	is greater than or equal to	$\mathbb{Q}$	the rational numbers
$<$	is less than	$\mathbb{R}$	the real numbers
$\leqslant$	is less than or equal to	$\mathbb{R}^+$	the positive real numbers
∞	infinity; infinitely large		excluding zero
$\Rightarrow$	implies	$[a, b]$	the interval $\{x : a \leqslant x \leqslant b\}$

A stroke through a symbol negates it, e.g. $\neq$ means 'is not equal to'

Abbreviations

$\parallel$	is parallel to	$-$ve	negative
$+$ve	positive	w.r.t.	with respect to

Useful Formulae

For a cone with base radius r, height h and slant height l

$$\text{volume} = \tfrac{1}{3}\pi r^2 h \qquad \text{curved surface area} = \pi r l$$

For a sphere of radius r

$$\text{volume} = \tfrac{4}{3}\pi r^3 h \qquad \text{surface area} = 4\pi r^2$$

For any pyramid with height h and base area a

$$\text{volume} = \tfrac{1}{3}ah$$

*Practical problems rarely have exact answers. Where numerical answers are given they are correct to two or three decimal places depending on their context, e.g. π is 3.142 correct to 3 d.p. and although we write $\pi = 3.142$ it is understood that this is not an exact value. We reserve the symbol $\approx$ for those cases where the approximation being made is part of the method used.

Computer Program References

Marginal symbols indicate a computer program which is helpful, programs being identified in the following manner,

Program No. 47 from *132 Short Programs for the Mathematics Classroom*

Super Graph

Instructions for Answering Multiple Choice Exercises

These exercises are included in each consolidation section. The questions are set in groups; the answering techniques are different for each group and are classified as follows:

TYPE I
These questions consist of a problem followed by several alternative answers, only *one* of which is correct.

Write down the letter corresponding to the correct answer.

TYPE II
In this type of question some information is given and is followed by a number of responses. *One or more* of these follow(s) directly and necessarily from the information given.

Write down the letter(s) corresponding to the correct response(s).
e.g. PQR is a triangle

A $\angle P + \angle Q + \angle R = 180°$
B PQ + QR is less than PR
C if $\angle P$ is obtuse, $\angle Q$ and $\angle R$ must both be acute.
D $\angle P = 90°$, $\angle Q = 45°$, $\angle R = 45°$

The correct responses are **A** and **C**.
B is definitely incorrect and **D** may or may not be true of triangle PQR, i.e. it does not follow directly and *necessarily* from the information given. Responses of this kind should not be regarded as correct.

TYPE III
A single statement is made. Write T if it is true and F if it is false.

CHAPTER 1

ALGEBRA

The ability to manipulate algebraic expressions is an essential base for any mathematics course beyond GCSE. Applying the processes involved needs to be almost as instinctive as the ability to manipulate simple numbers. This and the next two chapters present the facts and provide practice necessary for the development of these skills.

MULTIPLICATION OF ALGEBRAIC EXPRESSIONS

The multiplication sign is usually omitted, so that, for example,

$$2q \text{ means } 2 \times q$$

and $\qquad x \times y$ can be simplified to xy

Remember also that if a string of numbers and letters are multiplied, the multiplication can be done in any order, for example

$$2p \times 3q = 2 \times p \times 3 \times q$$

$$= 6pq$$

Powers can be used to simplify expressions such as $x \times x$,

i.e. $\qquad x \times x = x^2$

and $\qquad x \times x^2 = x \times x \times x = x^3$

But remember that a power refers only to the number or letter it is written above, for example

$$2x^2 \text{ means that } x \text{ is squared, but 2 is not.}$$

1

Example 1a _____

Simplify   (a) $(4pq)^2 \times 5$ (b) $\dfrac{ax^2}{y} \div \dfrac{x}{ay^2}$

(a) $(4pq)^2 \times 5 = 4pq \times 4pq \times 5$

$= 80p^2q^2$

(b) $\dfrac{ax^2}{y} \div \dfrac{x}{ay^2} = \dfrac{ax^{\cancel{2}}}{\cancel{y}} \times \dfrac{ay^{\cancel{2}}}{\cancel{x}}$

$= a^2xy$

_____

EXERCISE 1a

Simplify

1. $3 \times 5x$ **2.** $x \times 2x$ **3.** $(2x)^2$

4. $5p \times 2q$ **5.** $4x \times 2x$ **6.** $2pq \times 5pr$

7. $(3a)^2$ **8.** $7a \times 9b$ **9.** $8t \times 3st$

10. $2a^2 \times 4a$ **11.** $25x^2 \div 15x$ **12.** $12m^2 \div 6m$

13. $b^2 \times 4ab$ **14.** $25x^2y \div 5x$ **15.** $(7pq)^2 \times (2p)^2$

16. $\dfrac{22ab}{11b}$ **17.** $\dfrac{18ax^2}{3x}$ **18.** $\dfrac{36xy}{18y}$

19. $\dfrac{72ab^2}{40a^2b}$ **20.** $\dfrac{2}{5} \div \dfrac{1}{x}$ **21.** $\dfrac{x^2}{y} \div \dfrac{y}{x}$

ADDITION AND SUBTRACTION OF EXPRESSIONS

The *terms* in an algebraic expression are the parts separated by a plus or minus sign.

Like terms contain the same combination of letters; like terms can be added or subtracted.

For example, $2ab$ and $5ab$ are like terms and can be added,

i.e. $2ab + 5ab = 7ab$

Unlike terms contain different algebraic expressions; they cannot be added or subtracted. For example, ab and ac are unlike terms and $ab + ac$ cannot be simplified.

Example 1b _____

Simplify $5x - 3(4 - x)$

$$5x - 3(4 - x) = 5x - 12 + 3x$$
$$= 8x - 12$$

Note that $-3(4 - x)$ means 'take away 3 times everything inside the bracket':
remember that $(-3) \times (-x) = +3x$

_____

EXERCISE 1b

Simplify

1. $2x^2 - 4x + x^2$ **2.** $5a - 4(a + 3)$

3. $2y - y(x - y)$ **4.** $8pq - 9p^2 - 3pq$

5. $4xy - y(x - y)$ **6.** $x^3 - 2x^2 + x^2 - 4x + 5x + 7$

7. $t^2 - 4t + 3 - 2t^2 + 5t + 2$ **8.** $2(a^2 - b) - a(a + b)$

9. $3 - (x - 4)$ **10.** $5x - 2 - (x + 7)$
     Note that $-(x - 4)$ means $-1(x - 4)$

11. $3x(x + 2) + 4(3x - 5)$ **12.** $a(b - c) - c(a - b)$

13. $2cT(3 - T) + 5T(c - 11T)$ **14.** $x^2(x + 7) - 3x^3 + x(x^2 - 7)$

15. $(3y^2 + 4y - 2) - (7y^2 - 20y + 8)$ **16.** $6RS + 5RF - R(R + S)$

COEFFICIENTS

We can identify a particular term in an expression by using the letter,
or combination of letters, involved, for example

    $2x^2$ is 'the term in x^2'

    $3xy$ is 'the term in xy'

The number in front of the letters is called the *coefficient*, for example

    in the term $2x^2$, 2 is the coefficient of x^2

    in the term $3xy$, 3 is the coefficient of xy

If no number is written in front of a term, the coefficient is 1 or -1,
depending on the sign of the term.

Consider the expression $x^3 + 5x^2y - y^3$

the coefficient of x^3 is 1

the coefficient of x^2y is 5

the coefficient of y^3 is -1

There is no term in x^2, so the coefficient of x^2 is zero.

EXERCISE 1c

1. Write down the coefficient of x in $x^2 - 7x + 4$

2. What is the coefficient of xy^2 in the expression $y^3 + 2xy^2 - 7xy$?

3. For the expression $x^2 - 5xy - y^2$ write down the coefficient of
   (a) x^2 (b) xy (c) y^2

4. For the expression $x^3 - 3x + 7$ write down the coefficient of
   (a) x^3 (b) x^2 (c) x

EXPANSION OF TWO BRACKETS

Expanding an expression means multiplying it out.

To expand $(2x + 4)(x - 3)$ each term in the first bracket is multiplied by each term in the second bracket. To make sure that nothing is missed out, it is sensible to follow the same order every time. The order used in this book is:

$$(2x + 4)(x - 3) = 2x^2 - 6x + 4x - 12$$
$$= 2x^2 - 2x - 12$$

Use the next exercise to practice expanding and to develop the confidence to go straight to the simplified form.

EXERCISE 1d

Expand and simplify

1. $(x + 2)(x + 4)$ 2. $(x + 5)(x + 3)$

3. $(a + 6)(a + 7)$ 4. $(t + 8)(t + 7)$

5. $(s + 6)(s + 11)$ 6. $(2x + 1)(x + 5)$

7. $(5y + 3)(y + 5)$ 8. $(2a + 3)(3a + 4)$

9. $(7t + 6)(5t + 8)$ 10. $(11s + 3)(9s + 2)$

11. $(x - 3)(x - 2)$ 12. $(y - 4)(y - 1)$

13. $(a - 3)(a - 8)$ 14. $(b - 8)(b - 9)$

15. $(p - 3)(p - 12)$ 16. $(2y - 3)(y - 5)$

17. $(x - 4)(3x - 1)$ 18. $(2r - 7)(3r - 2)$

19. $(4x - 3)(5x - 1)$ 20. $(2a - b)(3a - 2b)$

21. $(x - 3)(x + 2)$ 22. $(a - 7)(a + 8)$

23. $(y + 9)(y - 7)$ 24. $(s - 5)(s + 6)$

25. $(q - 5)(q + 13)$ 26. $(2t - 5)(t + 4)$

27. $(x + 3)(4x - 1)$ 28. $(2q + 3)(3q - 5)$

29. $(x + y)(x - 2y)$ 30. $(s + 2t)(2s - 3t)$

Difference of Two Squares

Consider the expansion of $(x - 4)(x + 4)$,

$$(x - 4)(x + 4) = x^2 - 4x + 4x - 16$$
$$= x^2 - 16$$

EXERCISE 1e

Expand and simplify

1. $(x - 2)(x + 2)$ 2. $(5 - x)(5 + x)$

3. $(x + 3)(x - 3)$ 4. $(2x - 1)(2x + 1)$

5. $(x + 8)(x - 8)$ 6. $(x - a)(x + a)$

From Questions 1 to 6 it is clear that an expansion of the form $(ax + b)(ax - b)$ can be written down directly,

i.e. $$(ax + b)(ax - b) = a^2x^2 - b^2$$

Use this result to expand the following brackets.

7. $(x - 1)(x + 1)$ 8. $(3b + 4)(3b - 4)$

9. $(2y - 3)(2y + 3)$ 10. $(ab + 6)(ab - 6)$

11. $(5x + 1)(5x - 1)$ 12. $(xy + 4)(xy - 4)$

Squares

$(2x + 3)^2$ means $(2x + 3)(2x + 3)$

$$\therefore \quad (2x + 3)^2 = (2x + 3)(2x + 3)$$
$$= (2x)^2 + (2)(2x)(3) + (3)^2$$
$$= 4x^2 + 12x + 9$$

In general,
$$(ax + b)^2 = a^2x^2 + (2)(ax)(b) + b^2$$
$$= a^2x^2 + 2abx + b^2$$

and
$$(ax - b)^2 = a^2x^2 - 2abx + b^2$$

EXERCISE 1f

Use the results above to expand

1. $(x + 4)^2$
2. $(x + 2)^2$
3. $(2x + 1)^2$
4. $(3x + 5)^2$
5. $(2x + 7)^2$
6. $(x - 1)^2$
7. $(x - 3)^2$
8. $(2x - 1)^2$
9. $(4x - 3)^2$
10. $(5x - 2)^2$
11. $(3t - 7)^2$
12. $(x + y)^2$
13. $(2p + 9)^2$
14. $(3q - 11)^2$
15. $(2x - 5y)^2$

Important Expansions

The results from the last two sections should be memorised. They are summarised here.

$$(ax + b)^2 = a^2x^2 + 2abx + b^2$$
$$(ax - b)^2 = a^2x^2 - 2abx + b^2$$
$$(ax + b)(ax - b) = a^2x^2 - b^2$$

The next exercise contains a variety of expansions including some of the forms given above.

Example 1g

Expand $(4p + 5)(3 - 2p)$

$$(4p + 5)(3 - 2p) = (5 + 4p)(3 - 2p)$$
$$= 15 + 2p - 8p^2$$

EXERCISE 1g

Expand

1. $(2x - 3)(4 - x)$ 2. $(x - 7)(x + 7)$

3. $(6 - x)(1 - 4x)$ 4. $(7p + 2)(2p - 1)$

5. $(3p - 1)^2$ 6. $(5t + 2)(3t - 1)$

7. $(4 - p)^2$ 8. $(4t - 1)(3 - 2t)$

9. $(x + 2y)^2$ 10. $(4x - 3)(4x + 3)$

11. $(3x + 7)^2$ 12. $(R + 3)(5 - 2R)$

13. $(a - 3b)^2$ 14. $(2x - 5)^2$

15. $(7a + 2b)(7a - 2b)$ 16. $(3a + 5b)^2$

17. Write down the coefficients of x^2 and x in the expansion of
    (a) $(2x - 4)(3x - 5)$ (b) $(5x + 2)(3x + 5)$
    (c) $(2x - 3)(7x - 5)$ (d) $(9x + 1)^2$

FACTORISING QUADRATIC EXPRESSIONS

In the last four exercises, each bracket contained a *linear expression*, i.e. an expression that contained an x term and a number term.

An expression of the form $ax + b$, where a and b are numbers, is called a linear expression in x

When two linear expressions in x are multiplied, the result usually contains three terms: a term in x^2, a term in x and a number.

Expressions of this form, i.e. $ax^2 + bx + c$ where a, b and c are numbers and $a \neq 0$, are called *quadratic expressions in x*

Since the product of two linear brackets is quadratic, we might expect to be able to reverse this process. For instance, given a quadratic such as $x^2 - 5x + 6$, we could try to find two linear expressions in x whose product is $x^2 - 5x + 6$. To be able to do this we need to appreciate the relationship between what is inside the brackets and the resulting quadratic.

Consider the examples

$$(2x + 1)(x + 5) = 2x^2 + 11x + 5 \qquad [1]$$

$$(3x - 2)(x - 4) = 3x^2 - 14x + 8 \qquad [2]$$

$$(x - 5)(4x + 2) = 4x^2 - 18x - 10 \qquad [3]$$

The first thing to notice about the quadratic in each example is that

the coefficient of x^2 is the product of the coefficients of x in the two brackets,

the number is the product of the numbers in the two brackets,

the coefficient of x is the sum of the coefficients formed by multiplying the x term in one bracket by the number term in the other bracket.

The next thing to notice is the relationship between the signs.

Positive signs throughout the quadratic come from positive signs in both brackets, as in [1].

A positive number term and a negative coefficient of x in the quadratic come from a negative sign in each bracket, as in [2].

A negative number term in the quadratic comes from a negative sign in one bracket and a positive sign in the other, as in [3].

Examples 1h _____

1. Factorise $x^2 - 5x + 6$

The x term in each bracket is x as x^2 can only be $x \times x$
The sign in each bracket is $-$, so $x^2 - 5x + 6 = (x - \ )(x - \ )$
The numbers in the brackets could be 6 and 1 or 2 and 3
Checking the middle term tells us that the numbers must be 2 and 3

$$x^2 - 5x + 6 = (x - 2)(x - 3)$$

Mentally expanding the brackets checks that they are correct.

_____

2. Factorise $x^2 - 3x - 10$

The x term in each bracket is $x \ \Rightarrow \ x^2 - 3x - 10 = (x - \ )(x + \ )$
The numbers could be 10 and 1 or 5 and 2
Checking the middle term shows that they are 5 and 2

$$x^2 - 3x - 10 = (x - 5)(x + 2)$$

Mentally expanding the brackets confirms that they are correct.

_____

EXERCISE 1h

Factorise

1. $x^2 + 8x + 15$ 2. $x^2 + 11x + 28$ 3. $x^2 + 7x + 6$

4. $x^2 + 7x + 12$ 5. $x^2 - 10x + 9$ 6. $x^2 - 6x + 9$

7. $x^2 + 8x + 12$ 8. $x^2 - 9x + 8$ 9. $x^2 + 5x - 14$

10. $x^2 + x - 12$ 11. $x^2 - 4x - 5$ 12. $x^2 - 10x - 24$

13. $x^2 + 9x + 14$ 14. $x^2 - 2x + 1$ 15. $x^2 - 9$

16. $x^2 + 5x - 24$ 17. $x^2 + 4x + 4$ 18. $x^2 - 1$

19. $x^2 - 3x - 18$ 20. $x^2 + 10x + 25$ 21. $x^2 - 16$

22. $4 + 5x + x^2$ 23. $2x^2 - 3x + 1$ 24. $3x^2 + 4x + 1$

25. $9x^2 - 6x + 1$ 26. $6x^2 - x - 1$ 27. $9 + 6x + x^2$

28. $4x^2 - 9$ 29. $x^2 + 2ax + a^2$ 30. $x^2y^2 - 2xy + 1$

Harder Factorising

When the number of possible combinations of terms for the brackets increases, common sense considerations can help to reduce the possibilities.

For example, if the coefficient of x in the quadratic is odd, then there must be an even number and an odd number in the brackets.

Example 1i _____

Factorise $12 - x - 6x^2$

The x terms in the brackets could be $6x$ and x, or $3x$ and $2x$, one positive and the other negative.

The number terms could be 12 and 1 or 3 and 4 (not 6 and 2 because the coefficient of x in the quadratic is odd).

Now we try various combinations until we find the correct one.

$$12 - x - 6x^2 = (3 + 2x)(4 - 3x)$$

EXERCISE 1i

Factorise

1. $6x^2 + x - 12$ 2. $4x^2 - 11x + 6$ 3. $4x^2 + 3x - 1$

4. $3x^2 - 17x + 10$ 5. $4x^2 - 12x + 9$ 6. $3 - 5x - 2x^2$

7. $25x^2 - 16$ 8. $3 - 2x - x^2$ 9. $5x^2 - 61x + 12$

10. $9x^2 + 30x + 25$ 11. $3 + 2x - x^2$ 12. $12 + 7x - 12x^2$

13. $1 - x^2$ 14. $9x^2 + 12x + 4$ 15. $x^2 + 2xy + y^2$

16. $1 - 4x^2$ 17. $4x^2 - 4xy + y^2$ 18. $9 - 4x^2$

19. $36 + 12x + x^2$ 20. $40x^2 - 17x - 12$ 21. $7x^2 - 5x - 150$

22. $36 - 25x^2$ 23. $x^2 - y^2$ 24. $81x^2 - 36xy + 4y^2$

25. $49 - 84x + 36x^2$ 26. $25x^2 - 4y^2$ 27. $36x^2 + 60xy + 25y^2$

28. $4x^2 - 4xy - 3y^2$ 29. $6x^2 + 11xy + 4y^2$ 30. $49p^2q^2 - 28pq + 4$

Common Factors

Consider $4x^2 + 8x + 4$

$$4x^2 + 8x + 4 = 4(x^2 + 2x + 1)$$

The quadratic inside the bracket now has smaller coefficients and can be factorised more easily:

$$4x^2 + 8x + 4 = 4(x + 1)(x + 1)$$
$$= 4(x + 1)^2$$

Not all quadratics factorise.

Consider $3x^2 - x + 5$

The options we can try are $(3x - 5)(x - 1)$ [1]

$(3x - 1)(x - 5)$ [2]

From [1], $(3x - 5)(x - 1) = 3x^2 - 8x + 5$

From [2], $(3x - 1)(x - 5) = 3x^2 - 16x + 5$

As neither of the possible pairs of brackets expand to give $3x^2 - x + 5$, we conclude that $3x^2 - x + 5$ has no factors of the form $ax + b$ where a and b are integers.

Example 1j _____

Factorise $2x^2 - 8x + 16$

$$2x^2 - 8x + 16 = 2(x^2 - 4x + 8)$$

The possible brackets are $(x - 1)(x - 8)$ and $(x - 2)(x - 4)$
Neither pair expands to $x^2 - 4x + 8$, so there are no further factors.

_____

EXERCISE 1j

Factorise where possible

1. $x^2 + x + 1$ **2.** $2x^2 + 4x + 2$ **3.** $x^2 + 3x + 2$

4. $3x^2 + 12x - 15$ **5.** $x^2 + 4$ **6.** $x^2 - 4x - 6$

7. $x^2 + 3x + 1$ **8.** $2x^2 - 8x + 8$ **9.** $3x^2 - 3x - 6$

10. $2x^2 - 6x + 8$ **11.** $3x^2 - 6x - 24$ **12.** $x^2 - 4x - 12$

13. $x^2 + 1$ **14.** $4x^2 - 100$ **15.** $5x^2 - 25$

16. $7x^2 + x + 4$ **17.** $10x^2 - 39x - 36$ **18.** $x^2 + xy + y^2$

HARDER EXPANSIONS

Consider the product $(x - 2)(x^2 - x + 5)$

This expansion should be done in a systematic way.

First multiply each term of the quadratic by x, writing down the separate results as they are found. Then multiply each term of the quadratic by -2. Do not attempt to simplify at this stage.

$$(x - 2)(x^2 - x + 5)$$
$$= x^3 - x^2 + 5x - 2x^2 + 2x - 10$$

Now simplify

$$= x^3 - 3x^2 + 7x - 10$$

Example 1k _____

Expand $(x + 2)(2x - 1)(x + 4)$

First we expand the last two brackets.

$$
\begin{aligned}
(x + 2)(2x - 1)(x + 4) &= (x + 2)(2x^2 + 7x - 4) \\
&= 2x^3 + 7x^2 - 4x + 4x^2 + 14x - 8 \\
&= 2x^3 + 11x^2 + 10x - 8
\end{aligned}
$$

EXERCISE 1k

Expand and simplify

1. $(x - 2)(x^2 + x + 1)$ **2.** $(3x - 2)(x^2 - x - 1)$

3. $(2x - 1)(2x^2 - 3x + 5)$ **4.** $(x - 1)(x^2 - x - 1)$

5. $(2x + 3)(x^2 - 6x - 3)$ **6.** $(x + 1)(x + 2)(x + 3)$

7. $(x + 4)(x - 1)(x + 1)$ **8.** $(x - 2)(x - 3)(x + 1)$

9. $(x + 1)(2x + 1)(x + 2)$ **10.** $(x + 2)(x + 1)^2$

11. $(2x - 1)^2(x + 2)$ **12.** $(3x - 1)^3$

13. $(4x + 3)(x + 1)(x - 4)$ **14.** $(x - 1)(2x - 1)(2x + 1)$

15. $(2x + 1)(x + 2)(3x - 1)$ **16.** $(x + 1)^3$

17. $(x - 2)(x + 2)(x + 1)$ **18.** $(x + 3)(2x + 3)(x - 1)$

19. $(3x - 2)(2x + 5)(4x - 1)$ **20.** $2(x - 7)(2x + 3)(x - 5)$

21. Expand and simplify $(x - 2)^2(3x - 4)$. Write down the coefficients of x^2 and x.

22. Find the coefficients of x^3 and x^2 in the expansion of $(x - 4)(2x + 3)(3x - 1)$

23. Expand and simplify $(x + y)^3$

24. Expand and simplify $(x + y)^4$

PASCAL'S TRIANGLE

58

It is sometimes necessary to expand expressions such as $(x + y)^4$ but the multiplication is tedious when the power is three or more.
We now describe a far quicker way of obtaining such expansions.

Consider the following expansions,

$$(x + y)^1 = x + y$$
$$(x + y)^2 = x^2 + 2xy + y^2$$
$$(x + y)^3 = x^3 + 3x^2y + 3xy^2 + y^3$$
$$(x + y)^4 = x^4 + 4x^3y + 6x^2y^2 + 4xy^3 + y^4$$

The first thing to notice is that the powers of x and y in the terms of each expansion form a pattern. Looking at the expansion of $(x + y)^4$ we see that the first term is x^4 and then the power of x decreases by 1 in each succeeding term while the power of y increases by 1. For all the terms, the sum of the powers of x and y is 4 and the expansion ends with y^4. There is a similar pattern in the other expansions.

Now consider just the coefficients of the terms. Writing these as a triangular array gives

```
              1       1
          1       2       1
      1     3       3       1
  1       4     6       4       1
```

This array is called *Pascal's Triangle* and clearly it has a pattern.
Each row starts and ends with 1 and each other number is the sum of the two numbers in the row above it, as shown. When the pattern is known, Pascal's triangle can be written down to as many rows as needed. Using Pascal's triangle to expand $(x + y)^6$, for example, we go as far as row 6:

```
                1       1
            1       2       1
        1       3       3       1
    1       4       6       4       1
 1      5      10      10       5      1
1     6     15     20     15      6      1
```

We then use our knowledge of the pattern of the powers, together with row 6 of the array, to fill in the coefficients,

i.e. $(x + y)^6 = x^6 + 6x^5y + 15x^4y^2 + 20x^3y^3 + 15x^2y^4 + 6xy^5 + y^6$

The following worked examples show how expansions of other brackets can be found.

Examples 1I _____

1. Expand $(x + 5)^3$

From Pascal's triangle $(x + y)^3 = x^3 + 3x^2y + 3xy^2 + y^3$

Replacing y by 5 gives $(x + 5)^3 = x^3 + 3x^2(5) + 3x(5)^2 + (5)^3$

$$= x^3 + 15x^2 + 75x + 125$$

2. Expand $(2x - 3)^4$

From Pascal's triangle, $(x + y)^4 = x^4 + 4x^3y + 6x^2y^2 + 4xy^3 + y^4$

Replacing x by $2x$ and y by -3 gives

$(2x - 3)^4 = (2x)^4 + 4(2x)^3(-3) + 6(2x)^2(-3)^2 + 4(2x)(-3)^3 + (-3)^4$

$$= 16x^4 - 96x^3 + 216x^2 - 216x + 81$$

EXERCISE 1I

Expand

1. $(x + 3)^3$ 2. $(x - 2)^4$ 3. $(x + 1)^4$

4. $(2x + 1)^3$ 5. $(x - 3)^5$ 6. $(p - q)^4$

7. $(2x + 3)^3$ 8. $(x - 4)^5$ 9. $(3x - 1)^4$

10. $(1 + 5a)^4$ 11. $(2a - b)^6$ 12. $(2x - 5)^3$

MIXED EXERCISE 1

1. Find the coefficient of x in the expansion of $(3x - 7)(5x + 4)$

2. Expand $5(3x - 2)(3 - 7x)$

3. Write down the coefficient of y^2 in the expansion of $(2y + 9)^3$

4. Factorise $3x^2 - 9x + 6$

5. Write down the coefficient of x^3 in the expansion of $(x - 5)^5$

6. Factorise $4x^2 - 36$

7. Expand $x(2x - 1)^2$

8. Find the factors of $25 + x^2 - 10x$

9. Find the coefficient of x in the expansion of $(x - 4)(3 - x)$

10. Expand $3x(2x - 1)(2x + 1)$

11. Factorise $4x^2 - 12xy + 9y^2$

12. Find the coefficient of xy in the expansion of $(3x - y)(y - 4x)$

13. Write down the coefficient of x^2y in the expansion of $(2x - y)^3$

14. Factorise $6x^2 - 11xy - 10y^2$

15. Find the coefficient of a^2b^2 in the expansion of $(2a - b)^4$

CHAPTER 2

FRACTIONS

SIMPLIFICATION OF FRACTIONS

The value of a fraction is unaltered if *numerator* and *denominator* are multiplied or divided by the same number,

e.g.
$$\frac{3}{6} = \frac{1}{2} = \frac{2}{4} = \frac{7}{14} = \cdots$$

and
$$\frac{ax}{ay} = \frac{x}{y} = \frac{3x}{3y} = \frac{x(a+b)}{y(a+b)} = \cdots$$

A fraction can be simplified by multiplying or dividing *top* and *bottom* by a factor which is common to both.

When simplifying fractions, it is sensible first to get rid of any fractions in the numerator and/or denominator. Then factorise the numerator and denominator and look for any common factors.

Examples 2a

1. Simplify $\dfrac{2a^2 - 2ab}{6ab - 6b^2}$

$$\frac{2a^2 - 2ab}{6ab - 6b^2} = \frac{\cancel{2}a(a - b)}{\cancel{6}_3 b(a - b)}$$

$$= \frac{a}{3b}$$

16

2. Simplify $\dfrac{\frac{1}{2}x^2 - 2}{\frac{1}{4}y^2 + 3}$

$$\frac{\frac{1}{2}x^2 - 2}{\frac{1}{4}y^2 + 3} = \frac{2x^2 - 8}{y^2 + 12} \qquad \text{Multiplying top and bottom by 4}$$

$$= \frac{2(x^2 - 4)}{y^2 + 12}$$

$$= \frac{2(x - 2)(x + 2)}{y^2 + 12}$$

EXERCISE 2a

Simplify where possible.

1. $\dfrac{x - 2}{4x - 8}$

2. $\dfrac{2x + 4}{3x - 6}$

3. $\dfrac{2a + 8}{3a + 12}$

4. $\dfrac{3p - 3q}{5p - 5q}$

5. $\dfrac{x^2 + xy}{xy + y^2}$

6. $\dfrac{x - 3p}{2x + p}$

7. $\dfrac{a - 4}{a - 2}$

8. $\dfrac{x^2y + xy^2}{y^2 + \frac{2}{5}xy}$

9. $\dfrac{\frac{1}{3}a - b}{a + \frac{1}{6}b}$

10. $\dfrac{2x(b - 4)}{6x^2(b + 4)}$

11. $\dfrac{(x - 4)(x - 3)}{x^2 - 16}$

12. $\dfrac{4y^2 + 3}{y^2 - 9}$

13. $\dfrac{\frac{1}{3}(x - 3)}{x^2 - 9}$

14. $\dfrac{x^2 - x - 6}{2x^2 - 5x - 3}$

15. $\dfrac{(x - 2)(x + 2)}{x^2 + x - 2}$

16. $\dfrac{\frac{1}{2}(a + 5)}{a^2 - 25}$

17. $\dfrac{3p + 9q}{p^2 + 6pq + 9q^2}$

18. $\dfrac{a^2 + 2a + 4}{a^2 + 7a + 10}$

19. $\dfrac{x^2 + 2x + 1}{3x^2 + 12x + 9}$

20. $\dfrac{4(x - 3)^2}{(x + 1)(x^2 - 2x - 3)}$

MULTIPLICATION AND DIVISION

Fractions are multiplied by taking the product of the numerators and the product of the denominators,

e.g.
$$\frac{x}{a} \times \frac{y}{b} = \frac{x \times y}{a \times b} = \frac{xy}{ab}$$

To divide by a fraction, we multiply by the reciprocal of that fraction, for example

$$\frac{x}{a} \div \frac{y}{b} = \frac{x}{a} \times \frac{b}{y} = \frac{xb}{ay}$$

Example 2b _____

Simplify $\dfrac{2\pi x^2}{7y^2} \div 4\pi x$

$$\frac{2\pi x^2}{7y^2} \div 4\pi x = \frac{\cancel{2}\cancel{\pi} x^{\cancel{2}}}{7y^2} \times \frac{1}{\cancel{4\pi x}_2}$$

$$= \frac{x}{14y^2}$$

_____

EXERCISE 2b

Simplify

1. $\dfrac{4x}{y} \times \dfrac{x}{6y}$

2. $2st \times \dfrac{3t}{s^2}$

3. $\dfrac{4uv}{3} \div \dfrac{u}{2v}$

4. $\dfrac{4\pi r^2}{3} \div 2\pi r$

5. $x(2-x) \div \dfrac{2-x}{3}$

6. $\dfrac{3x^2}{2y} \bigg/ \dfrac{6xy}{9}$

7. $\dfrac{\pi x^3}{3} \div 8\pi x$

8. $\dfrac{1}{a^2 + ab} \bigg/ \dfrac{1}{2}$

9. $\dfrac{1}{x^2 - 1} \div \dfrac{1}{x - 1}$

10. $\left(\dfrac{1}{a}\right)^2 \times \tfrac{1}{2}a$

11. $(x + 1) \times \dfrac{1}{x^2 - 1}$

12. $\dfrac{x - 4}{x + 3} \div \dfrac{2(x - 4)}{3}$

13. $\dfrac{a^2}{3} \times \left(\dfrac{a}{3}\right)^2$

14. $\dfrac{x^2}{6} \bigg/ \left(\dfrac{x}{2}\right)^2$

15. $\dfrac{2r^3}{3} \times \left(\dfrac{1}{rs}\right)^2$

16. $\dfrac{3x^2}{2y} \times \dfrac{y}{y - 2}$

17. $\dfrac{ab}{c} \div \dfrac{ac}{b}$

18. $\dfrac{x^2 + 4x + 3}{5} \times \dfrac{10}{x + 1}$

19. $\dfrac{x^2 + x - 12}{3} \bigg/ (x + 4)$

20. $\dfrac{4x^2 - 9}{(x - 1)^2} \div \dfrac{2x + 3}{x(x - 1)}$

ADDITION AND SUBTRACTION OF FRACTIONS

Before fractions can be added or subtracted, they must be expressed with the same denominator, i.e. we have to find a common denominator. Then the numerators can be added or subtracted,

e.g. $$\frac{2}{p} + \frac{3}{q} = \frac{2q}{pq} + \frac{3p}{pq} = \frac{2q + 3p}{pq}$$

Example 2c _____

Simplify $x - \dfrac{1}{x}$

$$x - \frac{1}{x} = \frac{x}{1} - \frac{1}{x} = \frac{x^2}{x} - \frac{1}{x} = \frac{x^2 - 1}{x}$$

$$= \frac{(x - 1)(x + 1)}{x}$$

EXERCISE 2c

Simplify

1. $\dfrac{1}{a} - \dfrac{1}{b}$

2. $\dfrac{1}{3x} + \dfrac{1}{5x}$

3. $\dfrac{1}{p} - \dfrac{1}{q}$

4. $\dfrac{1}{2x} + \dfrac{3}{5x}$

5. $x + \dfrac{1}{x}$

6. $\dfrac{x}{y} - \dfrac{y}{x}$

7. $2p - \dfrac{1}{p}$

8. $\dfrac{x}{3} + \dfrac{x+1}{4}$

9. $\tfrac{1}{2}(x-1) + \tfrac{1}{3}(x+1)$

10. $\dfrac{x+2}{5} - \dfrac{2x-1}{3}$

11. $\dfrac{1}{\sin A} + \dfrac{1}{\sin B}$

12. $\dfrac{1}{\cos A} + \dfrac{1}{\sin A}$

13. $3x + \dfrac{1}{4x}$

14. $x - \dfrac{2}{2x+1}$

15. $x + 1 + \dfrac{1}{x+1}$

16. $1 + \dfrac{1}{x} + \dfrac{1}{2x}$

17. $1 - x + \dfrac{1}{x}$

18. $\dfrac{1}{n} + \dfrac{1}{n^2}$

19. $\dfrac{x}{a^2} + \dfrac{x}{b^2}$

20. $1 + \dfrac{1}{a} + \dfrac{1}{a+1}$

Example 2d

Simplify $\dfrac{2}{x+2} - \dfrac{x-4}{2x^2+x-6}$

$$\dfrac{2}{x+2} - \dfrac{x-4}{2x^2+x-6} = \dfrac{2}{x+2} - \dfrac{x-4}{(x+2)(2x-3)}$$

$$= \dfrac{2(2x-3)}{(x+2)(2x-3)} - \dfrac{x-4}{(x+2)(2x-3)}$$

$$= \dfrac{2(2x-3)-(x-4)}{(x+2)(2x-3)}$$

$$= \dfrac{4x-6-x+4}{(x+2)(2x-3)}$$

$$= \dfrac{3x-2}{(x+2)(2x-3)}$$

EXERCISE 2d

Simplify

1. $\dfrac{1}{x+1} + \dfrac{1}{x-1}$

2. $\dfrac{1}{x+1} + \dfrac{1}{x-2}$

3. $\dfrac{4}{x+2} + \dfrac{3}{x+3}$

4. $\dfrac{1}{x^2-1} + \dfrac{1}{x+1}$

5. $\dfrac{2}{a^2-1} - \dfrac{3}{a-1}$

6. $\dfrac{1}{x^2+2x+1} + \dfrac{1}{x+1}$

7. $\dfrac{3}{4x^2+4x+1} - \dfrac{2}{2x+1}$

8. $\dfrac{2}{x^2+5x+4} - \dfrac{3}{x+1}$

9. $\dfrac{4}{(x+1)^2} + \dfrac{2}{x+1}$

10. $\dfrac{3}{(x+2)^2} - \dfrac{1}{x+4}$

11. $\dfrac{1}{2(x-1)} + \dfrac{2}{3(x+4)}$

12. $\dfrac{7}{5(x+2)} - \dfrac{2}{x+4}$

13. $\dfrac{4}{3(x+2)} - \dfrac{3}{2(3x-5)}$

14. $\dfrac{3}{x+1} - \dfrac{2}{x-2} + \dfrac{4}{x+3}$

15. $\dfrac{1}{x+1} - \dfrac{2}{x+2} + \dfrac{3}{x+3}$

16. $\dfrac{x+2}{(x+1)^2} - \dfrac{1}{x}$

17. $\dfrac{4t}{t^2+2t+1} + \dfrac{3}{t+1}$

18. $\dfrac{2t}{t^2+1} - \dfrac{t^2+1}{t^2-1}$

19. $\dfrac{1}{y^2-x^2} + \dfrac{3}{y+x}$

20. $1 + \dfrac{1}{n} + \dfrac{1}{n+1} + \dfrac{1}{n+2}$

MIXED EXERCISE 2

1. Simplify (a) $\dfrac{x^2-9}{2x-6}$ (b) $\dfrac{1}{x^2-9} \div \dfrac{1}{x+3}$

2. Simplify (a) $\left(\dfrac{2p}{r}\right)^2 \times \dfrac{ar}{p^3}$ (b) $\dfrac{2p}{r} - \dfrac{3}{p}$

3. Simplify (a) $\dfrac{2n-4}{3} \div (n^2-4)$ (b) $\dfrac{1}{x+1} + \dfrac{1}{2x-1} + \dfrac{1}{x}$

4. Simplify (a) $x + \dfrac{3}{4x-1}$ (b) $\dfrac{x^2-9}{x^2-5x+6}$

5. Simplify (a) $\dfrac{2x}{x^2-1} \div \dfrac{x^2-2x}{x^2-2x+1}$

   (b) $\dfrac{2x}{x^2-1} + \dfrac{x^2-2x}{x^2-2x+1}$

6. Simplify (a) $1 + \dfrac{1}{2x} + \dfrac{1}{x-1}$ (b) $(x^2-1) \div \left(\dfrac{x+1}{x-1}\right)$

7. Simplify (a) $\dfrac{4x^2-25}{4x^2+20x+25}$ (b) $\dfrac{2t}{t^2+1} \div \dfrac{t^2-1}{t^2+1}$

8. Simplify (a) $\left(\dfrac{x-1}{x+1}\right)^2 \times (x^2-1)$ (b) $\dfrac{1}{a} + \dfrac{1}{b} + \dfrac{1}{c}$

CHAPTER 3

SURDS AND INDICES

SQUARE ROOTS

When we express a number as the product of two equal factors, that factor is called the *square root* of the number, for example

$$4 = 2 \times 2 \implies 2 \text{ is the square root of } 4$$

This is written $\qquad 2 = \sqrt{4}$

Now -2 is also a square root of 4, as $4 = -2 \times -2$ but we do *not* write $\sqrt{4} = -2$

The symbol $\sqrt{}$ is used *only for the positive square root.*

So, although $x^2 = 4 \implies x = \pm 2$, the only value of $\sqrt{4}$ is 2

The negative square root of 4 would be written as $-\sqrt{4}$ and, when both square roots are wanted, we write $\pm\sqrt{4}$

CUBE ROOTS

When a number can be expressed as the product of three equal factors, that factor is called the *cube root* of the number,

e.g. $\qquad 27 = 3 \times 3 \times 3 \qquad$ so 3 is the cube root of 27

This is written $\sqrt[3]{27} = 3$

OTHER ROOTS

The notation used for square and cube roots can be extended to represent fourth roots, fifth roots, etc,

e.g. $\qquad 16 = 2 \times 2 \times 2 \times 2 \implies \sqrt[4]{16} = 2$

and $\qquad 243 = 3 \times 3 \times 3 \times 3 \times 3 \implies \sqrt[5]{243} = 3$

In general, if a number, n, can be expressed as the product of p equal factors then each factor is called the pth root of n and is written $\sqrt[p]{n}$

23

RATIONAL NUMBERS

A number which is either an integer, or a fraction whose numerator and denominator are both integers, is called a *rational number*.

The square roots of certain numbers are rational,

e.g. $$\sqrt{9} = 3, \quad \sqrt{25} = 5, \quad \sqrt{\tfrac{4}{49}} = \tfrac{2}{7}$$

This is not true of all square roots however, e.g. $\sqrt{2}, \sqrt{5}, \sqrt{11}$ are not rational numbers. Such square roots can be given to as many decimal places as are required, for example

$$\sqrt{3} = 1.73 \quad \text{correct to 2 d.p.}$$
$$\sqrt{3} = 1.732\,05 \quad \text{correct to 5 d.p.}$$

but they can never be expressed exactly as a decimal. They are called *irrational numbers*.

The only way to give an exact answer when such irrational numbers are involved is to leave them in the form $\sqrt{2}, \sqrt{7}$ etc; in this form they are called *surds*. At this level of mathematics *answers should always be given exactly unless an approximate answer is asked for*, e.g. give your answer correct to 3 s.f.

Surds arise in many topics and the reader will find it necessary to be able to manipulate them.

Simplifying Surds

Consider $\sqrt{18}$
One of the factors of 18 is 9, and 9 has an exact square root,

i.e. $$\sqrt{18} = \sqrt{(9 \times 2)} = \sqrt{9} \times \sqrt{2}$$

But $\sqrt{9} = 3$, therefore $\sqrt{18} = 3\sqrt{2}$

$3\sqrt{2}$ is the simplest possible surd form for $\sqrt{18}$

Similarly $$\sqrt{\frac{2}{25}} = \frac{\sqrt{2}}{\sqrt{25}} = \frac{\sqrt{2}}{5}$$

EXERCISE 3a

Express in terms of the simplest possible surd.

1. $\sqrt{12}$ 2. $\sqrt{32}$ 3. $\sqrt{27}$ 4. $\sqrt{50}$
5. $\sqrt{200}$ 6. $\sqrt{72}$ 7. $\sqrt{162}$ 8. $\sqrt{288}$
9. $\sqrt{75}$ 10. $\sqrt{48}$ 11. $\sqrt{500}$ 12. $\sqrt{20}$

Multiplying Surds

Consider $(4 - \sqrt{5})(3 + \sqrt{2})$

The multiplication is carried out in the same way and order as when multiplying two linear brackets,

i.e.
$$(4 - \sqrt{5})(3 + \sqrt{2}) = (4)(3) + (4)(\sqrt{2}) - (3)(\sqrt{5}) - (\sqrt{5})(\sqrt{2})$$
$$= 12 + 4\sqrt{2} - 3\sqrt{5} - \sqrt{5}\sqrt{2}$$
$$= 12 + 4\sqrt{2} - 3\sqrt{5} - \sqrt{10}$$

In this example there are no like terms to collect but if the same surd occurs in each bracket the expansion can be simplified.

Examples 3b _____

1. Expand and simplify $(2 + 2\sqrt{7})(5 - \sqrt{7})$

$$(2 + 2\sqrt{7})(5 - \sqrt{7}) = (2)(5) - (2)(\sqrt{7}) + (5)(2\sqrt{7}) - (2\sqrt{7})(\sqrt{7})$$
$$= 10 - 2\sqrt{7} + 10\sqrt{7} - 14$$
$$= 8\sqrt{7} - 4$$

---

2. Expand and simplify $(4 - \sqrt{3})(4 + \sqrt{3})$

$$(4 - \sqrt{3})(4 + \sqrt{3}) = 16 + 4\sqrt{3} - 4\sqrt{3} - (\sqrt{3})(\sqrt{3})$$
$$= 16 - 3$$
$$= 13$$

Example 3b number 2 is a special case because the result is a single rational number. The reader will notice that the two given brackets were of the form $(x - a)(x + a)$, i.e. the factors of $a^2 - x^2$.

The product of any two brackets of the type $(p - \sqrt{q})(p + \sqrt{q})$ is, similarly, $p^2 - (\sqrt{q})^2 = p^2 - q$, which is always rational.

This property has an important application in a later section of this chapter.

EXERCISE 3b

Expand and simplify where this is possible.

1. $\sqrt{3}(2 - \sqrt{3})$

2. $\sqrt{2}(5 + 4\sqrt{2})$

3. $\sqrt{5}(2 + \sqrt{75})$

4. $\sqrt{2}(\sqrt{32} - \sqrt{8})$

5. $(\sqrt{3} + 1)(\sqrt{2} - 1)$

6. $(\sqrt{3} + 2)(\sqrt{3} + 5)$

7. $(\sqrt{5} - 1)(\sqrt{5} + 1)$

8. $(2\sqrt{2} - 1)(\sqrt{2} - 1)$

9. $(\sqrt{5} - 3)(2\sqrt{5} - 4)$

10. $(4 + \sqrt{7})(4 - \sqrt{7})$

11. $(\sqrt{6} - 2)^2$

12. $(2 + 3\sqrt{3})^2$

Multiply by a bracket which will make the product rational.

13. $(4 - \sqrt{5})$

14. $(\sqrt{11} + 3)$

15. $(2\sqrt{3} - 4)$

16. $(\sqrt{6} - \sqrt{5})$

17. $(3 - 2\sqrt{3})$

18. $(2\sqrt{5} - \sqrt{2})$

Rationalising a Denominator

A fraction whose denominator contains a surd is more awkward to deal with than one where a surd occurs only in the numerator.

There is a technique for transferring the surd expression from the denominator to the numerator; it is called *rationalising the denominator* (i.e. making the denominator into a rational number).

Examples 3c _____

1. Rationalise the denominator of $\dfrac{2}{\sqrt{3}}$

The square root in the denominator can be removed if we multiply it by another $\sqrt{3}$. If this is done we must, of course, multiply the numerator also by $\sqrt{3}$, otherwise the value of the fraction is changed.

$$\frac{2}{\sqrt{3}} = \frac{2\sqrt{3}}{(\sqrt{3})(\sqrt{3})} = \frac{2\sqrt{3}}{3}$$

2. Rationalise the denominator and simplify $\dfrac{3\sqrt{2}}{5-\sqrt{2}}$

We saw in Example 3b number 2, that a product of the type $(a-\sqrt{b})(a+\sqrt{b})$ is wholly rational so in this question we multiply numerator and denominator by $5+\sqrt{2}$

$$\frac{3\sqrt{2}}{5-\sqrt{2}} = \frac{3\sqrt{2}(5+\sqrt{2})}{(5-\sqrt{2})(5+\sqrt{2})}$$

$$= \frac{15\sqrt{2}+3(\sqrt{2})(\sqrt{2})}{25-(\sqrt{2})(\sqrt{2})}$$

$$= \frac{15\sqrt{2}+6}{23}$$

EXERCISE 3c

Rationalise the denominator, simplifying where possible.

1. $\dfrac{3}{\sqrt{2}}$ **2.** $\dfrac{1}{\sqrt{7}}$ **3.** $\dfrac{2}{\sqrt{11}}$

4. $\dfrac{3\sqrt{2}}{\sqrt{5}}$ **5.** $\dfrac{1}{\sqrt{27}}$ **6.** $\dfrac{\sqrt{5}}{\sqrt{10}}$

7. $\dfrac{1}{\sqrt{2}-1}$ **8.** $\dfrac{3\sqrt{2}}{5+\sqrt{2}}$ **9.** $\dfrac{2}{2\sqrt{3}-3}$

10. $\dfrac{5}{2-\sqrt{5}}$ **11.** $\dfrac{1}{\sqrt{7}-\sqrt{3}}$ **12.** $\dfrac{4\sqrt{3}}{2\sqrt{3}-3}$

13. $\dfrac{3-\sqrt{5}}{\sqrt{5}+1}$ **14.** $\dfrac{2\sqrt{3}-1}{4-\sqrt{3}}$ **15.** $\dfrac{\sqrt{5}-1}{\sqrt{5}-2}$

16. $\dfrac{3}{\sqrt{3}-\sqrt{2}}$ **17.** $\dfrac{3\sqrt{5}}{2\sqrt{5}+1}$ **18.** $\dfrac{\sqrt{2}+1}{\sqrt{2}-1}$

19. $\dfrac{2\sqrt{7}}{\sqrt{7}+2}$ **20.** $\dfrac{\sqrt{5}-1}{3-\sqrt{5}}$ **21.** $\dfrac{1}{\sqrt{11}-\sqrt{7}}$

22. $\dfrac{4-\sqrt{3}}{3-\sqrt{3}}$ **23.** $\dfrac{1-3\sqrt{2}}{3\sqrt{2}+2}$ **24.** $\dfrac{1}{3\sqrt{2}-2\sqrt{3}}$

25. $\dfrac{\sqrt{3}}{\sqrt{2}(\sqrt{6}-\sqrt{3})}$ **26.** $\dfrac{1}{\sqrt{3}(\sqrt{21}+\sqrt{7})}$ **27.** $\dfrac{\sqrt{2}}{\sqrt{3}(\sqrt{5}-\sqrt{2})}$

INDICES

Base and Index

In an expression such as 3^4, the *base* is 3 and the 4 is called the *power* or *index* (the plural is *indices*).

Working with indices involves using some properties which apply to any base, so we express these rules in terms of a general base a (i.e. a stands for any number).

Rule 1

Because a^3 means $a \times a \times a$ and a^2 means $a \times a$ it follows that

$$a^3 \times a^2 = (a \times a \times a) \times (a \times a) = a^5$$

i.e. $a^3 \times a^2 = a^{3+2}$

Similar examples with different powers all indicate the general rule that

$$a^p \times a^q = a^{p+q}$$

Rule 2

Now dealing with division we have

$$a^7 \div a^4 = \frac{\not a \times \not a \times \not a \times \not a \times a \times a \times a}{\not a \times \not a \times \not a \times \not a} = a^3$$

i.e. $a^7 \div a^4 = a^{7-4}$

Again this is just one example of the general rule

$$a^p \div a^q = a^{p-q}$$

When this rule is applied to certain fractions some interesting cases arise.

Consider $a^3 \div a^5$

$$\frac{a^3}{a^5} = \frac{\not a \times \not a \times \not a}{\not a \times \not a \times \not a \times a \times a} = \frac{1}{a^2}$$

But from Rule 2 we have

$$a^3 \div a^5 = a^{3-5} = a^{-2}$$

Therefore a^{-2} means $\dfrac{1}{a^2}$

In general
$$a^{-p} = \frac{1}{a^p}$$

i.e. a^{-p} means 'the reciprocal of a^p'

Now consider $a^4 \div a^4$

$$\frac{a^4}{a^4} = \frac{\cancel{a} \times \cancel{a} \times \cancel{a} \times \cancel{a}}{\cancel{a} \times \cancel{a} \times \cancel{a} \times \cancel{a}} = 1$$

From Rule 2, $\dfrac{a^4}{a^4} = a^{4-4} = a^0$

Therefore $a^0 = 1$

i.e. any base to the power zero is equal to 1

Rule 3

$$(a^2)^3 = (a \times a)^3$$
$$= (a \times a) \times (a \times a) \times (a \times a)$$
$$= a^6$$

i.e. $(a^2)^3 = a^{2 \times 3}$

In general
$$(a^p)^q = a^{pq}$$

Rule 4

This rule explains the meaning of a fractional index.

From the first rule we have

$$a^{1/2} \times a^{1/2} = a^{1/2 + 1/2} = a^1 = a$$

i.e. $a = a^{1/2} \times a^{1/2}$

But $a = \sqrt{a} \times \sqrt{a}$

Therefore $a^{1/2}$ means $\sqrt{a}$, i.e. the positive square root of a

Similarly $a^{1/3} \times a^{1/3} \times a^{1/3} = a^{1/3 + 1/3 + 1/3} = a^1 = a$

But $\sqrt[3]{a} \times \sqrt[3]{a} \times \sqrt[3]{a} = a$

Therefore $a^{1/3}$ means $\sqrt[3]{a}$, i.e. the cube root of a

In general $a^{1/p} = \sqrt[p]{a}$, i.e. the pth root of a

For a more general fractional index, $\frac{p}{q}$, the third rule shows that

$$a^{p/q} = (a^p)^{1/q} \quad \text{or} \quad (a^{1/q})^p$$

For example

$$a^{3/4} = (a^3)^{1/4} \quad \text{or} \quad (a^{1/4})^3$$
$$= \sqrt[4]{a^3} \quad \text{or} \quad (\sqrt[4]{a})^3$$

i.e. $a^{3/4}$ represents either 'the fourth root of a^3'
or 'the cube of the fourth root of a'

All the general rules can be applied to simplify a wide range of expressions containing indices *provided that the terms all have the same base.*

Examples 3d _____

1. Simplify (a) $\dfrac{2^3 \times 2^7}{4^3}$ (b) $(x^2)^7 \times x^{-3}$ (c) $\sqrt[3]{(a^4 b^5)} \times b^{1/3}/a$

(a) First we express all the terms to a base 2

$$\frac{2^3 \times 2^7}{4^3} = \frac{2^3 \times 2^7}{(2^2)^3}$$

$$= \frac{2^{3+7}}{2^{2 \times 3}}$$

$$= \frac{2^{10}}{2^6} = 2^4$$

(b) $(x^2)^7 \times x^{-3} = x^{2 \times 7} \times \dfrac{1}{x^3}$

$$= x^{14} \times \frac{1}{x^3} = x^{11}$$

(c) $\sqrt[3]{(a^4 b^5)} \times b^{1/3}/a = (a^{4/3})(b^{5/3})(b^{1/3})(a^{-1})$

$$= (a^{4/3 - 1})(b^{5/3 + 1/3})$$

$$= a^{1/3} b^2$$

2. Evaluate (a) $(64)^{-1/3}$ (b) $\left(\dfrac{25}{9}\right)^{-3/2}$

(a) $(64)^{-1/3} = \dfrac{1}{(64)^{1/3}} = \dfrac{1}{\sqrt[3]{64}} = \dfrac{1}{4}$

(b) $\left(\dfrac{25}{9}\right)^{-3/2} = \left(\dfrac{9}{25}\right)^{3/2} = \left(\sqrt{\dfrac{9}{25}}\right)^3 = \left(\dfrac{3}{5}\right)^3 = \dfrac{27}{125}$

Note that $\left(\dfrac{9}{25}\right)^{3/2}$ could have been expressed as $\sqrt{\left(\dfrac{9}{25}\right)^3}$ but this form involves *much* bigger numbers.

EXERCISE 3d

Simplify

1. $\dfrac{2^4}{2^2 \times 4^3}$

2. $4^{1/2} \times 2^{-3}$

3. $(3^3)^{1/2} \times 9^{1/4}$

4. $\dfrac{x^{1/3} \times x^{4/3}}{x^{-1/3}}$

5. $\dfrac{p^{1/2} \times p^{-3/4}}{p^{-1/4}}$

6. $(\sqrt{t})^3 \times (\sqrt{t^5})$

7. $(y^2)^{3/2} \times y^{-3}$

8. $(16)^{5/4} \div 8^{4/3}$

9. $\dfrac{y^{1/2}}{y^{-3/4}} \times \sqrt{(y^{1/2})}$

10. $x^2 \times x^{5/2} \div x^{-1/2}$

11. $\dfrac{y^{1/6} \times y^{-2/3}}{y^{1/4}}$

12. $(p^{1/3})^2 \times (p^2)^{1/3} \div \sqrt[3]{p}$

Evaluate

13. $\left(\dfrac{1}{3}\right)^{-1}$

14. $\left(\dfrac{1}{4}\right)^{5/2}$

15. $(8)^{-1/3}$

16. $\dfrac{1}{(16)^{-1/4}}$

17. $\left(\dfrac{1}{9}\right)^{-3/2}$

18. $\left(\dfrac{27}{8}\right)^{2/3}$

19. $\left(\dfrac{100}{9}\right)^{0}$

20. $\dfrac{1}{4^{-2}}$

21. $(0.64)^{-1/2}$

22. $\left(-\dfrac{1}{5}\right)^{-1}$

23. $(121)^{3/2}$

24. $\left(\dfrac{125}{27}\right)^{-1/3}$

25. $18^{1/2} \times 2^{1/2}$

26. $3^{-3} \times 2^0 \times 4^2$

27. $\dfrac{8^{1/2} \times 32^{1/2}}{(16)^{1/4}}$

28. $5^{1/3} \times 25^0 \times 25^{1/3}$

29. $27^{1/4} \times 3^{1/4} \times (\sqrt{3})^{-2}$

30. $\dfrac{9^{1/3} \times 27^{-1/2}}{3^{-1/6} \times 3^{-2/3}}$

MIXED EXERCISE 3

1. Simplify (a) $\sqrt{84}$ (b) $\sqrt{300}$ (c) $\sqrt{45}$

2. Expand and simplify
 (a) $\sqrt{3}(7 - 2\sqrt{3})$ (b) $\sqrt{2}(2\sqrt{2} + \sqrt{8})$

3. Expand and simplify   (a) $(3 + \sqrt{2})(4 - 2\sqrt{2})$ (b) $(\sqrt{5} - \sqrt{2})^2$

4. Multiply by a bracket that will make the product rational
 (a) $(7 - \sqrt{3})$ (b) $(2\sqrt{2} + 1)$ (c) $(\sqrt{7} - \sqrt{5})$

5. Rationalise the denominator and simplify where possible
 (a) $\dfrac{5}{\sqrt{7}}$ (b) $\dfrac{3}{\sqrt{13} - 2}$ (c) $\dfrac{4}{\sqrt{3} - \sqrt{2}}$ (d) $\dfrac{\sqrt{3} - 1}{\sqrt{3} + 1}$

6. Write down the value of
 (a) 2^5 (b) 3^3 (c) $243^{1/5}$

7. Simplify (a) $\dfrac{2^3 \times 4^{-2}}{2^{-1}}$ (b) $(x^3)^{-2} \times (x^2)^3$

8. Evaluate (a) $(64)^{-1/3}$ (b) $\left(\dfrac{49}{16}\right)^{-1/2}$ (c) $\left(\dfrac{8}{27}\right)^{3/2}$

9. Simplify (a) $8^{1/6} \times 2^0 \times 2^{-1/2}$ (b) $(\sqrt{5})^{-2} \times 75^{1/2} \times 25^{-1/4}$

10. What is the value of the missing number or index?
 (a) $9^{3/2} = 3^{\square}$ (b) $2^{\square} = 8^{-1/4}$ (c) $\dfrac{1}{16^{3/2}} = \square^{-3}$

CHAPTER 4

QUADRATIC EQUATIONS AND SIMULTANEOUS EQUATIONS

QUADRATIC EQUATIONS

When a quadratic expression has a particular value we have a quadratic equation, for example

$$2x^2 - 5x + 1 = 0$$

Using a, b and c to stand for any numbers, any quadratic equation can be written in the general form

$$ax^2 + bx + c = 0$$

Solution by Factorising

Consider the quadratic equation $x^2 - 3x + 2 = 0$

The quadratic expression on the left-hand side can be factorised,

i.e. $$x^2 - 3x + 2 = (x - 2)(x - 1)$$

Therefore the given equation becomes

$$(x - 2)(x - 1) = 0 \qquad\qquad [1]$$

Now if the product of two quantities is zero then one, or both, of those quantities must be zero.

Applying this fact to equation [1] gives

$$x - 2 = 0 \quad \text{or} \quad x - 1 = 0$$

i.e. $$x = 2 \quad \text{or} \quad x = 1$$

This is the solution of the given equation.
The values 2 and 1 are called the *roots* of that equation.

This method of solution can be used for any quadratic equation in which the quadratic expression factorises.

Example 4a

Find the roots of the equation $x^2 + 6x - 7 = 0$

$$x^2 + 6x - 7 = 0$$
$$\Rightarrow \quad (x - 1)(x + 7) = 0$$
$$\therefore \quad x - 1 = 0 \quad \text{or} \quad x + 7 = 0$$
$$\therefore \quad x = 1 \quad \text{or} \quad x = -7$$

The roots of the equation are 1 and -7

EXERCISE 4a

Solve the equations.

1. $x^2 + 5x + 6 = 0$
2. $x^2 + x - 6 = 0$
3. $x^2 - x - 6 = 0$
4. $x^2 + 6x + 8 = 0$
5. $x^2 - 4x + 3 = 0$
6. $x^2 + 2x - 3 = 0$
7. $2x^2 + 3x + 1 = 0$
8. $4x^2 - 9x + 2 = 0$
9. $x^2 + 4x - 5 = 0$
10. $x^2 + x - 72 = 0$

Find the roots of the equations.

11. $x^2 - 2x - 3 = 0$
12. $x^2 + 5x + 4 = 0$
13. $x^2 - 6x + 5 = 0$
14. $x^2 + 3x - 10 = 0$
15. $x^2 - 5x - 14 = 0$
16. $x^2 - 9x + 14 = 0$

Rearranging the Equation

The terms in a quadratic equation are not always given in the order $ax^2 + bx + c = 0$. When they are given in a different order they should be rearranged into the standard form.

For example

$$x^2 - x = 4 \quad \text{becomes} \quad x^2 - x - 4 = 0$$
$$3x^2 - 1 = 2x \quad \text{becomes} \quad 3x^2 - 2x - 1 = 0$$
$$x(x - 1) = 2 \quad \text{becomes} \quad x^2 - x = 2 \quad \Rightarrow \quad x^2 - x - 2 = 0$$

It is usually best to collect the terms on the side where the x^2 term is positive, for example

$$2 - x^2 = 5x \quad \text{becomes} \quad 0 = x^2 + 5x - 2$$

i.e. $$x^2 + 5x - 2 = 0$$

Losing a Solution

Quadratic equations sometimes have a common factor containing the unknown quantity. It is very tempting in such cases to divide by the common factor, but doing this results in the loss of part of the solution, as the following example shows.

First solution

$$x^2 - 5x = 0$$

$$x(x - 5) = 0$$

$\therefore \quad x = 0 \quad \text{or} \quad x - 5 = 0$

$\Rightarrow x = 0 \text{ or } 5$

Second solution

$$x^2 - 5x = 0$$

$$x - 5 = 0 \quad \text{(Dividing by } x\text{)}$$

$\therefore \quad x = 5$

The solution $x = 0$ has been lost.

Although dividing an equation by a numerical common factor is correct and sensible, dividing by a common factor containing the unknown quantity results in the loss of a solution.

Examples 4b

1. Solve the equation $4x - x^2 = 3$

$$4x - x^2 = 3$$

$\Rightarrow$ $$0 = x^2 - 4x + 3$$

$\Rightarrow$ $$x^2 - 4x + 3 = 0$$

$\Rightarrow$ $$(x - 3)(x - 1) = 0$$

$\Rightarrow$ $$x - 3 = 0 \quad \text{or} \quad x - 1 = 0$$

$\Rightarrow$ $$x = 3 \quad \text{or} \quad x = 1$$

2. Find the roots of the equation $x^2 = 3x$

$$x^2 = 3x$$
$$\Rightarrow \qquad x^2 - 3x = 0$$
$$\Rightarrow \qquad x(x - 3) = 0$$
$$\Rightarrow \qquad x = 0 \quad \text{or} \quad x - 3 = 0$$
$$\Rightarrow \qquad x = 0 \quad \text{or} \quad x = 3$$

Therefore the roots are 0 and 3

EXERCISE 4b

Solve the equations.

1. $x^2 + 10 - 7x = 0$ **2.** $15 - x^2 - 2x = 0$

3. $x^2 - 3x = 4$ **4.** $12 - 7x + x^2 = 0$

5. $2x - 1 + 3x^2 = 0$ **6.** $x(x + 7) + 6 = 0$

7. $2x^2 - 4x = 0$ **8.** $x(4x + 5) = -1$

9. $2 - x = 3x^2$ **10.** $6x^2 + 3x = 0$

11. $x^2 + 6x = 0$ **12.** $x^2 = 10x$

13. $x(4x + 1) = 3x$ **14.** $20 + x(1 - x) = 0$

15. $x(3x - 2) = 8$ **16.** $x^2 - x(2x - 1) + 2 = 0$

17. $x(x + 1) = 2x$ **18.** $4 + x^2 = 2(x + 2)$

19. $x(x - 2) = 3$ **20.** $1 - x^2 = x(1 + x)$

Solution by Completing the Square

When there are no obvious factors, another method is needed to solve the equation. One such method involves adding a constant to the x^2 term and the x term, to make a perfect square. This technique is called *completing the square*.

Consider $\qquad\qquad\qquad x^2 - 2x$

Adding 1 gives $\qquad\qquad\quad x^2 - 2x + 1$

Now $x^2 - 2x + 1 = (x - 1)^2$ which is a perfect square.

Adding the number 1 was not a guess, it was found by using the fact that

$$x^2 + 2ax + \boxed{a^2} = (x + a)^2$$

We see from this that the number to be added is always
$$\text{(half the coefficient of } x)^2$$

Hence $x^2 + 6x$ requires 3^2 to be added to make a perfect square,

i.e. $\qquad\qquad\qquad x^2 + 6x + 9 = (x + 3)^2$

To complete the square when the coefficient of x^2 is not 1, we first take out the coefficient of x^2 as a factor,

e.g. $\qquad\qquad\qquad 2x^2 + x = 2(x^2 + \tfrac{1}{2}x)$

Now we add $(\tfrac{1}{2} \times \tfrac{1}{2})^2$ inside the bracket, giving

$$2(x^2 + \tfrac{1}{2}x + \tfrac{1}{16}) = 2(x + \tfrac{1}{4})^2$$

Take extra care when the coefficient of x^2 is negative

e.g. $\qquad\qquad\qquad -x^2 + 4x = -(x^2 - 4x)$

Then $\qquad\qquad -(x^2 - 4x + 4) = -(x - 2)^2$

$\therefore \qquad\qquad\quad -x^2 + 4x - 4 = -(x - 2)^2$

Examples 4c

1. Solve the equation $x^2 - 4x - 2 = 0$, giving the solution in surd form.

$$x^2 - 4x - 2 = 0$$

No factors can be found so we isolate the two terms with x in,

i.e. $\qquad\qquad\qquad x^2 - 4x = 2$

Add $\{\tfrac{1}{2} \times (-4)\}^2$ to *both* sides

i.e. $\qquad\qquad\qquad x^2 - 4x + 4 = 2 + 4$

$\Rightarrow \qquad\qquad\qquad (x - 2)^2 = 6$

$\therefore \qquad\qquad\qquad x - 2 = \pm\sqrt{6}$

$\therefore \qquad\qquad x = 2 + \sqrt{6} \quad \text{or} \quad x = 2 - \sqrt{6}$

2. Find in surd form the roots of the equation $2x^2 - 3x - 3 = 0$

$$2x^2 - 3x - 3 = 0$$
$$2x^2 - 3x = 3$$
$$2(x^2 - \tfrac{3}{2}x) = 3$$
$$x^2 - \tfrac{3}{2}x = \tfrac{3}{2}$$
$$x^2 - \tfrac{3}{2}x + \tfrac{9}{16} = \tfrac{3}{2} + \tfrac{9}{16}$$
$$(x - \tfrac{3}{4})^2 = \tfrac{33}{16}$$

$\therefore \qquad x - \tfrac{3}{4} = \pm\sqrt{\tfrac{33}{16}} = \pm\tfrac{1}{4}\sqrt{33}$

$\therefore \qquad x = \tfrac{3}{4} \pm \tfrac{1}{4}\sqrt{33}$

The roots of the equation are $\tfrac{1}{4}(3 + \sqrt{33})$ and $\tfrac{1}{4}(3 - \sqrt{33})$

EXERCISE 4c

Add a number to each expression so that the result contains a perfect square.

1. $x^2 - 4x$ **2.** $x^2 + 2x$ **3.** $x^2 - 6x$

4. $x^2 + 10x$ **5.** $2x^2 - 4x$ **6.** $x^2 + 5x$

7. $3x^2 - 48x$ **8.** $x^2 + 18x$ **9.** $2x^2 - 40x$

10. $x^2 + x$ **11.** $3x^2 - 2x$ **12.** $2x^2 + 3x$

Solve the equations by completing the square, giving the solutions in surd form.

13. $x^2 + 8x = 1$ **14.** $x^2 - 2x - 2 = 0$ **15.** $x^2 + x - 1 = 0$

16. $2x^2 + 2x = 1$ **17.** $x^2 + 3x + 1 = 0$ **18.** $2x^2 - x - 2 = 0$

19. $x^2 + 4x = 2$ **20.** $3x^2 + x - 1 = 0$ **21.** $2x^2 + 4x = 7$

22. $x^2 - x = 3$ **23.** $4x^2 + x - 1 = 0$ **24.** $2x^2 - 3x - 4 = 0$

The Formula for Solving a Quadratic Equation

Solving a quadratic equation by completing the square is rather tedious. If the method is applied to a general quadratic equation, a formula can be derived which can then be used to solve any particular equation.

Using a, b and c to represent any numbers we have the general quadratic equation

$$ax^2 + bx + c = 0$$

Using the method of completing the square for this equation gives

$$ax^2 + bx = -c$$

i.e.
$$a\left(x^2 + \frac{b}{a}x\right) = -c$$

$\Rightarrow$
$$x^2 + \frac{b}{a}x = -\frac{c}{a}$$

$\therefore$
$$x^2 + \frac{b}{a}x + \left(\frac{b}{2a}\right)^2 = \left(\frac{b}{2a}\right)^2 - \frac{c}{a}$$

$\therefore$
$$\left(x + \frac{b}{2a}\right)^2 = \frac{b^2}{4a^2} - \frac{c}{a} = \frac{b^2 - 4ac}{4a^2}$$

$\Rightarrow$
$$x + \frac{b}{2a} = \pm\sqrt{\frac{(b^2 - 4ac)}{4a^2}}$$

$\Rightarrow$
$$x = -\frac{b}{2a} \pm \frac{\sqrt{(b^2 - 4ac)}}{2a}$$

i.e.
$$x = \frac{-b \pm \sqrt{(b^2 - 4ac)}}{2a}$$

Example 4d

Find, by using the formula, the roots of the equation
$2x^2 - 7x - 1 = 0$ giving them correct to 3 decimal places.

$$2x^2 - 7x - 1 = 0$$

Comparing with $ax^2 + bx + c = 0$ gives $a = 2$, $b = -7$, $c = -1$

$$x = \frac{-b \pm \sqrt{(b^2 - 4ac)}}{2a}$$

$$= \frac{7 \pm \sqrt{\{49 - 4(2)(-1)\}}}{4}$$

Therefore, in surd form,
$$x = \frac{7 \pm \sqrt{57}}{4}$$

Correct to 3 d.p. the roots are 3.637 and -0.137

EXERCISE 4d

Solve the equations by using the formula. Give the solutions in surd form.

1. $x^2 + 4x + 2 = 0$

2. $2x^2 + x - 2 = 0$

3. $x^2 + 5x + 1 = 0$

4. $2x^2 - x - 4 = 0$

5. $x^2 + 1 = 4x$

6. $2x^2 - x = 5$

7. $1 + x - 3x^2 = 0$

8. $3x^2 = 1 - x$

9. $5 + 2x = x^2$

10. $5x^2 - 1 = x$

Find, correct to 3 d.p., the roots of the equations.

11. $5x^2 + 9x + 2 = 0$

12. $2x^2 - 7x + 4 = 0$

13. $4x^2 - 7x - 1 = 0$

14. $3x = 5 - 4x^2$

15. $4x^2 + 3x = 5$

16. $1 = 5x - 5x^2$

17. $8x - x^2 = 1$

18. $x^2 - 3x = 1$

19. $2x^2 - x = 2$

20. $x^2 + 1 = 8x$

SIMULTANEOUS EQUATIONS

When only one unknown quantity has to be found, only one equation is needed to provide a solution.

If two unknown quantities are involved in a problem we need two equations connecting them. Then, between the two equations we can eliminate one of the unknowns, producing just one equation containing just one unknown. This is then ready for solution.

The reader will already have met a variety of techniques used to eliminate one of the unknowns from a pair of linear equations.

If one of the equations is linear and the other is quadratic, one unknown can be isolated in the linear equation and then substituted in the quadratic equation.

Example 4e _____

Solve the equations $\qquad x - y = 2$

$$2x^2 - 3y^2 = 15$$

$$x - y = 2 \qquad\qquad [1]$$

$$2x^2 - 3y^2 = 15 \qquad\qquad [2]$$

Equation 1 is linear so we use it for the substitution, i.e. $x = y + 2$

Substituting $y + 2$ for x in [2] gives

$$2(y + 2)^2 - 3y^2 = 15$$

$\Rightarrow \qquad\qquad 2(y^2 + 4y + 4) - 3y^2 = 15$

$\Rightarrow \qquad\qquad 2y^2 + 8y + 8 - 3y^2 = 15$

Collecting terms on the side where y^2 is positive gives

$$0 = y^2 - 8y + 7$$

$\Rightarrow \qquad\qquad 0 = (y - 7)(y - 1)$

$\therefore \qquad\qquad y = 7 \quad \text{or} \quad 1$

Now we use $x = y + 2$ to find corresponding values of x

y	7	1
x	9	3

$\therefore \qquad$ either $\qquad\qquad x = 9$ and $y = 7$

$\qquad\quad$ or $\qquad\qquad\quad x = 3$ and $y = 1$

Note that the values of x and y must be given in *corresponding pairs*.

It is incorrect to write the answer as $y = 7$ or 1 and $x = 9$ or 3

because $\begin{cases} y = 7 & \text{with} \quad x = 3 \\ y = 1 & \text{with} \quad x = 9 \end{cases}$ are *not* solutions

EXERCISE 4e

Solve the following pairs of equations.

1. $x^2 + y^2 = 5$

$y - x = 1$

2. $y^2 - x^2 = 8$

$x + y = 2$

3. $3x^2 - y^2 = 3$

$2x - y = 1$

4. $y = 4x^2$

$y + 2x = 2$

5. $y^2 + xy = 3$

$2x + y = 1$

6. $x^2 - xy = 14$

$y = 3 - x$

7. $xy = 2$

$x + y - 3 = 0$

8. $2x - y = 2$

$x^2 - y = 5$

9. $y - x = 4$

$y^2 - 5x^2 = 20$

10. $x + y^2 = 10$

$x - 2y = 2$

11. $4x + y = 1$

$4x^2 + y = 0$

12. $3xy - x = 0$

$x + 3y = 2$

13. $x^2 + 4y^2 = 2$

$2y + x + 2 = 0$

14. $x + 3y = 0$

$2x + 3xy = 1$

15. $3x - 4y = 1$

$6xy = 1$

16. $x^2 + 4y^2 = 2$

$x + 2y = 2$

17. $xy = 9$

$x - 2y = 3$

18. $4x + y = 2$

$4x + y^2 = 8$

19. $1 + 3xy = 0$

$x + 6y = 1$

20. $x^2 - xy = 0$

$x + y = 1$

21. $xy + y^2 = 2$

$2x + y = 3$

22. $xy + x = -3$

$2x + 5y = 8$

PROPERTIES OF THE ROOTS OF A QUADRATIC EQUATION

A number of interesting facts can be observed by examining the formula used for solving a quadratic equation, especially when it is written in the form

$$x = -\frac{b}{2a} \pm \frac{\sqrt{(b^2 - 4ac)}}{2a}$$

The Sum of the Roots

The separate roots are

$$-\frac{b}{2a} + \frac{\sqrt{(b^2 - 4ac)}}{2a} \quad \text{and} \quad -\frac{b}{2a} - \frac{\sqrt{(b^2 - 4ac)}}{2a}$$

When the roots are added, the terms containing the square root disappear giving

$$\text{sum of roots} = -\frac{b}{a}$$

This fact is very useful as a check on the accuracy of roots that have been calculated.

The Nature of the Roots

In the formula there are two terms. The first of these, $-\frac{b}{2a}$, can always be found for any values of a and b.

The second term however, i.e. $\frac{\sqrt{(b^2 - 4ac)}}{2a}$, is not so straightforward as there are three different cases to consider.

1) If $b^2 - 4ac$ is positive, its square root can be found and, whether it is a whole number, a fraction or a decimal, it is a number of the type we are familiar with – it is called a *real* number.

The two square roots, i.e. $\pm \sqrt{(b^2 - 4ac)}$ have different (or distinct) values giving two different real values of x
So the equation has *two different real roots*.

2) If $b^2 - 4ac$ is zero then its square root also is zero and
$x = -\dfrac{b}{2a} - \dfrac{\sqrt{(b^2 - 4ac)}}{2a}$ gives

$$x = -\frac{b}{2a} + 0 \quad \text{and} \quad x = -\frac{b}{2a} - 0$$

i.e. there is just one value of x that satisfies the equation.

An example of this case is $x^2 - 2x + 1 = 0$

From the formula we get $x = -\dfrac{(-2)}{2} \pm 0$

i.e. $\qquad\qquad\qquad\qquad x = 1 \text{ or } 1$

By factorising we can see that there are two equal roots,

i.e. $\qquad\qquad\qquad (x - 1)(x - 1) = 0$

$\Rightarrow \qquad\qquad\qquad x = 1 \quad \text{or} \quad x = 1$

This type of equation can be said to have a *repeated root*.

3) If $b^2 - 4ac$ is negative we cannot find its square root because there is no real number whose square is negative. In this case the equation has *no real roots*.

From these three considerations we see that the roots of a quadratic equation can be

either	real and different
or	real and equal
or	not real

and that it is the value of $b^2 - 4ac$ which determines the nature of the roots, i.e.

Condition	Nature of Roots
$b^2 - 4ac > 0$	Real and different
$b^2 - 4ac = 0$	Real and equal
$b^2 - 4ac < 0$	Not real

Sometimes it matters only that the roots are real, in which case the first two conditions can be combined to give

if $b^2 - 4ac \geqslant 0$, the roots are real.

Examples 4f _____

1. Determine the nature of the roots of the equation $x^2 - 6x + 1 = 0$

$$x^2 - 6x + 1 = 0$$

$a = 1, \ b = -6, \ c = 1$

$$b^2 - 4ac = (-6)^2 - 4(1)(1) = 32$$

$b^2 - 4ac > 0$ so the roots are real and different.

_____

2. If the roots of the equation $2x^2 - px + 8 = 0$ are equal, find the value of p.

$$2x^2 - px + 8 = 0$$

$a = 2, \ b = -p, \ c = 8$

The roots are equal so $b^2 - 4ac = 0$,

i.e. $$(-p)^2 - 4(2)(8) = 0$$

$\Rightarrow$ $$p^2 - 64 = 0$$

$\Rightarrow$ $$p^2 = 64$$

$\therefore$ $$p = \pm 8$$

_____

3. Prove that the equation $(k - 2)x^2 + 2x - k = 0$ has real roots whatever the value of k

$$(k - 2)x^2 + 2x - k = 0$$

$a = k - 2, \ b = 2, \ c = -k$

$$b^2 - 4ac = 4 - 4(k - 2)(-k)$$

$$= 4 + 4k^2 - 8k$$

$$= 4k^2 - 8k + 4$$

$$= 4(k^2 - 2k + 1) = 4(k - 1)^2$$

Now $(k - 1)^2$ cannot be negative whatever the value of k, so $b^2 - 4ac$ cannot be negative. Therefore the roots are always real.

_____

EXERCISE 4f

Without solving the equation, write down the sum of its roots.

1. $x^2 - 4x - 7 = 0$ 2. $3x^2 + 5x + 1 = 0$

3. $2 + x - x^2 = 0$ 4. $3x^2 - 4x - 2 = 0$

5. $x^2 + 3x + 1 = 0$ 6. $7 + 2x - 5x^2 = 0$

Without solving the equation, determine the nature of its roots.

7. $x^2 - 6x + 4 = 0$ 8. $3x^2 + 4x + 2 = 0$

9. $2x^2 - 5x + 3 = 0$ 10. $x^2 - 6x + 9 = 0$

11. $4x^2 - 12x - 9 = 0$ 12. $4x^2 + 12x + 9 = 0$

13. $x^2 + 4x - 8 = 0$ 14. $x^2 + ax + a^2 = 0$

15. $x^2 - ax - a^2 = 0$ 16. $x^2 + 2ax + a^2 = 0$

17. If the roots of $3x^2 + kx + 12 = 0$ are equal, find k

18. If $x^2 - 3x + a = 0$ has equal roots, find a

19. The roots of $x^2 + px + (p - 1) = 0$ are equal. Find p

20. Prove that the roots of the equation $kx^2 + (2k + 4)x + 8 = 0$ are real for all values of k

21. Show that the equation $ax^2 + (a + b)x + b = 0$ has real roots for all values of a and b

22. Find the relationship between p and q if the roots of the equation $px^2 + qx + 1 = 0$ are equal.

Summary

Methods for solving quadratic equations.

1) Collect the terms in the order $ax^2 + bx + c = 0$, then factorise the left-hand side.

2) Arrange in the form $ax^2 + bx = -c$, then complete the square on the left-hand side, adding the appropriate number to *both* sides.

3) Use the formula $x = \dfrac{-b \pm \sqrt{(b^2 - 4ac)}}{2a}$

Note. Roots that are not rational should be given in surd form (i.e. the exact form) unless an approximate form (such as correct to 3 s.f.) is specifically asked for.

Properties of Roots

$b^2 - 4ac > 0 \quad \Rightarrow \quad$ real different roots

$b^2 - 4ac = 0 \quad \Rightarrow \quad$ real equal roots

$b^2 - 4ac \geqslant 0 \quad \Rightarrow \quad$ real roots

$b^2 - 4ac < 0 \quad \Rightarrow \quad$ no real roots

Sum of roots $= -\dfrac{b}{a}$

LOSING SOLUTIONS

It has already been shown that a solution is lost if an equation is divided by a *common factor containing the unknown quantity*.

There is another situation where a valid solution *may* be overlooked.

The Infinite Solution

Consider the equation $t(t - 1) = t^2 + 2$
This gives $t^2 - t = t^2 + 2$
and $t = -2$ appears to be the only solution.

Suppose, however, that the value of t can be very large indeed, so large that t approaches infinity (we write, $t \to \infty$). Then, in the equation $t(t - 1) = t^2 + 2$, we see that

> 1 is so small compared with the value of t
> that $t(t - 1)$ is very nearly equal to t^2

also   2 is so small compared with the value of t^2
that $t^2 + 2$ is very nearly equal to t^2

That is, for very large values of t, $t(t - 1)$ is nearly equal to $t^2 + 2$ and, the larger t becomes the more nearly is the equation satisfied. Therefore, in an equation where the squared terms cancel, we must always *consider* a solution of the type $t \to \infty$

In real problems the unknown quantity usually represents something specific and in most cases it could not possibly have an infinitely large value.

There are cases, however, when the infinite solution is meaningful. The reader will meet some of these later on and will probably have met one already, i.e. if the unknown quantity, t, represents the tangent of an angle, then $t \to \infty$ gives an angle of $90°$

MIXED EXERCISE 4

In each Question from 1 to 10
(a) write down the value of $-b/a$
(b) use any suitable method to find the roots of the equation, giving any irrational roots in surd form.
(c) find the sum of the roots and check that it is equal to the answer to (a).

1. $x^2 - 5x - 6 = 0$
2. $x^2 - 6x - 5 = 0$
3. $2x^2 + 3x = 1$
4. $5 - 3x^2 = 4x$
5. $x(2 - x) = 1$
6. $4x^2 - 3 = 11x$
7. $(x - 1)(x + 2) = 1$
8. $x^2 + 4x + 4 = 16$
9. $x^2 + 2x = 2$
10. $2(x^2 + 2) = x(x - 4)$

In Questions 11 to 16, solve the equations giving *all possible* solutions.

11. $x(x - 2) = 0$
12. $x(x + 3) = 4$
13. $x^2 + x + 8 = x(x + 5)$
14. $x^2 + 5x + 2 = 2(2x + 1)$
15. $x(x - 5) = 2(x + 5)$
16. $2x(x + 3) = x(2x - 1) + 7$

17. Determine the nature of the roots of the equations
    (a) $x^2 + 3x + 7 = 0$
    (b) $3x^2 - x - 5 = 0$
    (c) $ax^2 + 2ax + a = 0$
    (d) $2 + 9x - x^2 = 0$

18. For what values of p does the equation $px^2 + 4x + (p - 3) = 0$ have equal roots?

19. Show that the equation $2x^2 + 2(p + 1)x + p = 0$ always has real roots.

20. The equation $x^2 + kx + k = 1$ has equal roots. Find k

In Questions 21 and 22 solve the pair of equations. (Choose your substitution carefully, to keep the amount of squaring to a minimum.)

21. $2x^2 - y^2 = 7$
    $x + y = 9$
22. $2x = y - 1$
    $x^2 - 3y + 11 = 0$

23. Use the formula to solve the equation $3x^2 - 17x + 10 = 0$
    (a) Are the roots of the equation rational or irrational?
    (b) What does your answer to (a) tell you about the LHS of the equation?

CONSOLIDATION A

SUMMARY

TERMS AND COEFFICIENTS

In an algebraic expression, terms are separated by plus or minus signs. An individual term is identified by the combination of letters involved. The coefficient of a term is the number in the term, e.g. $\textcircled{2}\,x^2y$

EXPANSION OF BRACKETS

Important results are

$$(ax + b)^2 = a^2x^2 + 2abx + b^2$$

$$(ax - b)^2 = a^2x^2 - 2abx + b^2$$

$$(ax + b)(ax - b) = a^2x^2 - b^2$$

PASCAL'S TRIANGLE

$$
\begin{array}{ccccccccc}
 & & & & 1 & & 1 & & \\
 & & & 1 & & 2 & & 1 & \\
 & & 1 & & 3 & & 3 & & 1 \\
 & 1 & & 4 & & 6 & & 4 & & 1 \\
\end{array}
$$

The 1st, 2nd, 3rd, ... rows in this array give the coefficients in the expansion of $(1 + x)^1$, $(1 + x)^2$, $(1 + x)^3$, ...

INDICES

$$a^n \times a^m = a^{n+m}$$

$$a^n \div a^m = a^{n-m}$$

$$(a^n)^m = a^{nm}$$

$$\sqrt[n]{a} = a^{1/n}$$

$$a^0 = 1$$

$$a^{n/m} = \sqrt[m]{a^n} = (\sqrt[m]{a})^n$$

49

QUADRATIC EQUATIONS

The general quadratic equation is $ax^2 + bx + c = 0$

The roots of this equation can be found by

    factorising when this is possible,

or completing the square,

or by using the formula $x = \dfrac{-b \pm \sqrt{(b^2 - 4ac)}}{2a}$

When $b^2 - 4ac > 0$, the roots are real and different.

When $b^2 - 4ac = 0$, the roots are real and equal.

When $b^2 - 4ac < 0$, the roots are not real.

MULTIPLE CHOICE EXERCISE A

TYPE I

1. The roots of the equation $x^2 - 3x + 2 = 0$ are

    A 2, 1   **B** $-2, -1$ **C** $-3, 2$ **D** $0, \frac{2}{3}$ **E** not real

2. The coefficient of xy in the expansion of $(x - 3y)(2x + y)$ is

    A 1   **B** 6   **C** 5   **D** 0   **E** -5

3. $\dfrac{1 - \sqrt{2}}{1 + \sqrt{2}}$ is equal to

    A 1   **B** -1 **C** $3 - \sqrt{2}$ **D** $1 - \frac{2}{3}\sqrt{2}$ **E** $2\sqrt{2} - 3$

4. Expanding $(1 + \sqrt{2})^3$ gives

    A $3 + 3\sqrt{2}$ **C** $1 + 3\sqrt{2}$ **E** $1 + 2\sqrt{2}$

    B $7 + 5\sqrt{2}$ **D** $3 + \sqrt{6}$

5. If $x^2 + px + 6 = 0$ has equal roots and $p > 0$; or p is

    A $\sqrt{48}$ **B** 0   **C** $\sqrt{6}$ **D** 3   **E** $\sqrt{24}$

6. If $x^2 + 4x + p \equiv (x + q)^2 + 1$, the values of p and q are

    A $p = 5, q = 2$ **D** $p = -1, q = 5$

    B $p = 1, q = 2$ **E** $p = 0, q = -1$

    C $p = 2, q = 5$

7. $\dfrac{p^{-1/2} \times p^{3/4}}{p^{-1/4}}$ simplifies to

    A 1   **B** $p^{-1/2}$ **C** $p^{3/4}$ **D** p **E** $p^{1/2}$

8. In the expansion of $(a - 2b)^3$ the coefficient of b^2 is

   A $-2a^2$ **B** $-8a$ **C** $12a$ **D** $-4a$ **E** -12

TYPE II

9. If $y = 2x - 1$ and $xy = 3$ then

   A   there is only one value of y that satisfies both equations
   B   there are two values of x that satisfy both equations
   C $x = 1$ and $y = 3$ satisfy one of the equations.

10. When $(3 - 5x)^4$ is expanded

   A   the coefficient of x^4 is 1
   B   the coefficient of x is -540
   C   there are four terms after all simplification.

11. $f(x) = x^2 - 2x + 2$

   A $f(x) = (x - 1)^2 + 1$
   B $f(1) = 0$
   C $f(x) = 0$ has equal roots.

12. $f(x) = 2x^2 + 3x - 2$

   A $f(x)$ can be expressed in the form $(x + a)^2 + b$
   B   the equation $f(x) = 0$ has two real distinct roots
   C $x + 2$ is a factor of $f(x)$

13. $\dfrac{2\sqrt{3} - 2}{2\sqrt{3} + 2}$

   A   can be expressed as a fraction with a rational denominator
   B   is an irrational number
   C   is equal to -1

TYPE III

14. If $x - a$ is a factor of $x^2 + px + q$, the equation $x^2 + px + q = 0$ has a root equal to a

15. $10x^6 \div x^{1/2} = 10x^3$

16. In the expansion of $(1 + x)^6$ the coefficient of x is 6

17. Values of A and B can be found such that

   $3x^2 - 2x + 7 \equiv (x - A)^2 + B$

MISCELLANEOUS EXERCISE A

1. Find the value of x when   (a) $3^x = 3\sqrt{3}$ (b) $5^x = 125\sqrt{5}$

2. Show that the roots of the equation $x^2 + kx - 2 = 0$ are real and different for all values of k

3. Express $3x^2 + 6x - 4$ in the form $a[(x + b)^2 + c]$

4. Find the coefficient of x^3 in the expansion of $(3 - 2x)^4$

5. Find the value of a for which $x - 1$ is a factor of $2x^2 - 3x + a$

6. Find the value of k for which the equation $x^2 - 9x + k$ has equal roots.

7. Find the values of p and q for which
$$x^2 - 4x + p \equiv (x - q)^2 + 4$$

8. Given that k is a real constant such that $0 < k < 1$, show that the roots of the equation
$$kx^2 + 2x + (1 - k) = 0$$
are (a) always real
     (b) always negative.   (U of L)

9. Find the values of p and q for which
$$2x^2 + px + 3 = 2[(x - 1)^2 + q]$$
Hence show that there are no real values of x for which
$$2[(x - 1)^2 + q] = 0$$

10. Find the values of a and b for which
$$(2x - a)^3 = 8x^3 + bx^2 + 6a^2x - 27$$

11. Find the value of c for which $x = 2$ is a root of the equation $3x^2 - 4x + c = 0$

12. Find both solutions of the equation $x + 1 = \dfrac{20}{x}$ (O/C, SU & C)

13. Find all the solutions that satisfy both the equations $x^2 + y^2 = 9$ and $x - 2y + 3 = 0$

14. Determine the range of values of k for which the quadratic equation $x^2 + 2kx + 4k - 3 = 0$ has two real and distinct roots.   (WJEC)

CHAPTER 5

STRAIGHT LINE GEOMETRY

PROOF

This chapter contains some geometric facts and definitions that will be needed later in this course.

Up to now, many rules have been based on investigating a few particular cases. For example the reader may have accepted that the sum of the interior angles in any triangle is 180°, only because the measured angles in some specific triangles had this property. This fact may be reinforced by our not being able to find a triangle whose angles have a different sum but that does not rule out the possibility that such a triangle exists.

It is no longer satisfactory to assume that a fact is *always* true without *proving that it is*, because as a mathematics course progresses, one fact is often used to produce another. Hence it is very important to distinguish between a 'fact' that is assumed from a few particular cases and one that has been *proved* to be true, as results deduced from an assumption cannot be reliable.

A proof deals with a general case, e.g. a triangle in which the sides and angles are not specified. The formal statement of a proved result is called a *theorem*.

PROOF THAT THE ANGLES OF A TRIANGLE ADD UP TO 180°

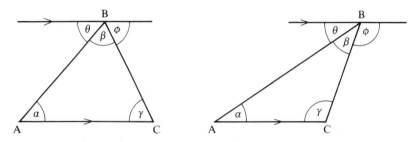

Let ABC be any triangle.

Draw a straight line through B parallel to AC.

Using the notation on the diagram, and the fact that alternate angles are equal, we have

$$\alpha = \theta \quad \text{and} \quad \gamma = \phi$$

At B, $\theta + \beta + \phi = 180°$ (angles on a straight line)

∴ $\alpha + \beta + \gamma = 180°$

We have proved the general case and hence can now be certain that, for *all* triangles *the angles of a triangle add up to 180°*

Proof that in any triangle, an exterior angle is equal to the sum of the two interior opposite angles

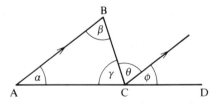

Let ABC be any triangle, with AC produced to D.

Draw a line through C parallel to AB.

Using the notation on the diagram,

        $\theta = \beta$ (alternate $\angle$s)   and $\phi = \alpha$ (corresponding $\angle$s)

∴ $\theta + \phi = \beta + \alpha$

i.e. $\angle BCD = \angle BAC + \angle CBA$

DEFINITIONS

Division of a Line in a Given Ratio

A point P is said to divide a line AB *internally* if P is between A and B.

Further, if P divides AB internally in the ratio $p:q$, then

$$AP:PB = p:q$$

If a point P is on AB produced, or on BA produced, then P is said to divide a line AB *externally*.

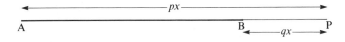

Further, if P divides AB externally in the ratio $p:q$ then
$AP:PB = p:q$

Examples 5a _____

1. A line AB of length 15 cm is divided internally by P in the ratio $2:3$. Find the length of PB.

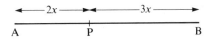

As P divides AB internally in the ratio $2:3$, P is between A and B and is nearer to A than to B.

AB is divided into 5 portions of which PB is 3 portions. If one portion is x cm, then

$$AB = 5x = 15$$

$\Rightarrow \qquad\qquad\qquad x = 3$

$\therefore \qquad\qquad PB = 3x = 9$

i.e. PB is 9 cm long.

_____

2. A line AB of length 15 cm is divided externally by P in the ratio 2:3. Find the length of PB.

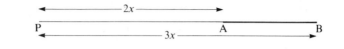

As AP:PB = 2:3, P is nearer to A than to B, so P is on BA produced.

$$AB = 3x - 2x = 15$$

$$\Rightarrow \qquad\qquad\qquad x = 15$$

$$\therefore \qquad\qquad PB = 3x = 45$$

i.e. PB is 45 cm long.

EXERCISE 5a

1. A line AB of length 12 cm is divided internally by P in the ratio 1:5. Find the length of AP.

2. A line AB of length 18 cm is divided externally by P in the ratio 5:6. Find the length of AP.

3. The point D divides a line PQ internally in the ratio 3:4. If PQ is of length 35 cm, find the length of DQ.

4. A line LM is divided externally by a point T in the ratio 5:7 Find the length of MT if   (a) LM = 24 cm   (b) LM = 2x cm.

5. P is a point on a line AB of length 12 cm and AP = 5 cm. Find the ratio in which P divides AB.

6. AB is a line of length 16 cm and M is a point on AB produced such that AM = 24 cm. Find the ratio in which M divides AB.

7. AB is a line of length x units and P is a point on AB such that AP is of length y units. Find, in terms of x and y, the ratio in which P divides AB.

8. PQ is a line of length a units. It is divided externally in the ratio $n:m$ by a point L, where $n > m$. Find, in terms of a, n and m, the length of QL.

9. ST is a line of length a units and L is a point on ST produced such that TL is of length b units. Find, in terms of a and b, the ratio in which L divides ST.

THE INTERCEPT THEOREM

> A straight line drawn parallel to one side of a triangle divides the
> other two sides in the same ratio.

i.e.

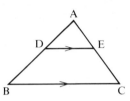

if DE is parallel to BC then AD/DB = AE/EC

Proof

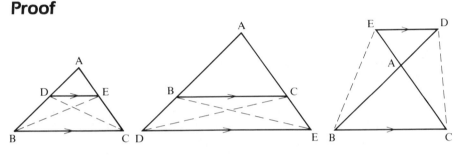

Triangle ABC is any triangle and DE is parallel to BC.

Three different positions for DE must be considered, as shown in the
diagrams. The proof that follows applies to all three cases. This proof
uses the fact that the area of a triangle can be found by multiplying
half the base by the height.

Area $\triangle$ DEC = area $\triangle$ DEB

($\triangle$s have same base DE and equal heights)

∴   area $\triangle$ AED : area $\triangle$ DEB = area $\triangle$ AED : area $\triangle$ DEC

Now area $\triangle$ AED : area $\triangle$ DEB = AD : DB

($\triangle$s have the same heights)

and   area $\triangle$ AED : area $\triangle$ DEC = AE : EC

($\triangle$s have the same heights)

Therefore AD : DB = AE : EC

The converse of this theorem is also true, i.e. if a line divides two sides
of a triangle in the same ratio, then it is parallel to the third side of the
triangle.

PYTHAGORAS' THEOREM AND ITS CONVERSE

Pythagoras' theorem is familiar and very useful. Here is a reminder.

> In any right-angled triangle, the square on the hypotenuse is equal to the sum of the squares on the other two sides.

i.e.

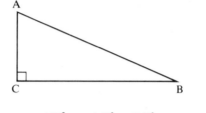

$$AB^2 = AC^2 + BC^2$$

The converse of this theorem is less well known, but equally useful. It states that

> if the square on one side of a triangle is equal to the sum of the squares on the other two sides, then the angle opposite the first side is a right-angle.

EXERCISE 5b

1. In triangle ABC, DE is parallel to BC.
If AD = 2 cm, DB = 3 cm and AE = 2.5 cm, find EC.

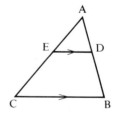

2. In triangle LMN, ST is parallel to MN.
If LM = 9 cm, LS = 4 cm and LT = 2 cm, find
(a) TN   (b) LN.

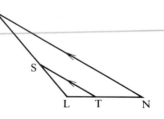

3. In the diagram, ED is parallel
to AB.
AC = 3 cm, CE = 1.5 cm
and CD = 2 cm. Find BC.

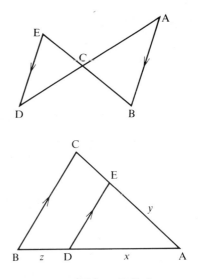

4. Using the measurements given in
the diagram find the length of
CE in terms of x, y and z

5. ABC is a triangle and a line PQ is drawn parallel to BC, but on
the opposite side of A from BC. PQ cuts BA produced at P and
cuts CA produced at Q.
AC = 2.5 cm, AQ = 1 cm and AP = 0.5 cm. Find
(a) the length of AB   (b) the ratio in which P divides AB.

6. Determine whether or not a triangle is right-angled if the lengths of
its sides are
(a) 3, 5, 4   (b) 2, 1, $\sqrt{3}$ (c) 2, 2, $\sqrt{8}$
(d) $5x$, $12x$, $13x$ (e) 7, 5, 12   (f) $\sqrt{2}$, $\sqrt{3}$, 1

SIMILAR TRIANGLES

If one triangle is an enlargement of another triangle, then the two
triangles are *similar*.

This means that the three angles of one triangle are equal to the three
angles of the other triangle *and* that the corresponding sides of the two
triangles are in the same ratio.

However, to prove that triangles are similar it is necessary only to
show that *one* of these conditions is satisfied because the other one
follows, i.e.

if two triangles contain the same angles then their corresponding
sides are in the same ratio.

Proof

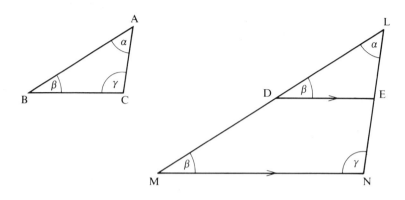

The angles of triangle ABC are equal to the angles of triangle LMN.

It follows that we can find a point D on LM such that LD = AB and a point E on LN such that LE = AC.

Joining DE, it is clear that △LDE and △ABC are identical, i.e. congruent.

∴ ∠ABC = ∠LDE

These are corresponding angles with respect to DE and MN.

∴ DE is parallel to MN

The intercept theorem then tells us that DE divides LM and LN in the same ratio, i.e.

$$LD:LM = LE:LN$$

⇒ $$AB:LM = AC:LN$$

The converse of this theorem is also true, i.e.

if two triangles are such that their corresponding sides are in the same ratio, then corresponding pairs of angles are equal.

It is left to the reader to prove this in Question 11 in the next exercise.

SIMILAR FIGURES

Two figures are similar if one figure is an enlargement of the other. This means that their corresponding sides are in the same ratio and that the angles in one figure are equal to the corresponding angles in the other figure.

To show that figures other than triangles are similar, both the side and the angle property have to be proved. In the case of triangles, we have seen that it is necessary only to show that one of these conditions is satisfied to prove the triangles similar, because the other condition follows, i.e.

> two triangles are similar if we can show
> either that the angles of the triangles are equal
> or that the corresponding sides of the triangles are in the same ratio

THE ANGLE BISECTOR THEOREM

Another useful fact concerning triangles and ratios is

> the line bisecting an angle of a triangle divides the side opposite to that angle in the ratio of the sides containing the angle.

e.g. if AD bisects $\angle A$, then $BD : DC = AB : AC$

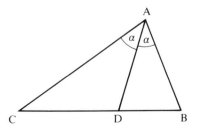

Proof

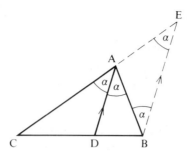

In △ABC, AD bisects the angle at A.

Drawing BE parallel to DA to cut CA produced at E, we have

$$\angle BEA = \angle DAC \quad \text{(corresponding angles)}$$

and $$\angle EBA = \angle BAD \quad \text{(alternate angles)}$$

∴ △BEA is isosceles   ⇒   EA = AB

In △BCE the intercept theorem gives

$$BD:DC = EA:AC$$

⇒ $$BD:DC = AB:AC$$

ALTITUDES AND MEDIANS

We end this chapter with a couple of definitions.
The line drawn from a vertex of a triangle, perpendicular to the opposite side, is called an *altitude*, for example

AD is the altitude through A,
and   BE is the altitude through B.

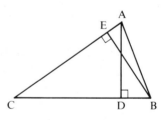

A *median* of a triangle is the line joining a vertex to the midpoint of the opposite side, for example

XP is the median through X,
and   YQ is the median through Y.

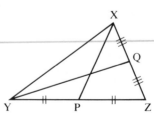

Example 5c _____

In △ABC, AB = 4 cm, AC = 3 cm and ∠A = 90°. The bisector of ∠A cuts BC at D. Find the length of BD.

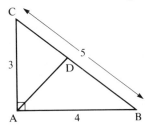

From Pythagoras' theorem, BC = 5 cm.

From the angle bisector theorem, BD : DC = AB : AC = 4 : 3

∴   BD : BC = 4 : 7

⇒   BD = $\frac{4}{7}$ × 5

                                                = $2\frac{6}{7}$

∴ BD is $2\frac{6}{7}$ cm long.

_____

EXERCISE 5c

1. XYZ is a triangle with a right angle at X, XW is the altitude from X. Show that triangles XYZ, XWZ and XYW are all similar.

2. In △ABC, AB = 6 cm, AC = 5 cm and BC = 7 cm. BD is the median from B to AC and BE is the bisector of ∠B to AC. Find the length of DE.

3. Triangles PQR and XYZ are such that ∠P = ∠X and ∠Q = ∠Z. XY = 3 cm, YZ = 4 cm, PQ = 7 cm and PR = 12 cm. Find the lengths of XZ and QR.

4.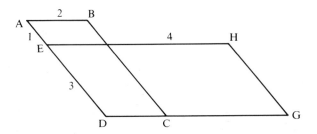

ABCD and EDGH are parallelograms. Prove that they are similar.

5. PQR is a triangle in which PQ = 4 cm, PR = 3 cm and QR = 6 cm. T is a point outside the triangle on the side of QR, and ∠RQT = ∠PRQ and ∠QRT = ∠QPR. Find the lengths of QT and RT.

6. ABC is any triangle and equilateral triangles ABD and ACE are drawn on the sides AB and AC respectively. The bisector of ∠BAC meets BC at F such that BF:FC = 3:2. Find the ratio of the areas of the two equilateral triangles.

7. D and E are two points on the side BC of △ABC such that AD is the bisector of ∠BAC and AE is an altitude of the triangle. If ∠ACB = 40° and ∠ABC = 60° find ∠DAE.

8. In triangle ABC, AB = 24 cm, BC = 7 cm and AC = 25 cm. Show that △ABC is right-angled. The bisector of ∠BAC meets BC at D. Find the lengths of BD, DC and AD.

9. AB and CD are two lines that intersect at E. AC and DB are parallel. Show that triangles ACE and EDB are similar.

10. Triangle ABC has a right-angle at B and BE is an altitude of the triangle. BC = 5 cm and BE = 4 cm. Calculate the length of EC and of AC.

11. (Proof that if the corresponding sides of two triangles are in the same ratio, then the triangles contain the same angles.)

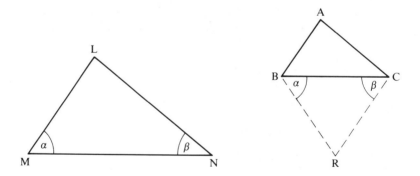

Triangles LMN and ABC are such that
LM:AB = MN:BC = LN:AC.
R is a point such that ∠CBR = ∠LMN and ∠BCR = ∠LNM.
Prove that LM:BR = LM:AB and hence that BR = AB.
Similarly show that CR = AC. *Hence* show that △LMN and △ABC have equal angles.

CHAPTER 6

COORDINATE GEOMETRY

LOCATION OF A POINT IN A PLANE

Graphical methods lend themselves particularly well to the investigation of the geometrical properties of many curves and surfaces. At this stage we will restrict ourselves to rectilinear plane figures (i.e. two dimensional figures bounded by straight lines). To begin, we need a simple and unambiguous way of describing the position of a point on a graph.

Consider the problem of describing the location of a city, London say.

There are many ways in which this can be done, but they all require reference to at least one known place and known directions. This is called a *system* or *frame of reference*. Within this frame of reference, two measurements are needed to locate the city precisely. These measurements are called coordinates.

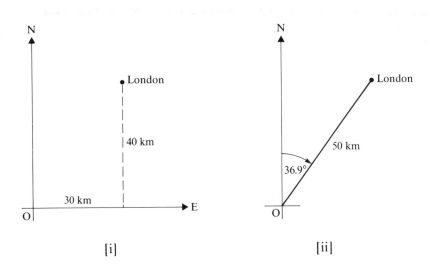

[i]   [ii]

The position of London is described in two alternative ways in the diagrams above.

In [i] the frame of reference comprises a fixed point O and the directions due east and due north from O. The coordinates of London are 30 km east of O and 40 km north of O.

In [ii] the frame of reference comprises a fixed point O and the direction due north from O. The coordinates of London are 50 km from O and a bearing of 036.9°

The system used at this level for graphical work is based on the first of the two practical systems described above.

CARTESIAN COORDINATES

This system of reference uses a fixed point O, called *the origin*, and a pair of perpendicular lines through O. One of these lines is drawn horizontally and is called the *x*-axis. The other line is drawn vertically and is called the *y*-axis.

The coordinates of a point P are the directed distances of P from O parallel to the axes.

A positive coordinate is a distance measured in the positive direction of the axis and a negative coordinate is a distance in the opposite direction.

The coordinates are given as an ordered pair (a, b) with the x-*coordinate* or *abscissa* first and the y-*coordinate* or *ordinate* second.

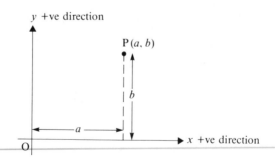

The diagram opposite represents the points whose Cartesian coordinates are $(7, 4)$ and $(-3, -1)$

These points are referred to in future simply as the points $(7, 4)$ and $(-3, -1)$

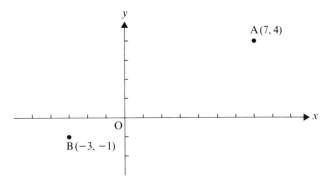

EXERCISE 6a

1. Represent on a diagram the points whose coordinates are
   (a) $(1, 6)$ (b) $(0, 5)$ (c) $(-4, 0)$ (d) $(-3, -2)$ (e) $(3, -4)$

2. Two adjacent corners of a square are the points $(3, 5)$ and $(3, -1)$. What could the coordinates of the other two corners be?

3. The two opposite corners of a square are $(-2, -3)$ and $(3, 2)$. Write down the coordinates of the other two corners.

COORDINATE GEOMETRY

Coordinate geometry is the name given to the graphical analysis of geometric properties. For this analysis we need to refer to three types of points:

1) fixed points whose coordinates are known, e.g. the point $(1, 2)$

2) fixed points whose coordinates are not known numerically. These points are referred to as (x_1, y_1), (x_2, y_2), ... etc. or (a, b), etc.

3) points which are not fixed. We call these general points and we refer to them as (x, y), (X, Y), etc.

It is conventional to use the letters A, B, C, ... for fixed points and the letters P, Q, R, ... for general points.

It is also conventional to graduate the axes using identical scales. This avoids distorting the shape of figures.

THE LENGTH OF A LINE JOINING TWO POINTS

Consider the line joining the points A(1, 2) and B(3, 4)

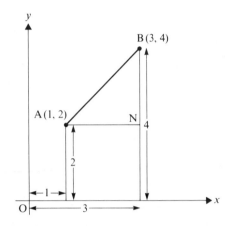

The length of the line joining A and B can be found by using Pythagoras' theorem, i.e.

$$AB^2 = AN^2 + BN^2$$
$$= (3 - 1)^2 + (4 - 2)^2$$
$$= 8$$

Therefore $AB = \sqrt{8} = 2\sqrt{2}$

In the same way the length of the line joining any two points $A(x_1, y_1)$ and $B(x_2, y_2)$ can be found.

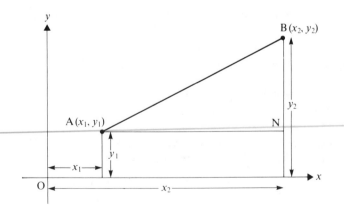

From the diagram, $AB^2 = AN^2 + BN^2$

$$= (x_2 - x_1)^2 + (y_2 - y_1)^2$$

$\Rightarrow$ $\qquad AB = \sqrt{[(x_2 - x_1)^2 + (y_2 - y_1)^2]}$

i.e.   the length of the line joining $A(x_1, y_1)$ to $B(x_2, y_2)$ is given by

$$AB = \sqrt{[(x_2 - x_1)^2 + (y_2 - y_1)^2]}$$

This formula still holds when some, or all, of the coordinates are negative. This is illustrated in the next worked example.

Examples 6b

1. Find the length of the line joining $A(-2, 2)$ to $B(3, -1)$

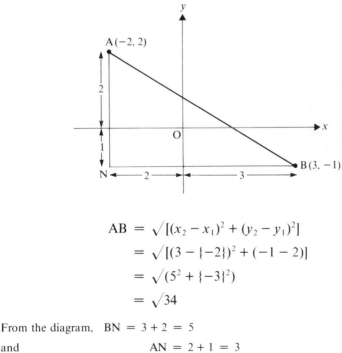

$$AB = \sqrt{[(x_2 - x_1)^2 + (y_2 - y_1)^2]}$$
$$= \sqrt{[(3 - \{-2\})^2 + (-1 - 2)]}$$
$$= \sqrt{(5^2 + \{-3\}^2)}$$
$$= \sqrt{34}$$

From the diagram, $BN = 3 + 2 = 5$

and $\qquad\qquad\qquad\qquad AN = 2 + 1 = 3$

$\Rightarrow$ $\qquad\qquad\qquad\qquad AB^2 = 5^2 + 3^2 = 34$

$\Rightarrow$ $\qquad\qquad\qquad\qquad AB = \sqrt{34}$

This confirms that the formula used above is valid when some of the coordinates are negative.

THE MIDPOINT OF THE LINE JOINING TWO GIVEN POINTS

Consider the line joining the points A(1, 1) and B(5, 3)

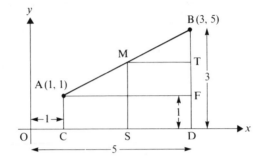

Using the intercept theorem, we see that if M is the midpoint of AB then S is the midpoint of CD.

Therefore the x-coordinate of M is given by OS, where

$$OS = OC + \tfrac{1}{2}CD = 1 + \tfrac{1}{2}(5 - 1) = 3$$

Similarly, T is the midpoint of BF, so the y-coordinate of M is given by SM (= DT), where

$$DT = DF + \tfrac{1}{2}FB = 1 + \tfrac{1}{2}(3 - 1) = 2$$

Therefore M is the point (3, 2)

In general, if $A(x_1, y_1)$ and $B(x_2, y_2)$ are two points, then the coordinates of M, the midpoint of AB, can be found in the same way.

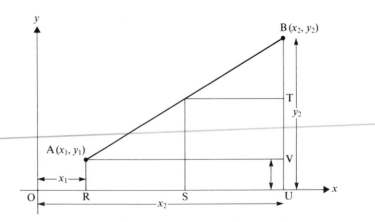

At M, $\qquad x = OS = OR + \frac{1}{2}RU$

$$= x_1 + \tfrac{1}{2}(x_2 - x_1) = \tfrac{1}{2}(x_1 + x_2)$$

and $\qquad y = SM = UT = UV + \frac{1}{2}BV$

$$= y_1 + \tfrac{1}{2}(y_2 - y_1) = \tfrac{1}{2}(y_1 + y_2)$$

Note that the coordinates of M are the average of the coordinates of A and B. Hence

> the coordinates of the midpoint of the line joining $A(x_1, y_1)$ and $B(x_2, y_2)$ are $[\tfrac{1}{2}(x_1 + x_2), \tfrac{1}{2}(y_1 + y_2)]$

The next worked example shows that this formula holds when some of the coordinates are negative.

Examples 6b (continued) _____

2. Find the coordinates of the midpoint of the line joining $A(-3, -2)$ and $B(1, 3)$.

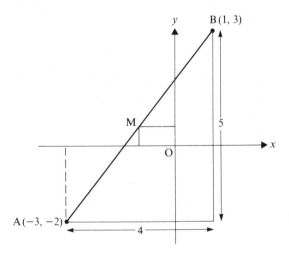

The coordinates of M are $[\tfrac{1}{2}(x_1 + x_2), \tfrac{1}{2}(y_1 + y_2)]$

$$= [\tfrac{1}{2}(-3 + 1), \tfrac{1}{2}(-2 + 3)] = (-1, \tfrac{1}{2})$$

Alternatively, from the diagram, M is half-way from A to B horizontally and vertically, i.e.

at M $\qquad x = -3 + \frac{1}{2}(4) = -1 \quad$ and $\quad y = -2 + \frac{1}{2}(5) = \frac{1}{2}$

This confirms that the formula works when some of the coordinates are negative.

EXERCISE 6b

1. Find the length of the line joining
   (a) $A(1, 2)$ and $B(4, 6)$ (b) $C(3, 1)$ and $D(2, 0)$
   (c) $J(4, 2)$ and $K(2, 5)$

2. Find the coordinates of the midpoints of the lines joining the points in Question 1.

3. Find (i) the length, (ii) the coordinates of the midpoint of the line, joining
   (a) $A(-1, -4)$, $B(2, 6)$ (b) $S(0, 0)$, $T(-1, -2)$
   (c) $E(-1, -4)$, $F(-3, -2)$

4. Find the distance from the origin to the point $(7, 4)$

5. Find the length of the line joining the point $(-3, 2)$ to the origin.

6. Find the coordinates of the midpoint of the line from the point $(4, -8)$ to the origin.

7. Show, by using Pythagoras' Theorem, that the lines joining $A(1, 6)$, $B(-1, 4)$ and $C(2, 1)$ form a right-angled triangle.

8. A, B and C are the points $(7, 3)$, $(-4, 1)$ and $(-3, -2)$ respectively.
   (a) Show that $\triangle ABC$ is isosceles.
   (b) Find the midpoint of BC.
   (c) Find the area of $\triangle ABC$.

9. The vertices of a triangle are $A(0, 2)$, $B(1, 5)$ and $C(-1, 4)$
   Find
   (a) the perimeter of the triangle
   (b) the coordinates of D such that AD is a median of $\triangle ABC$
   (c) the length of AD.

10. Show that the lines OA and OB are perpendicular where A and B are the points $(4, 3)$ and $(3, -4)$ respectively.

11. M is the midpoint of the line joining A to B. The coordinates of A and M are $(5, 7)$ and $(0, 2)$ respectively. Find the coordinates of B.

GRADIENT

The gradient of a straight line is a measure of its slope with respect to the x-axis. Gradient is defined as

the increase in y divided by the increase in x between one point and another point on the line.

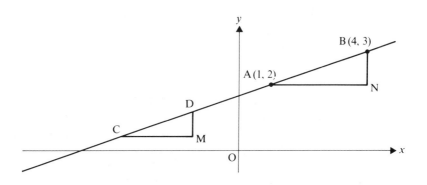

Consider the line passing through A$(1, 2)$ and B$(4, 3)$.

From A to B,   the increase in y is 1
                     the increase in x is 3

Therefore the gradient of AB is $\frac{1}{3}$.

Now NB measures the increase in the y-coordinate and AN measures the increase in the x-coordinate, so the gradient can be written as $\dfrac{\text{NB}}{\text{AN}}$.

If C and D are any other two points on the line then $\triangle$ABN and $\triangle$CDM are similar, so

$$\frac{\text{NB}}{\text{AN}} = \frac{\text{MD}}{\text{CM}} = \frac{1}{3}$$

i.e.

the gradient of a line may be found from *any* two points on the line.

Now consider the line through the points A(2, 3) and B(6, 1)

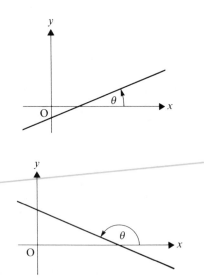

Moving from A to B

$$\frac{\text{increase in } y}{\text{increase in } x} = \frac{-2}{4} = -\frac{1}{2}$$

Alternatively, moving from B to A

$$\frac{\text{increase in } y}{\text{increase in } x} = \frac{2}{-4} = -\frac{1}{2}$$

This shows that it does not matter in which order the two points are considered, provided that they are considered in the *same* order when calculating the increases in x and in y.

From these two examples we see that the gradient of a line may be positive or negative.

A positive gradient indicates an 'uphill' slope with respect to the positive direction of the x-axis, i.e. the line makes an acute angle with the positive sense of the x-axis.

A negative gradient indicates a 'downhill' slope with respect to the positive direction of the x-axis, i.e. the line makes an obtuse angle with the positive sense of the x-axis.

In general,

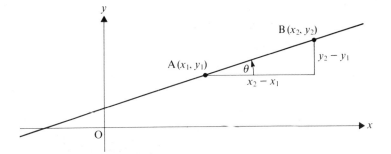

the gradient of the line passing through $A(x_1, y_1)$ and $B(x_2, y_2)$ is

$$\frac{\text{the increase in } y}{\text{the increase in } x} = \frac{y_2 - y_1}{x_2 - x_1}$$

As the gradient of a straight line is the increase in y divided by the increase in x from one point on the line to another,

gradient measures the increase in y per unit increase in x, i.e. the rate of increase of y with respect to x.

PARALLEL LINES

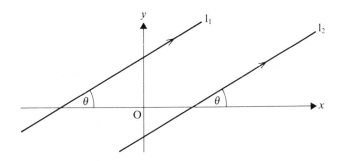

If l_1 and l_2 are parallel lines, they are equally inclined to the positive direction of the x-axis, i.e.

parallel lines have equal gradients.

PERPENDICULAR LINES

Consider the perpendicular lines AB and CD whose gradients are m_1 and m_2 respectively.

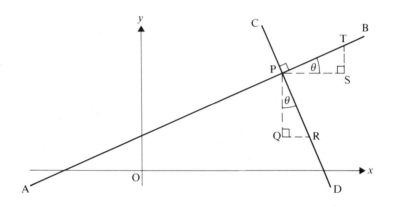

If AB makes an angle θ with the x-axis then CD makes an angle θ with the y-axis. Therefore triangles PQR and PST are similar.

Now the gradient of AB is $\dfrac{ST}{PS} = m_1$

and the gradient of CD is $\dfrac{-PQ}{QR} = m_2$,   i.e. $\dfrac{PQ}{QR} = -m_2$

But $\dfrac{ST}{PS} = \dfrac{QR}{PQ}$ ($\triangle$s PQR and PST are similar)

therefore $m_1 = -\dfrac{1}{m_2}$ or $m_1 m_2 = -1$

i.e.

the product of the gradients of perpendicular lines is -1, or, if one line has gradient m, any line perpendicular to it has gradient $-\dfrac{1}{m}$

Example 6c _____

Determine, by comparing gradients, whether the following three points are collinear (i.e. lie on the same straight line).

$$A(\tfrac{2}{3}, 1), \ B(1, \tfrac{1}{2}), \ C(4, -4)$$

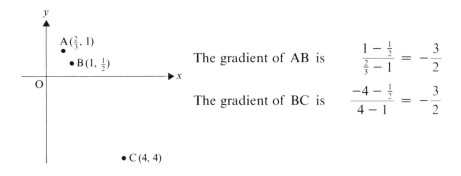

The gradient of AB is $\dfrac{1 - \tfrac{1}{2}}{\tfrac{2}{3} - 1} = -\dfrac{3}{2}$

The gradient of BC is $\dfrac{-4 - \tfrac{1}{2}}{4 - 1} = -\dfrac{3}{2}$

As the gradients of AB and BC are the same, A, B and C are collinear.

The diagram, although not strictly necessary, gives a check that the answer is reasonable.

_____

EXERCISE 6c

1. Find the gradient of the line through the pair of points.
   (a) $(0, 0), \ (1, 3)$ (b) $(1, 4), \ (3, 7)$ (c) $(5, 4), \ (2, 3)$
   (d) $(-1, 4), \ (3, 7)$ (e) $(-1, -3), \ (-2, 1)$ (f) $(-1, -6), \ (0, 0)$
   (g) $(-2, 5), \ (1, -2)$ (h) $(3, -2), \ (-1, 4)$ (i) $(h, k), \ (0, 0)$

2. Determine whether the given points are collinear.
   (a) $(0, -1), \ (1, 1), \ (2, 3)$ (b) $(0, 2), \ (2, 5), \ (3, 7)$
   (c) $(-1, 4), \ (2, 1), \ (-2, 5)$ (d) $(0, -3), \ (1, -4), \ (-\tfrac{1}{2}, -\tfrac{5}{2})$

3. Determine whether AB and CD are parallel, perpendicular or neither.
   (a) $A(0, -1), \ B(1, 1), \ C(1, 5), \ D(-1, 1)$
   (b) $A(1, 1), \ B(3, 2), \ C(-1, 1), \ D(0, -1)$
   (c) $A(3, 3), \ B(-3, 1), \ C(-1, -1), \ D(1, -7)$
   (d) $A(2, -5), \ B(0, 1), \ C(-2, 2), \ D(3, -7)$
   (e) $A(2, 6), \ B(-1, -9), \ C(2, 11), \ D(0, 1)$

PROBLEMS IN COORDINATE GEOMETRY

This chapter ends with a miscellaneous selection of problems on coordinate geometry. A clear and reasonably accurate diagram showing all the given information will often suggest the most direct method for solving a particular problem.

Example 6d _____

The vertices of a triangle are the points A(2, 4), B(1, −2) and C(−2, 3) respectively. The point H(a, b) lies on the altitude through A. Find a relationship between a and b

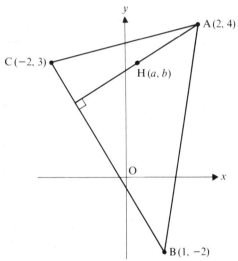

As H is on the altitude through A,   AH is perpendicular to BC.

The gradient of AH is $\dfrac{4 - b}{2 - a}$,

the gradient of BC is $\dfrac{3 - (-2)}{-2 - 1} = -\dfrac{5}{3}$

The product of the gradients of perpendicular lines is −1

Therefore $\left[\dfrac{4 - b}{2 - a}\right]\left[-\dfrac{5}{3}\right] = -1$

$\Rightarrow$ $\dfrac{-20 + 5b}{6 - 3a} = -1$

$\Rightarrow$ $5b = 3a + 14$

EXERCISE 6d

1. $A(1, 3)$, $B(5, 7)$, $C(4, 8)$, $D(a, b)$ form a rectangle ABCD. Find a and b

2. The triangle ABC has its vertices at the points $A(1, 5)$, $B(4, -1)$ and $C(-2, -4)$
   (a) Show that $\triangle ABC$ is right-angled.
   (b) Find the area of $\triangle ABC$.

3. Show that the point $(-\frac{32}{3}, 0)$ is on the altitude through A of the triangle whose vertices are $A(1, 5)$, $B(1, -2)$ and $C(-2, 5)$

4. Show that the triangle whose vertices are $(1, 1)$, $(3, 2)$, $(2, -1)$ is isosceles.

5. Find, in terms of a and b, the length of the line joining (a, b) and $(2a, 3b)$

6. The point $(1, 1)$ is the centre of a circle whose radius is 2. Show that the point $(1, 3)$ is on the circumference of this circle.

7. A circle, radius 2 and centre the origin, cuts the x-axis at A and B and cuts the positive y-axis at C. *Prove* that $\angle ACB = 90°$

8. Find in terms of p and q, the coordinates of the midpoint of the line joining $C(p, q)$ and $D(q, p)$. Hence show that the origin is on the perpendicular bisector of the line CD.

9. The point (a, b) is on the circumference of the circle of radius 3 whose centre is at the point $(2, 1)$. Find a relationship between a and b

10. ABCD is a quadrilateral where A, B, C and D are the points $(3, -1)$, $(6, 0)$, $(7, 3)$ and $(4, 2)$. Prove that the diagonals bisect each other at right angles and hence find the area of ABCD.

11. The vertices of a triangle are at the points $A(a, 0)$, $B(0, b)$ and $C(c, d)$ and $\angle B = 90°$. Find a relationship between a, b, c and d

12. A point $P(a, b)$ is equidistant from the y-axis and from the point $(4, 0)$. Find a relationship between a and b

CHAPTER 7

OBTUSE ANGLES, SINE AND COSINE FORMULAE

TRIGONOMETRIC RATIOS OF ACUTE ANGLES

The sine, cosine and tangent of an acute angle in a right-angled triangle are defined in terms of the sides of the triangle as follows

$$\cos A = \frac{\text{adjacent}}{\text{hypotenuse}}$$

$$\sin A = \frac{\text{opposite}}{\text{hypotenuse}}$$

$$\tan A = \frac{\text{opposite}}{\text{adjacent}}$$

If any of these trig ratios is given as a fraction, the lengths of two of the sides of the right-angled triangle can be marked. Then the third side can be calculated by using Pythagoras' theorem.

Example 7a _____

Given that $\sin A = \frac{3}{5}$ find $\cos A$ and $\tan A$

Because $\sin A = \dfrac{\text{opp}}{\text{hyp}}$, we can draw a right-angled triangle with the side opposite to angle A of length 3 units and a hypotenuse of length 5 units.

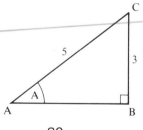

Applying Pythagoras' theorem to $\triangle ABC$ gives

$$(AB)^2 + (BC)^2 = (AC)^2$$

i.e. $(AB)^2 + 3^2 = 5^2$

$\Rightarrow$ $(AB)^2 = 25 - 9 = 16$

$\Rightarrow$ $AB = 4$

Then $\cos A = \dfrac{\text{adj}}{\text{hyp}} = \dfrac{4}{5}$

and $\tan A = \dfrac{\text{opp}}{\text{adj}} = \dfrac{3}{4}$

EXERCISE 7a

If any of the square roots in this exercise are not integers, leave them in surd form.

1. If $\tan A = \frac{12}{5}$ find $\sin A$ and $\cos A$

2. Given that $\cos X = \frac{4}{5}$ find $\tan X$ and $\sin X$

3. If $\sin P = \frac{40}{41}$ find $\cos P$ and $\tan P$

4. Tan $A = 1$ Find $\sin A$ and $\cos A$

5. If $\cos Y = \frac{2}{3}$ find $\sin Y$ and $\tan Y$

6. Given that $\sin A = \frac{1}{2}$ what is $\cos A$? Use your calculator to find the size of angle A

7. If $\sin X = \frac{2}{25}$ and $\tan X = \frac{7}{?}$ find $\cos X$

8. If $\sin X = \frac{3}{5}$ find $\cos X$ and hence calculate $\cos^2 X + \sin^2 X$

9. Repeat Question 8 with $\sin X = \frac{1}{2}$

10. If $\sin X = \frac{p}{q}$ find $\cos X$ in terms of p and q
    Hence find the value of $\sin^2 X + \cos^2 X$

TRIGONOMETRIC RATIOS OF OBTUSE ANGLES

The Cosine of an Obtuse Angle

Clearly the definition in the preceeding paragraph is restricted to acute angles, so if we want to work with the cosine of an obtuse angle we need a broader definition.

First however we will examine the values given by a calculator for the cosines of angles from 0 to 180°

θ	0	30°	45°	60°	90°	120°	135°	150°	180°
$\cos \theta$ (to 2 d.p.)	1	0.87	0.71	0.50	0	−0.50	−0.71	−0.87	−1

Plotting these figures on graph paper gives a shape which is called a cosine curve. Note that θ, the symbol used for the angle, is the most commonly used symbol for a varying angle.

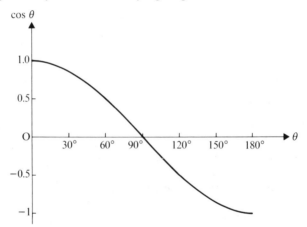

From the graph or the table it can be seen that

$$\cos 60° = 0.5$$

and $\qquad \cos 120° = -0.5$

i.e. $\qquad \cos 120° = -\cos 60° \qquad (120° + 60° = 180°)$

also $\qquad \cos 45° = 0.71$

and $\qquad \cos 135° = -0.71$

i.e. $\qquad \cos 135° = -\cos 45° \qquad (135° + 45° = 180°)$

The reader can find many more pairs of angles where the relationship is

$$\cos \theta = -\cos (180° - \theta)$$

The Sine of an Obtuse Angle

If we start by listing, and plotting, the values given by a calculator for the sines of angles from 0 to 180°, we have

θ	0	30°	45°	60°	90°	120°	135°	150°	180°
sin θ (to 2 d.p.)	0	0.5	0.71	0.87	1	0.87	0.71	0.5	0

and

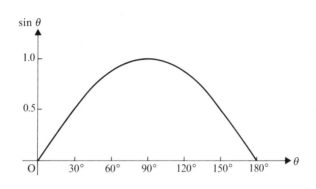

This graph is called a sine curve; notice that it looks symmetrical about a vertical line through 90°

Again relationships can be observed between the sines of pairs of angles, for example

$$\sin 30° = 0.5$$

and $\qquad \sin 150° = 0.5$

i.e. $\qquad \sin 150° = \sin 30° \qquad (150° + 30° = 180°)$

also $\qquad \sin 60° = 0.87$

and $\qquad \sin 120° = 0.87$

i.e. $\qquad \sin 120° = \sin 60° \qquad (120° + 60° = 180°)$

This time we have

$$\sin \theta = \sin (180° - \theta)$$

The Tangent of an Obtuse Angle

Plotting the values of $\tan\theta$ given by a calculator for angles from 0 to 180° we have

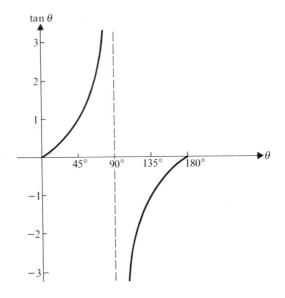

This time observation shows that

$$\tan\theta = -\tan(180° - \theta)$$

Collecting these results together we have

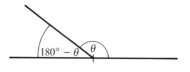

$$\sin\theta = \sin(180° - \theta)$$
$$\cos\theta = -\cos(180° - \theta)$$
$$\tan\theta = -\tan(180° - \theta)$$

Note that, so far, it just *looks as if* these results are true and, at this level of study, we should look for a more general explanation of these relationships. This is given in Chapter 17 where the trigonometric ratios of any angle are defined.

Examples 7b _____

1. If $\sin \theta = \frac{1}{5}$ find two possible values for θ

As given by a calculator, the angle with a sine of 0.2 is 11.5°

But $\sin \theta = \sin(180° - \theta)$ so $\sin 11.5° = \sin(180° - 11.5°)$
if $\sin \theta = \frac{1}{5}$, two values of θ are 11.5° and 168.5°

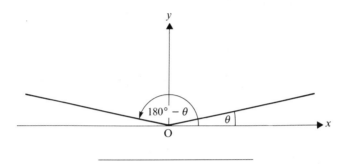

2. Use the information in the diagram to find $\cos \theta$ and $\tan \theta$

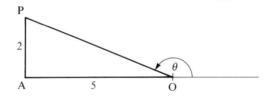

In $\triangle OPA$ $\qquad\qquad OP^2 = 4 + 25 = 29$ (Pythagoras)

and $\qquad\qquad\qquad A\widehat{O}P = (180° - \theta)$

$$\cos(180° - \theta) = \frac{OA}{OP} = \frac{5}{\sqrt{29}}$$

$$\cos \theta = -\cos(180° - \theta) = -\frac{5}{\sqrt{29}}$$

$$\tan \theta = -\tan(180° - \theta) = -\frac{AP}{OA} = -\frac{2}{5}$$

EXERCISE 7b

In each question from 1 to 4, find $\sin \theta$, $\cos \theta$ and $\tan \theta$, giving unknown lengths in surd form when necessary.

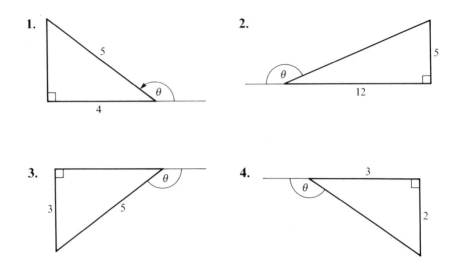

1.

2.

3.

4.

In each question from 5 to 16, find X where X is an angle from 0 to 180°

5. $\sin X = \sin 80°$ **6.** $\tan X = -\tan 120°$

7. $\cos X = -\cos 75°$ **8.** $\sin X = \sin 128°$

9. $\tan 45° = -\tan X$ **10.** $\cos 30° = -\cos X$

11. $\sin X = \sin 81°$ **12.** $-\cos 123 = \cos X$

13. $\sin 90° = \sin X$ **14.** $\tan X = -\tan 100°$

15. $\cos 91° = -\cos X$ **16.** $\cos 0 = -\cos X$

The unknown angles in questions 17 to 23 are in the range 0 to 180°

17. If $\sin A = \frac{3}{5}$ find two possible values of $\cos(180° - A)$

18. If $\cos X = -\frac{12}{13}$ find $\sin X$

19. If $\sin \theta = \frac{4}{5}$ find, to the nearest degree, two possible values of θ

20. Given that $\sin A = 0.5$ and $\cos A = -0.8660$, find $\angle A$

21. Find the angle P if $\cos P = -0.7071$ and $\sin P = 0.7071$

22. Is there an angle X for which
(a) $\cos X = 0$ and $\sin X = 1$
(b) $\sin X = 0$ and $\cos X = 1$
(c) $\cos X = 0$ and $\sin X = -1$
(d) $\tan X = 0$ and $\sin X = 0$?

23. If $\cos A = -\cos B$, what is the relationship between $\angle A$ and $\angle B$

24. Draw a diagram to show the angle T for which $\tan T = \frac{3}{4}$.
Draw on your diagram an angle with a tangent of $-\frac{3}{4}$.

25. Within the range $0 < \theta < 180°$, are there angles for which $\sin \theta$ and $\cos \theta$ are
(a) both positive   (b) both negative?

FINDING UNKNOWN SIDES AND ANGLES IN A TRIANGLE

Triangles are involved in many practical measurements (e.g. surveying) so it is important to be able to make calculations from limited data about a triangle.

Although a triangle has three sides and three angles, it is not necessary to know all of these in order to define a particular triangle. If enough information about a triangle is known, the remaining sides and angles can be calculated. This is called *solving* the triangle and it requires the use of one of a number of formulae.

The two relationships that are used most frequently are the sine rule and the cosine rule.

When working with a triangle ABC the side opposite to $\angle A$ is denoted by a, the side opposite to $\angle B$ by b and so on.

THE SINE RULE

In a triangle ABC, $\dfrac{a}{\sin A} = \dfrac{b}{\sin B} = \dfrac{c}{\sin C}$

This relationship can be proved quite simply and the proof is given here for students who are interested.

Proof

Consider a triangle ABC in which there is no right angle.

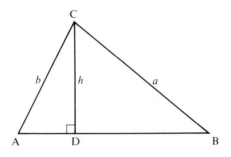

A line drawn from C, perpendicular to AB, divides triangle ABC into two right-angled triangles, CDA and CDB.

In $\triangle$CDA $\qquad \sin A = h/b \ \Rightarrow \ h = b \sin A$

In $\triangle$CDB $\qquad \sin B = h/a \ \Rightarrow \ h = a \sin B$

Therefore $\qquad\qquad a \sin B = b \sin A$

i.e. $\qquad\qquad\qquad \dfrac{a}{\sin A} = \dfrac{b}{\sin B}$

We could equally well have divided $\triangle$ABC into two right-angled triangles by drawing the perpendicular from A to BC (or from B to AC). This would have led to a similar result,

i.e. $\qquad\qquad\qquad \dfrac{b}{\sin B} = \dfrac{c}{\sin C}$

By combining the two results we produce the sine rule,

$$\frac{a}{\sin A} = \frac{b}{\sin B} = \frac{c}{\sin C}$$

Note that this proof is equally valid when $\triangle ABC$ contains an obtuse angle.

Suppose that $\angle A$ is obtuse.

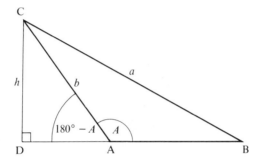

This time $h = b \sin(180° - A)$ but, as $\sin(180° - A) = \sin A$, we see that once again $h = b \sin A$

In all other respects the proof given above is unaltered, showing that the sine rule applies to any triangle.

Using the Sine Rule

$$\frac{a}{\sin A} = \frac{b}{\sin B} = \frac{c}{\sin C}$$

This rule is made up of three separate fractions, only two of which can be used at a time. We select the two which contain three known quantities and only one unknown.

Note that, when the sine rule is being used to find an unknown angle, it is more conveniently written in the form

$$\frac{\sin A}{a} = \frac{\sin B}{b} = \frac{\sin C}{c}$$

Examples 7c _____

1. In △ABC, BC = 5 cm, A = 43° and B = 61°. Find the length of AC.

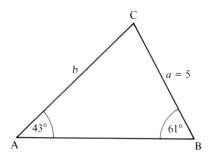

∠A, ∠B and b are known and b is required, so the two fractions we select from the sine rule are

$$\frac{a}{\sin A} = \frac{b}{\sin B}$$

i.e.
$$\frac{5}{\sin 43°} = \frac{b}{\sin 61°}$$

⇒
$$b = \frac{5 \sin 61°}{\sin 43°} = 6.412$$

Therefore AC = 6.41 cm correct to 3 s.f.

_____

2. In ABC, AC = 17 cm, ∠A = 105° and ∠B = 33°. Find AB.

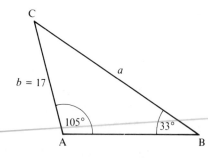

The two sides involved are b and c, so before the sine rule can be used we must find C.

$$∠A + ∠B + ∠C = 180° ⇒ ∠C = 42°$$

Now from the sine rule we can use

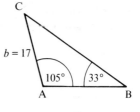

$$\frac{b}{\sin B} = \frac{c}{\sin C}$$

$\Rightarrow$ $$\frac{17}{\sin 33°} = \frac{c}{\sin 42°}$$

i.e. $$c = \frac{17 \times 0.6691}{0.5446} = 20.88$$

Therefore AB $=$ 20.9 cm correct to 3 s.f.

EXERCISE 7c

1. In $\triangle ABC$, AB $=$ 9 cm, $\angle A = 51°$ and $\angle C = 39°$
   Find BC.

2. In $\triangle XYZ$, $\angle X = 27°$, YZ $=$ 6.5 cm and $\angle Y = 73°$
   Find ZX.

3. In $\triangle PQR$, $\angle R = 52°$, $\angle Q = 79°$ and PR $=$ 12.7 cm.
   Find PQ.

4. In $\triangle ABC$, AC $=$ 9.1 cm, $\angle A = 59°$ and $\angle B = 62°$
   Find BC.

5. In $\triangle DEF$, DE $=$ 174 cm, $\angle D = 48°$ and $\angle F = 56°$
   Find EF.

6. In $\triangle XYZ$, $\angle X = 130°$, $\angle Y = 21°$ and XZ $=$ 53 cm.
   Find YZ.

7. In $\triangle PQR$, $\angle Q = 37°$, $\angle R = 101°$ and PR $=$ 4.3 cm.
   Find PQ.

8. In $\triangle ABC$, BC $=$ 73 cm, $\angle A = 54°$ and $\angle C = 99°$
   Find AB.

9. In $\triangle LMN$, LN $=$ 637 cm, $\angle M = 128°$ and $\angle N = 46°$
   Find LM.

10. In $\triangle XYZ$, XY $=$ 92 cm, $\angle X = 59°$ and $\angle Y = 81°$
    Find XZ.

11. In △PQR, ∠Q = 64°, ∠R = 38° and PR = 15 cm.
    Find QR.

12. In △ABC, AB = 24 cm, ∠A = 132° and ∠C = 22°
    Find AC.

13. In △XYZ, ∠X = 49°, XY = 98 cm and ∠Z = 100°
    Find XZ.

14. In △ABC, AB = 10 cm, BC = 9.1 cm and AC = 17 cm.
    Can you use the sine rule to find ∠A? If you answer YES, write
    down the two parts of the sine rule that you would use. If you
    answer NO, give your reason.

The Ambiguous Case

Consider a triangle specified by two sides and one angle.

If the angle is between the two sides there is only one possible triangle,
e.g.

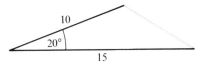

If, however, the angle is not between the two given sides it is sometimes
possible to draw two triangles from the given data.

Consider, for example, a triangle ABC in which ∠A = 20°,
$b = 10$ and $a = 8$

The two triangles with this specification are shown in the diagram; in
one of them B is an acute angle, while in the other one, B is obtuse.

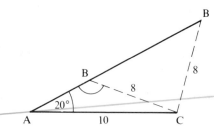

Therefore, when two sides and an angle of a triangle are given,

it is essential to check whether the obtuse angle is possible.

The following worked examples illustrate this special case.

Examples 7d _____

1. In the triangle ABC, find C given that AB = 5 cm, BC = 3 cm and A = 35°

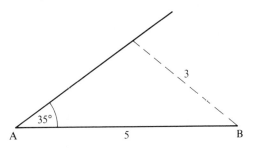

We know a, c and $\angle A$ so the sine rule can be used to find $\angle C$.

As we are looking for an angle, the form we use is

$$\frac{\sin A}{a} = \frac{\sin C}{c} \quad \Rightarrow \quad \frac{\sin 35°}{3} = \frac{\sin C}{5}$$

Hence $\qquad\qquad \sin C = \dfrac{5 \times 0.5736}{3} = 0.9560$

One angle whose sine is 0.9560 is 73° but there is also an obtuse angle with the same sine, i.e. 107°

If C = 107, then A + C = 107 + 35 = 142
⇒   B = 180 − 142 = 38

So in this case $\angle C = 107°$ *is* an acceptable solution and we have two possible triangles.

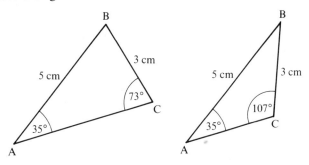

Therefore $\angle C$ is either 73° or 107°

_____

The reader should *not* assume that there are *always* two possible angles when the sine rule is used to find a second angle in a triangle. The next example shows that this is not so.

Examples 7d (continued)

2. In the triangle XYZ, $\angle Y = 41°$, $XZ = 11$ cm and $YZ = 8$ cm.
Find $\angle X$.

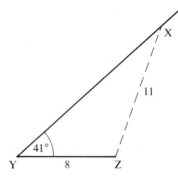

Using the part of the sine rule that involves x, y, $\angle X$ and $\angle Y$ we have

$$\frac{\sin X}{x} = \frac{\sin Y}{y} \quad \Rightarrow \quad \frac{\sin X}{8} = \frac{\sin 41°}{11}$$

Hence $\qquad \sin X = \dfrac{8 \times 0.6561}{11} = 0.4771$

The two angles with a sine of 0.4771 are $28°$ and $152°$

Checking to see whether $152°$ is a possible value for $\angle X$ we see that

$\angle X + \angle Y = 152° + 41° = 193°$

This is greater than $180°$, so it is not possible for the angle X to have the value $152°$.

In this case then, there is only one possible triangle containing the given data, i.e. the triangle in which $\angle X = 28°$

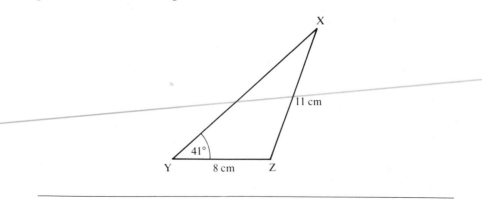

It is interesting to notice how the two different situations that arose in Examples 1 and 2 above can be illustrated by the construction of the triangles with the given data.

When AB = 5 cm, BC = 3 cm and ∠A = 35° we have

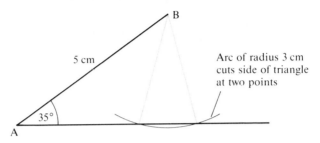

When XZ = 11 cm, YZ = 8 cm and ∠Y = 41° we have

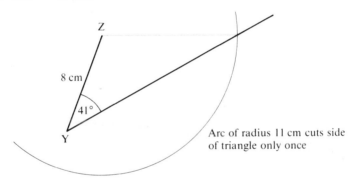

EXERCISE 7d

In each of the following questions, find the angle indicated by the question mark, giving two values in those cases where there are two possible triangles. Illustrate your solution to each question.

	AB	BC	CA	∠A	∠B	∠C
1.		2.9 cm	6.1 cm	?	40°	
2.	5.7 cm		2.3 cm		20°	?
3.	21 cm	36 cm		29.5°		?
4.		2.7 cm	3.8 cm	?	54°	
5.	4.6 cm		7.1 cm		?	33°
6.	9 cm	7 cm		?		40°

THE COSINE RULE

When solving a triangle, the sine rule cannot be used unless the data given includes one side and the angle opposite to that side. If, for example, a, b and C are given then in the sine rule we have

$$\frac{\textcircled{a}}{\sin A} = \frac{\textcircled{b}}{\sin B} = \frac{c}{\sin \textcircled{C}}$$

and it is clear that no pair of fractions contains only one unknown quantity.

Some other method is therefore needed in such circumstances and the one we use is called the *cosine rule*. This rule states that

$$a^2 = b^2 + c^2 - 2bc \cos A$$

The proof of the cosine rule is given here for readers who are interested.

Proof

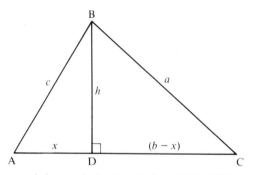

Let ABC be a non-right-angled triangle in which BD is drawn perpendicular to AC. Taking x as the length of AD, the length of CD is $(b - x)$. Then, using h as the length of BD, we can use Pythagoras' theorem to find h in each of the right-angled triangles BDA and BDC, i.e.

$$h^2 = c^2 - x^2 \quad \text{and} \quad h^2 = a^2 - (b - x)^2$$

Therefore $\quad c^2 - x^2 = a^2 - (b - x)^2$

$\Rightarrow \quad c^2 - x^2 = a^2 - b^2 + 2bx - x^2$

$\Rightarrow \quad a^2 = b^2 + c^2 - 2bx$

But $x = c \cos A$

Therefore $\quad a^2 = b^2 + c^2 - 2bc \cos A$

The proof is equally valid for an obtuse-angled triangle, as is shown opposite.

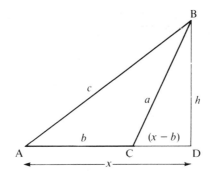

In this case the length of CD is $(x - b)$, so in $\triangle BCD$ we have

$$h^2 = a^2 - (x - b)^2 = a^2 - x^2 + 2bx - b^2$$

This is identical to the expression found for h^2 above.

The remainder of the proof above is unchanged so we have now proved that, in *any* triangle,

$$a^2 = b^2 + c^2 - 2bc \cos A$$

When the altitude is drawn from A or from C similar expressions for the other sides of a triangle are obtained, i.e.

$$b^2 = c^2 + a^2 - 2ca \cos B$$

and

$$c^2 = a^2 + b^2 - 2ab \cos C$$

Examples 7e

1. In $\triangle ABC$, $BC = 7\,cm$, $AC = 9\,cm$ and $C = 61°$. Find AB.

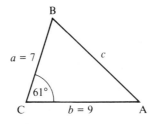

Using the cosine rule, starting with c^2, we have

$$c^2 = a^2 + b^2 - 2ab \cos C$$

$\Rightarrow \qquad c^2 = 7^2 + 9^2 - (2)(7)(9)(0.4848)$

$\Rightarrow \qquad c = 8.302$

Hence $AB = 8.30\,cm$ correct to 3 s.f.

2. XYZ is a triangle in which $\angle Y = 121°$, XY $= 14$ cm and YZ $= 26.9$ cm. Find XZ

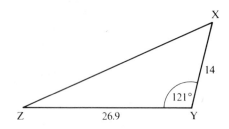

Using $y^2 = z^2 + x^2 - 2zx \cos Y$ gives

$$y^2 = (14)^2 + (26.9)^2 - (2)(14)(26.9)(-0.5150)$$

Note that, because Y is an obtuse angle, it has a negative cosine. Extra care therefore has to be taken with the sign of the term $-2zx \cos Y$. The best way to avoid mistakes is to enclose the cosine in brackets as shown.

Hence $\qquad y^2 = 1307.51 \quad \Rightarrow \quad y = 36.16$

Therefore XZ $= 36.2$ cm correct to 3 s.f.

EXERCISE 7e

In each question use the data given for $\triangle PQR$ to find the length of the third side.

	PQ	QR	RP	P	Q	R
1.		8 cm	4.6 cm			39°
2.	11.7 cm		9.2 cm	75°		
3.	29 cm	37 cm			109°	
4.		2.1 cm	3.2 cm			97°
5.	135 cm		98 cm	48°		
6.	4.7 cm	8.1 cm			138°	
7.		44 cm	62 cm			72°
8.	19.4 cm		12.6 cm	167°		

Using the Cosine Rule to Find an Angle

So far the cosine rule has been used only to find an unknown side of a triangle. When we want to find an unknown angle, it is advisable to rearrange the formula to some extent.

The version of the cosine rule that starts with c^2,

i.e. $c^2 = a^2 + b^2 - 2ab \cos C$

can be written as $2ab \cos C = a^2 + b^2 - c^2$

and further as $\cos C = \dfrac{a^2 + b^2 - c^2}{2ab}$

The reader should find this last form quite easy to remember if it is noted that the side opposite to the angle being found, c^2 in this case, appears only once as the last term in the formula. Some readers however may prefer to work from the basic cosine formula for all calculations, carrying out any necessary manipulation in each problem as it arises.

Examples 7f

1. If, in $\triangle ABC$, $a = 9$, $b = 16$ and $c = 11$, find, to the nearest degree, the largest angle in the triangle.

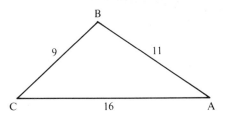

The largest angle in a triangle is opposite to the longest side, so in this question we are looking for angle B and we use

$$\cos B = \frac{c^2 + a^2 - b^2}{2ca}$$

$$= \frac{121 + 81 - 256}{(2)(11)(9)}$$

$$= -0.2727$$

The negative sign shows that $\angle B$ is obtuse.

Hence $B = 106°$ and this is the largest angle in $\triangle ABC$.

2. The sides a, b, c of a triangle ABC are in the ratio $3:6:5$ Find the smallest angle in the triangle.

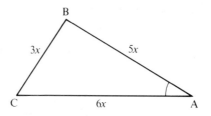

The actual lengths of the sides are not necessarily 3, 6 and 5 units so we represent them by $3x$, $6x$ and $5x$. The smallest angle is A (opposite to the smallest side).

$$\cos A = \frac{b^2 + c^2 - a^2}{2bc}$$

$$= \frac{36x^2 + 25x^2 - 9x^2}{60x^2}$$

$$= \frac{52}{60}$$

$$= 0.8667$$

Therefore the smallest angle in $\triangle ABC$ is $30°$

EXERCISE 7f

1. In $\triangle XYZ$, $XY = 34\,cm$, $YZ = 29\,cm$ and $ZX = 21\,cm$. Find the smallest angle in the triangle.

2. In $\triangle PQR$, $PQ = 1.3\,cm$, $QR = 1.8\,cm$ and $RP = 1.5\,cm$. Find $\angle Q$.

3. In $\triangle ABC$, $AB = 51\,cm$, $BC = 37\,cm$ and $CA = 44\,cm$. Find $\angle A$.

4. Find the largest angle in $\triangle XYZ$ given that $x = 91$, $y = 77$ and $z = 43$

5. What is the size of (a) the smallest, (b) the largest angle in $\triangle ABC$ if $a = 13$, $b = 18$ and $c = 7$?

6. In $\triangle PQR$ the sides PQ. QR and RP are in the ratio $2:1:2$ Find $\angle P$.

7. ABCD is a quadrilateral in which AB $=$ 5 cm, BC $=$ 8 cm, CD $=$ 11 cm, DA $=$ 9 cm and angle ABC $=$ 120° Find the length of AC and the size of the angle ADC.

GENERAL TRIANGLE CALCULATIONS

If three independent facts are given about the sides and/or angles of a triangle and further facts are required, a choice must be made between using the sine rule or the cosine rule for the first step.

As the sine rule is easier to work out, it is preferred to the cosine rule whenever the given facts make this possible, i.e. whenever an angle and the opposite side are known. (Remember that if two angles are given, then the third angle is also known.)

The cosine rule is used only when the sine rule is not suitable and it is never necessary to use it more than once in solving a triangle.

Suppose, for example, that the triangle PQR given below is to be solved.

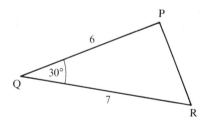

Only one angle is known and the side opposite to it is not given. We must therefore use the cosine rule first to find the length of PR.

Once we know q as well as $\angle Q$, the sine rule can be used to find either of the remaining angles, the third angle then following from the sum of the angles in the triangle.

EXERCISE 7g

Each of the following questions refers to a triangle ABC. Fill in the blank spaces in the table.

	∠A	∠B	∠C	a	b	c
1.		80°	50°			68 cm
2.			112°	15.7 cm	13 cm	
3.	41°	69°		12.3 cm		
4.	58°				131 cm	87 cm
5.		49°	94°		206 cm	
6.	115°		31°			21 cm
7.	59°	78°		17 cm		
8.		48°	80°		31.3 cm	
9.	77°				19 cm	24 cm
10.		125°		14 cm		20 cm

11. A tower stands on level ground. From a point P on the ground, the angle of elevation of the top of the tower is 26° Another point Q is 3 m vertically above P and from this point the angle of elevation of the top of the tower is 21° Find the height of the tower.

12. A survey of a triangular field, bounded by straight fences, found the three sides to be of lengths 100 m, 80 m and 65 m. Find the angles between the boundary fences.

MIXED EXERCISE 7

1. Find the value, between 0° and 180°, of ∠A if
   (a) $\cos A = -\cos 64°$ (b) $\sin 94° = \sin A$

2. If ∠X is acute and $\sin X = \frac{7}{25}$, find $\cos(180° - X)$

3. Given that $\sin A = \frac{5}{8}$, find $\tan A$ in surd form if
   (a) ∠A is acute   (b) ∠A is obtuse

4. Find, in surd form, $\sin \theta$ and $\cos \theta$, given

(a)   (b)

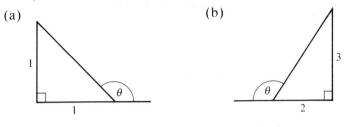

5. Given that $\sin X = \frac{12}{13}$ and X is obtuse, find $\cos X$.

6. In $\triangle ABC$, $BC = 11$ cm, $\angle B = 53°$ and $\angle A = 76°$, find AC.

7. In $\triangle PQR$, $p = 3$, $q = 5$ and $R = 69°$, find r

8. In $\triangle XYZ$, $XY = 8$ cm, $YZ = 7$ cm and $ZX = 10$ cm, find $\angle Y$.

9. In $\triangle ABC$, $AB = 7$ cm, $BC = 6$ cm and $\angle A = 44°$, find all possible values of $\angle ACB$.

10. Find the angles of a triangle whose sides are in the ratio $2:4:5$

11. Use the cosine formula, $\cos A = \dfrac{b^2 + c^2 - a^2}{2bc}$, to show that

   (a) $\angle A$ is acute if $a^2 < b^2 + c^2$
   (b) $\angle A$ is obtuse if $a^2 > b^2 + c^2$

CHAPTER 8

TRIANGLES

THE AREA OF A TRIANGLE

The simplest way to find the area of a triangle is to use the formula

$$\text{Area} = \tfrac{1}{2}\,\text{base} \times \text{perpendicular height}$$

Clearly this is of immediate use only when the perpendicular height is known. It can be adapted, however, to cover other cases.

Consider the triangle shown below, in which b, c, and A are known

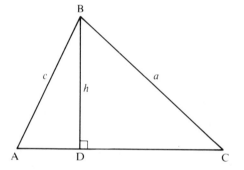

The line BD, drawn from B perpendicular to AC, is the height, h, of the triangle, so the area of the triangle is $\tfrac{1}{2}bh$

In the triangle ADB, $\sin A = \dfrac{h}{c} \;\Rightarrow\; h = c \sin A$

Therefore the area of triangle ABC is

$$\tfrac{1}{2}bc \sin A$$

Drawing the perpendicular heights from A to C give similar expressions, i.e.

$$\text{Area of triangle ABC} = \tfrac{1}{2}ab \sin C = \tfrac{1}{2}ac \sin B$$

Each of these formulae can be expressed in the 'easy to remember' form

$$\text{Area} = \tfrac{1}{2}\,\text{product of two sides} \times \text{sine of included angle}$$

Example 8a _____

Find the area of triangle PQR, given that P = 65°, Q = 79° and PQ = 30 cm.

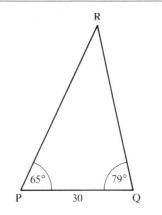

The given facts do not include two sides and the included angle so we must first find another side.   To do this the sine rule can be used and we need angle R.

$$\angle R = 180° - 65° - 79° = 36°$$

From the sine rule, $\dfrac{p}{\sin P} = \dfrac{r}{\sin R}$

$\Rightarrow$ $\qquad\qquad\qquad\qquad p = \dfrac{30 \times \sin 65°}{\sin 36°} = 46.26$

i.e. QR = 46.3 cm (correct to 3 s.f.).

Now we can use area PQR $= \frac{1}{2}pr \sin Q$

$\qquad \frac{1}{2}pr \sin Q = \frac{1}{2} \times 46.26 \times 30 \times \sin 79 = 681.2$

So the area of triangle PQR is 681 cm^2 (corr to 3 s.f.).

_____

EXERCISE 8a

Find the area of each triangle given in Questions 1 to 5.

1. $\triangle$XYZ; XY = 180 cm, YZ = 145 cm, $\angle$Y = 70°

2. $\triangle$ABC; AB = 75 cm, AC = 66 cm, $\angle$A = 62°

3. $\triangle$PQR; QR = 69 cm, PR = 49 cm, $\angle$R = 85°

4. $\triangle$XYZ; x = 30, y = 40, $\angle$Z = 49°

5. $\triangle$PQR; p = 9, r = 11, $\angle$Q = 120°

6. In triangle ABC, AB = 6 cm, BC = 7 cm and CA = 9 cm. Find $\angle$A and the area of the triangle.

7. $\triangle$PQR is such that $\angle P = 60°$, $\angle R = 50°$ and QR = 12 cm. Find PQ and the area of the triangle.

8. In $\triangle$XYZ, XY = 150 cm, YZ = 185 cm and the area is 11 000 cm². Find $\angle Y$ and XZ.

9. The area of triangle ABC is 36.4 cm². Given that AC = 14 cm and $\angle A = 98°$, find AB.

PROBLEMS

Many practical problems which involve distances and angles can be illustrated by a diagram. Often, however, this diagram contains too many lines, dimensions, etc. to be clear enough to work from. In these cases we can draw a second figure by extracting a triangle (or triangles) in which three facts about sides and/or angles are known. The various methods given in Chapter 7 can then be used to analyse this triangle and so to solve the problem.

Examples 8b

1. Two boats, P and Q, are 300 m apart. The base, A, of a lighthouse is in line with PQ. From the top, B, of the lighthouse the angles of depression of P and Q are found to be 35° and 48°. Write down the values of the angles BQA, PBQ and BPQ and find, correct to the nearest metre, the height of the lighthouse.

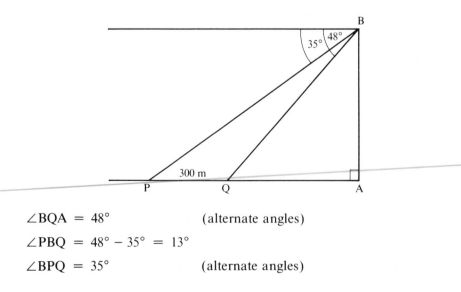

$\angle$BQA $= 48°$ (alternate angles)

$\angle$PBQ $= 48° - 35° = 13°$

$\angle$BPQ $= 35°$ (alternate angles)

Now we can extract △PBQ, knowing two angles and a side.

From the sine rule,

$$\frac{p}{\sin P} = \frac{b}{\sin B}$$

$$\therefore \qquad p = \frac{(300)(\sin 35°)}{\sin 13°}$$

$$= 764.9$$

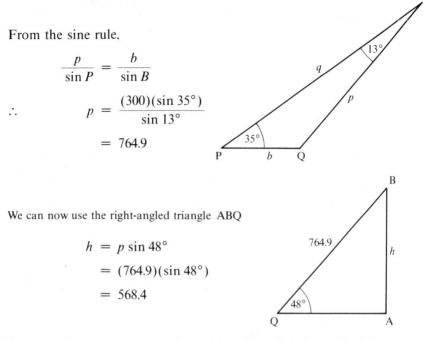

We can now use the right-angled triangle ABQ

$$h = p \sin 48°$$

$$= (764.9)(\sin 48°)$$

$$= 568.4$$

Correct to the nearest metre the height of the light-house is 568 m

2. A traveller pitches camp in a desert. He knows that there is an oasis in the distance, but cannot see it. Wishing to know how far away it is, he measures 250 m due north from his starting point, A, to a point B where he can see the oasis, O, and finds that its bearing is 276°. He then measures a further 250 m due north to point C from which the bearing of the oasis is 260°. Find how far from the oasis he has camped.

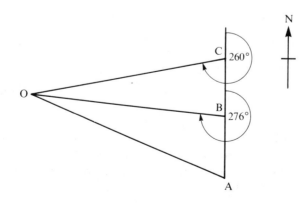

$\angle OCB = 260° - 180° = 80°$ and $\angle OBC = 360° - 276° = 84°$

As two angles and a side are known, $\triangle OBC$ can be used.

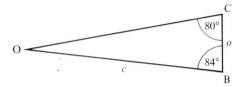

$$\angle BOC = 180° - 80° - 84° = 16°$$

From the sine rule, $\dfrac{c}{\sin C} = \dfrac{o}{\sin O}$

$\Rightarrow$ $c = \dfrac{250 \times \sin 80°}{\sin 16°} = 893.2$

Now in $\triangle ABO$, $\angle ABO = 276° - 180° = 96°$ and we also know OB and AB. As two sides and the included angle are known, it is the cosine rule that must be used.

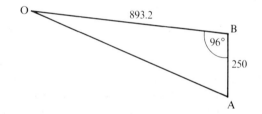

$$OA^2 = OB^2 + AB^2 - 2 \times OB \times AB \times \cos ABO$$
$$= (893.2)^2 + (250)^2 - 2 \times 893.2 \times 250 \times \cos 96°$$
$$= 906\,989$$

$\Rightarrow$ $OA = 952.4$

To the nearest metre the initial distance from the oasis was 952 m

---

EXERCISE 8b

1. In a quadrilateral PQRS, PQ $=$ 6 cm, QR $=$ 7 cm, RS $=$ 9 cm, $\angle PQR = 115°$ and $\angle PRS = 80°$. Find the length of PR. Considering it as split into two separate triangles, find the area of the quadrilateral PQRS.

2. A light aircraft flies from an airfield, A, a distance of 50 km on a bearing of 049° to a town, B. The pilot then changes course and flies on a bearing of 172° to a landing strip, C, 68 km from B. How far is the landing strip from the airfield?

3. In a surveying exercise, P and Q are two points on land which is inaccessible. To find the distance PQ, a line AB of length 300 metres is marked out so that P and Q are on opposite sides of AB. The directions of P and Q relative to the line AB are then measured and are shown in the diagram. Calculate the length of PQ. (Hint. Find AP and AQ.)

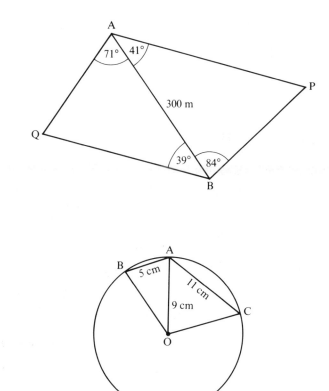

4.

AB, of length 5 cm, and AC, of length 11 cm, are two chords of a circle with centre O and radius 9 cm. Find each of the angles BAO and CAO and hence calculate the area of the triangle ABC.

5.

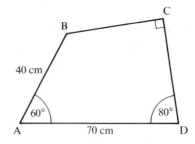

The diagram shows the cross section of a beam of length 2 m. Calculate

(a) the length of BD

(b) the angle ADB

(c) the length of CD

(d) the area of the cross section

(e) the volume of the beam.

THREE-DIMENSIONAL PROBLEMS

One of the difficulties which many people experience with this topic, arises when attempting to illustrate a three-dimensional situation on a two-dimensional diagram. The following hints may help in producing a clear representation of the 3-D problem from which appropriate calculations can be made.

1) Vertical lines should be drawn vertically on the page.

2) Lines in the East–West direction should be drawn horizontally on the page. North–South lines are shown as inclined at an acute angle to the East direction.

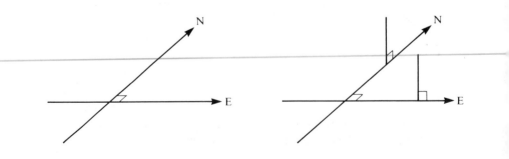

3) All angles that are 90° in three dimensions should be marked as right angles on the diagram, particularly those that do not *appear* to be 90°

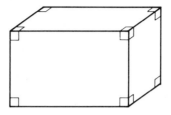

4) Perspective drawing is rarely used, so parallel lines are drawn parallel in the diagram.

5) When viewing a 3-D object, some of its sides are usually not visible. It is helpful to indicate these by broken lines.

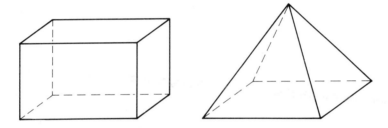

6) In a situation involving two points in the foreground and an object in the background it is usually clearer to draw the object *between* the two points.

e.g.

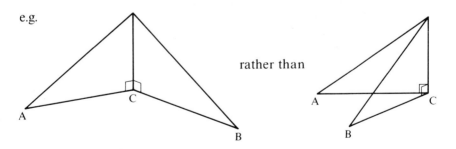

rather than

7) It is often helpful to draw a separate diagram showing each individual triangle in which calculations are needed.

The following facts and definitions should also be known.

1) Two non-parallel planes meet in a line called the common line.

2) A line that is perpendicular to a plane is also perpendicular to every line in that plane

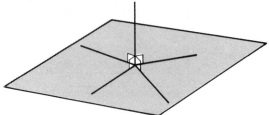

and if a line is perpendicular to two non-parallel lines in a plane, then it is perpendicular to the plane.

3) The angle between a line and a plane is defined as the angle between that line and its projection on the plane. (Its projection can be thought of as the shadow of the line cast on the plane by a beam of light shining at right-angles to the plane.)

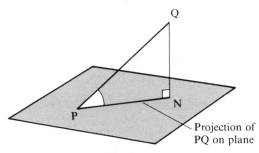

Projection of PQ on plane

4) The angle between two planes is defined as follows. From any point A, on the common line of two planes P_1 and P_2, lines AB and AC are drawn, one in each plane, perpendicular to the common line. Then angle BAC is the angle between the two planes.

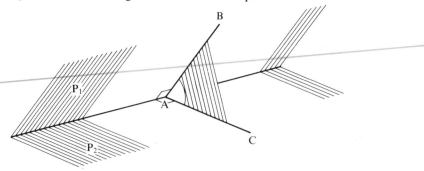

5) If, in fact 4, one of the planes, P_2 say, is horizontal, then AB is called *a line of greatest slope* of the plane P_1

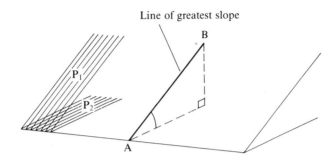

Line of greatest slope

Examples 8c _____

1. The diagram shows a cube of side 6 cm. M is the midpoint of AB. Find

   (a) the length of MC   (b) the length of MR

   (c) to the nearest degree, the angle between MR and the plane ABCD.

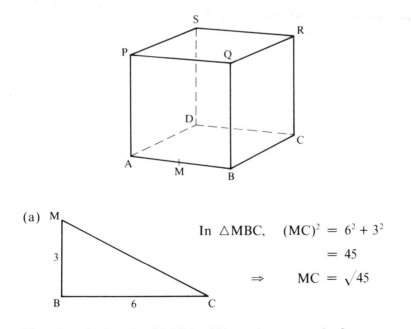

(a) In $\triangle$MBC, $\quad (MC)^2 = 6^2 + 3^2$

$\qquad\qquad\qquad\qquad\qquad = 45$

$\qquad\qquad \Rightarrow \qquad MC = \sqrt{45}$

Therefore the length of MC is 6.71 cm (correct to 3.s.f).

(b)

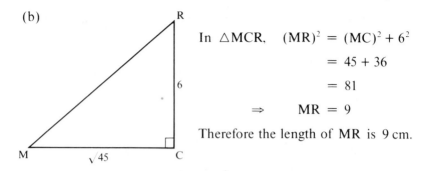

In $\triangle$MCR, $(MR)^2 = (MC)^2 + 6^2$

$= 45 + 36$

$= 81$

$\Rightarrow \quad MR = 9$

Therefore the length of MR is 9 cm.

To find the angle between RM and the plane ABCD, we need the projection of RM on the plane. As RC is perpendicular to ABCD, it follows that CM is the projection of RM on that plane. So the angle we are looking for is RMC.

(c)

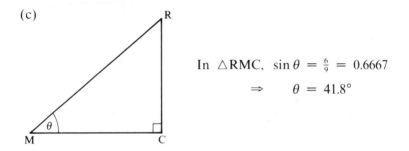

In $\triangle$RMC, $\sin\theta = \frac{6}{9} = 0.6667$

$\Rightarrow \quad \theta = 41.8°$

Therefore, to the nearest degree, the angle between RM and the plane ABCD is 42°

2. The base AB of an isosceles triangle ABC is horizontal. The plane containing the triangle is inclined to the horizontal at 54° If the angle ACB is 48°, find the angle between AC and the horizontal plane.

AB is the line common to the horizontal plane and the plane of the triangle and M is its midpoint. Because the triangle is isosceles, CM is perpendicular to AB. So CM is a line of greatest slope and therefore makes an angle of 54° with its projection, MD, on the horizontal plane.

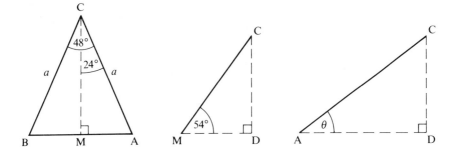

Let the length of AC and BC be a

In △CMA, $CM = a \cos 24°$

In △CDM, $CD = CM \sin 54°$

$$= (a \cos 24°)(\sin 54°)$$

The angle between AC and the horizontal plane is the angle between AC and its projection AD on that plane. If this angle is θ then

$$\sin \theta = \frac{CD}{CA} = \frac{(a)(\cos 24°)(\sin 54°)}{a}$$

$\Rightarrow$ $\theta = 47.7°$

EXERCISE 8c

1.

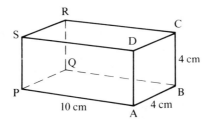

In the cuboid shown above, ABCD is a square of side 4 cm and PA = 10 cm. Find the length of

(a) AC   (b) AS   (c) AQ   (d) a diagonal of the cuboid.

2.

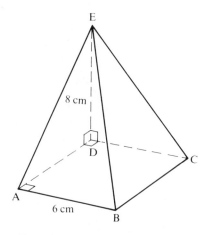

The pyramid ABCDE has a square base of side 6 cm. E is 8 cm vertically above D. Calculate

(a) the lengths of AD, BD and BE

(b) the angle between AE and the plane ABCD

(c) the angle between BE and the plane ABCD

(d) the angle between the planes EBC and ABCD.

3.

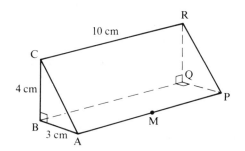

Given the triangular prism in the diagram, in which M is the midpoint of AP, find the following lengths and angles.

(a) RM   (b) RA   (c) QA

(d) the angle between RA and the plane ABQP

(e) the angle between RM and the plane ABQP.

4. Given a regular tetrahedron (i.e. a pyramid where each face is an equilateral triangle), find the cosine of the angle between two of the faces.

5.

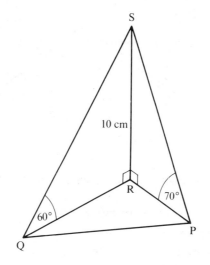

Three points, **P**, **Q** and **R** lie in a plane. The line **RS** is perpendicular to the plane and is of length 10 cm. If angle **SPR** = 70°, angle **SQR** = 60° and **PQ** = 7 cm, calculate each of the angles in triangle **PQR**.

6. Find the angle between two diagonals of a cube.

7.

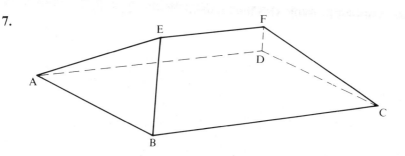

The diagram shows a roof whose base is a rectangle of length 15 m and width 9 m. Each end face is an isosceles triangle inclined to the horizontal at an angle, α, and each long face is a trapezium inclined to the horizontal at an angle β
If $\tan \alpha = 2$ and $\tan \beta = \frac{4}{3}$, calculate

(a) the height of the ridge (**EF**) above the base

(b) the length of the ridge

(c) the angle between **AE** and the horizontal

(d) the total surface area of the roof.

8.

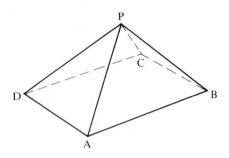

The diagram shows a solid figure in which ABCD is a horizontal rectangle. AB = 13 cm, BC = 8 cm, AP = DP = 9 cm and BP = CP = 7 cm. Calculate

(a) the length of AC

(b) the height of P above the plane ABCD

(c) the angle between AP and the horizontal

(d) the angle between the faces APB and ABCD.

9. An aircraft is noted simultaneously by three observers, A, B and C, stationed in a horizontal straight line. AB and BC are each 200 m and the noted angles of elevation of the aircraft from A and C are 25° and 40° respectively. What is the height of the aircraft? Find also the angle of elevation of the aircraft from B.

10.

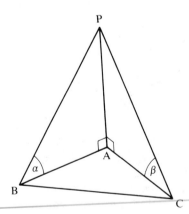

ABC is a horizontal triangle in which BC = 10 m. P is a point 12 m vertically above A. The angles of elevation of P from B and C are α and β, where $\tan \alpha = 1$ and $\tan \beta = \frac{6}{7}$. Find the angle between the planes PBC and ABC.

Harder Problems

Certain problems in three dimensions are rather more demanding than those seen up to now. An example of such a problem is given below and, in the following Mixed Exercise, Questions 8 to 10 are also a little harder. It is recommended that they be used at a later date for revision.

Example 8d

A, B and C are points on a horizontal line such that $AB = 60$ m and $BC = 30$ m. The angles of elevation, from A, B and C respectively, of the top of a clock tower are α, β, and γ, where $\tan \alpha = \frac{1}{13}$, $\tan \beta = \frac{1}{15}$ and $\tan \gamma = \frac{1}{20}$. The foot of the tower is at the same level as A, B and C. Find the height of the tower.

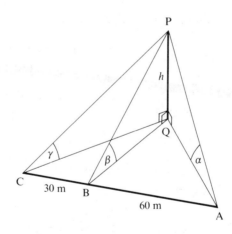

If the height of the tower, PQ, is h then

$$h = QA \tan \alpha = QB \tan \beta = QC \tan \gamma$$

i.e.

$$h = \frac{QA}{13} = \frac{QB}{15} = \frac{QC}{20}$$

$\Rightarrow \quad QA = 13h, \quad QB = 15h, \quad QC = 20h$

Now considering the base triangle ABCQ, we have

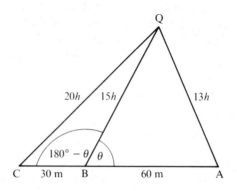

Using the cosine rule in $\triangle ABQ$ gives

$$\cos\theta = \frac{(60)^2 + (15h)^2 - (13h)^2}{2(60)(15h)}$$

and using the cosine rule in $\triangle CBQ$ gives

$$\cos(180° - \theta) = -\cos\theta = \frac{(30)^2 + (15h)^2 - (20h)^2}{2(30)(15h)}$$

$$\therefore \qquad \frac{(60)^2 + (15h)^2 - (13h)^2}{2(60)(15h)} = \frac{(20h)^2 - (30)^2 - (15h)^2}{2(30)(15h)}$$

$$\Rightarrow \qquad 3600 + 56h^2 = 2(175h^2 - 900)$$

$$\Rightarrow \qquad 5400 = 294h^2$$

Hence $\qquad\qquad\qquad h = 4.285$

The height of the clock tower is 4.29 m (correct to 3 s.f.).

MIXED EXERCISE 8

1. In $\triangle PQR$, $PQ = 11\,cm$, $PR = 14\,cm$ and $QPR = 100°$.
   Find the area of the triangle.

2. The area of ABC is $9\,cm^2$. If $AB = AC = 6\,cm$, find $\sin A$.
   Are there two possible triangles? Give a reason for your answer.

3.

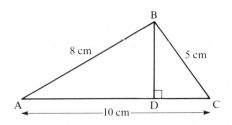

Given the information in the diagram,

(a) find $\angle ABC$

(b) find the area of $\triangle ABC$

(c) *hence* find the length of BD.

4. Triangle PQR, in which PRQ = 120°, lies in a horizontal plane and X is a point 6 cm vertically above R. If XQR = 45° and XPR = 60°, find the lengths of the sides of $\triangle PQR$. Find also the area of this triangle.

5.

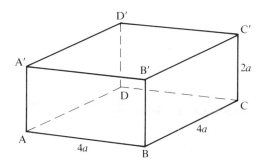

Given the cuboid shown in the diagram, find

(a) the angle between AC and the plane ABB′A′

(b) the angle between the planes ACD′ and ABCD.

6. Two rectangular panels, ABCD and ABEF, each measure 1.5 m by 2 m. They are hinged along the edge AB which is 2 m long. If the angle between their planes is 60°, find the angle between the diagonals AC and AE.

7. A river running due east has straight parallel banks. A vertical post stands with its base, P, on the north bank of the river. On the south bank are two surveyors, A who is to the east, and B who is to the west of the post. A and B are at a distance $\frac{2}{7}a$ apart and the angle APB is 150° The angles of elevation from A and B of the top, Q, of the post are 45° and 30° Find, in terms of a, the width of the river and the height of the post.

The remaining questions are a little more demanding.

8. ABCD is a tetrahedron whose horizontal base ABC is an equilateral triangle. The angle between each pair of slant edges is θ where $\tan\theta = \frac{5}{12}$ and the length of these edges is a. Find the height of D above ABC.

9. Two lamp standards are each of height h m. They are d m apart on level ground where a man of height t m is also standing. Each light casts a shadow of the man on the ground. Prove that, no matter where the man stands, the distance between the ends of his two shadows is $\dfrac{dt}{h-t}$ m. (Hint. Look for similar triangles.)

10. The roof of a south-facing house slopes down at an angle α to the horizontal. A gulley at the end of the roof is in the direction θ east of north. If the gulley is inclined to the horizontal at an angle β, show that $\tan\beta = \tan\alpha\cos\theta$

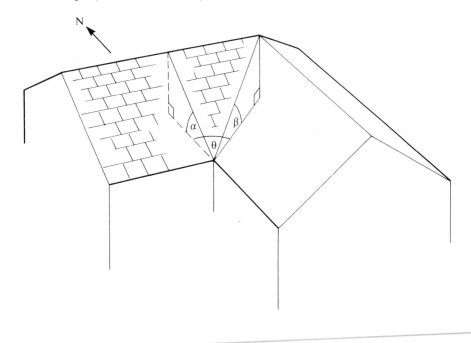

CHAPTER 9

STRAIGHT LINES

THE MEANING OF EQUATIONS

The Cartesian frame of reference provides a means of defining the position of any point in a plane. This plane is called the xy-plane.

In general x and y are independent variables. This means that they can each take any value independently of the value of the other unless some restriction is placed on them.

Consider the case when the value of x is restricted to 2

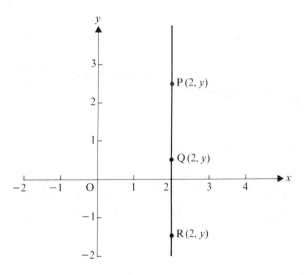

As the value of y is not restricted, the condition above gives a set of points which form a straight line parallel to the y-axis and passing through P, Q and R as shown. Therefore the condition that x is equal to 2, i.e. $x = 2$, defines the line through P, Q and R, i.e.

in the context of the xy-plane, the equation $x = 2$ defines the line shown in the diagram. Further, $x = 2$ is called *the equation of this line* and we can refer briefly to *the line $x = 2$*

Now consider the set of points for which the condition is $x > 2$

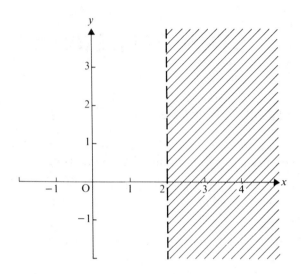

All the points to the right of the line $x = 2$ have an x-coordinate that is greater than 2

So the inequality $x > 2$ defines the shaded region of the xy-plane shown above.
Similarly, the inequality $x < 2$ defines the region left unshaded in the diagram.

Note that the region defined by $x > 2$ does not include the line $x = 2$

When a region does not include a boundary line this is drawn as a broken line. When the points on a boundary *are* included in a region, this boundary is drawn as a solid line.

Example 9a

Draw a sketch of the region of the xy-plane defined by the inequalities $0 \leqslant x \leqslant 2$ and $0 < y < 4$

The relationship $0 \leqslant x \leqslant 2$ contains two inequalities which must be considered separately, i.e. $x \geqslant 0$ and $x \leqslant 2$. Similarly, $0 < y < 4$ contains two relationships, i.e. $y > 0$ and $y < 4$. The required region is found by considering each inequality in turn and shading the unwanted region. This leaves the required region clear.

$x \geqslant 0$ defines both the line $x = 0$
(i.e. the y-axis) and the region to the
right of the y-axis $(x > 0)$ so we
shade the region to the left of the
y-axis.

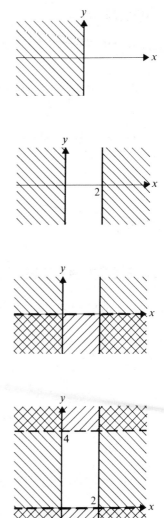

$x \leqslant 2$ defines both the line $x = 2$
and the region to the left of the line.

$y > 0$ defines the region above the
x-axis, but does not include the x-axis.

$y < 4$ defines the region below the
line $y = 4$ but does not include the
line $y = 4$

Therefore the unshaded region in the bottom figure, including the
solid boundary lines but not the broken ones, is the set of points that
satisfy all the given inequalities.

Note that in this book, we use the convention of shading unwanted regions when
dealing with more than one inequality in a plane. This convention is not universal
and some problems ask for such a diagram to be drawn with the required region
shaded. In this case we recommend that the reader follows the procedure used in
the worked example and then redraws the diagram to shade the required region.

EXERCISE 9a

1. Draw a sketch showing the lines defined by the equations $x = 5$, $x = -3$, $y = 0$, $y = 6$

2. Draw a sketch showing the lines defined by the equations $y = -3$, $y = -10$, $x = 7$, $x = -5$

Draw a sketch showing the region of the xy-plane defined by the following inequalities.

3. $x > 3$ 4. $y \leqslant 2$ 5. $x \geqslant -8$

6. $0 < x < 5$ 7. $-1 < y < 4$ 8. $-2 \leqslant x < 2$

9. $0 \leqslant x \leqslant 5,\ -1 \leqslant y \leqslant 3$ 10. $x \leqslant -3,\ x \geqslant 4,\ y < -2$

THE EQUATION OF A STRAIGHT LINE

A straight line may be defined in many ways; for example, a line passes through the origin and has a gradient of $\frac{1}{2}$.

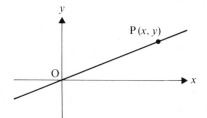

The point $P(x, y)$ is on this line if and only if the gradient of OP is $\frac{1}{2}$

In terms of x and y, the gradient of OP is $\dfrac{y}{x}$, so the statement above can be written in the form

$\quad\quad$ $P(x, y)$ is on the line if and only if $\quad \dfrac{y}{x} = \dfrac{1}{2}$, i.e. $2y = x$

Therefore the coordinates of points on the line satisfy the relationship $2y = x$, and the coordinates of points that are not on the line do not satisfy this relationship.

$\quad\quad$ $2y = x$ is called the equation of the line.

The equation of a line (straight or curved) is a relationship between the x and y coordinates of all points on the line and which is not satisfied by any other point in the plane.

Examples 9b _____

1. Find the equation of the line through the points $(1, -2)$ and $(-2, 4)$.

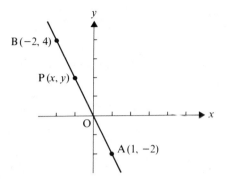

$P(x, y)$ is on the line if and only if the gradient of PA is equal to the gradient of AB (or PB).

The gradient of PA is $\dfrac{y - (-2)}{x - 1} = \dfrac{y + 2}{x - 1}$

The gradient of AB is $\dfrac{-2 - 4}{1 - (-2)} = -2$

Therefore the coordinates of P satisfy the equation $\dfrac{y + 2}{x - 1} = -2$

i.e. $y + 2x = 0$

_____

Consider the more general case of the line whose gradient is m and which cuts the y-axis at a directed distance c from the origin. Note that c is called the *intercept* on the y-axis.

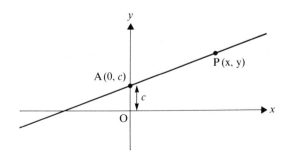

Now $P(x, y)$ is on this line if and only if the gradient of AP is m

Therefore the coordinates of P satisfy the equation $\dfrac{y - c}{x - 0} = m$

i.e. $y = mx + c$

This is the *standard form* for the equation of a straight line.

> An equation of the form $y = mx + c$ represents a straight line with gradient m and intercept c on the y-axis.

Because the value of m and/or c may be fractional, this equation can be rearranged and expressed as $ax + by + c = 0$, i.e.

> $$ax + by + c = 0$$

where a, b and c are constants, is the equation of a straight line.

Note that in this form c is *not* the intercept.

Examples 9b (continued)

2. Write down the gradient of the line $3x - 4y + 2 = 0$ and find the equation of the line through the origin which is perpendicular to the given line.

Rearranging $3x - 4y + 2 = 0$ in the standard form gives

$$y = \tfrac{3}{4}x + \tfrac{1}{2}$$

Comparing with $y = mx + c$ we see that the gradient (m) of the line is $\tfrac{3}{4}$ (and the intercept on the y-axis is $\tfrac{1}{2}$).

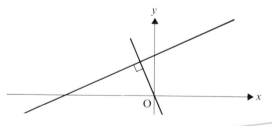

The gradient of the perpendicular line is given by $-\dfrac{1}{m} = -\dfrac{4}{3}$

and it passes through the origin so the intercept on the y-axis is 0

Therefore its equation is $y = -\tfrac{4}{3}x + 0$

$\Rightarrow$ $4y + 3x = 0$

3. Sketch the line $x - 2y + 3 = 0$

This line can be located accurately in the xy-plane when we know two points on the line. We will use the intercepts on the axes as these can be found easily (i.e. $x = 0 \Rightarrow y = \frac{3}{2}$ and $y = 0 \Rightarrow x = -3$).

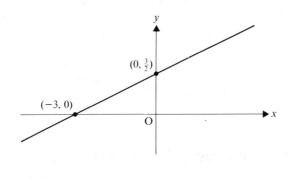

Notice that the diagrams in the worked examples are sketches, not accurate plots, but they show reasonably accurately the position of the lines in the plane.

EXERCISE 9b

1. Write down the equation of the line through the origin and with gradient

(a) 2   (b) -1 (c) $\frac{1}{3}$ (d) $-\frac{1}{4}$ (e) 0   (f) ∞

Draw a sketch showing all these lines on the same set of axes.

2. Write down the equation of the line passing through the given point and with the given gradient.

(a) $(0, 1)$, $\frac{1}{2}$ (b) $(0, 0)$, $-\frac{2}{3}$ (c) $(-1, -4)$, 4

Draw a sketch showing all these lines on the same set of axes.

3. Write down the equation of the line passing through the points

(a) $(0, 0)$, $(2, 1)$ (b) $(1, 4)$, $(3, 0)$ (c) $(-1, 3)$, $(-4, -3)$

4. Write down the equation of the line passing through the origin and perpendicular to

(a) $y = 2x + 3$ (b) $3x + 2y - 4 = 0$ (c) $x - 2y + 3 = 0$

5. Write down the equation of the line passing through $(2, 1)$ and perpendicular to

(a) $3x + y - 2 = 0$ (b) $2x - 4y - 1 = 0$

Draw a sketch showing all four lines on the same set of axes.

6. Write down the equation of the line passing through $(3, -2)$ and parallel to

(a) $5x - y + 3 = 0$ (b) $x + 7y - 5 = 0$

7. $A(1, 5)$ and $B(4, 9)$ are two adjacent vertices of a square. Find the equation of the line on which the side BC of the square lies. How long are the sides of this square?

Formulae for Finding the Equation of a Line

Straight lines play a major role in graphical analysis and it is important to be able to find their equations easily. This section gives two formulae derived from the commonest ways in which a straight line is defined.

The appropriate formula can then be used to write down the equation of a particular line.

The equation of a line with gradient m and passing through the point (x_1, y_1)

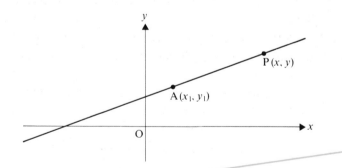

$P(x, y)$ is a point on the line if and only if the gradient of AP is m

i.e.

$$\frac{y - y_1}{x - x_1} = m$$

$\Rightarrow$

$$y - y_1 = m(x - x_1) \qquad [1]$$

The equation of the line passing through (x_1, y_1) and (x_2, y_2)

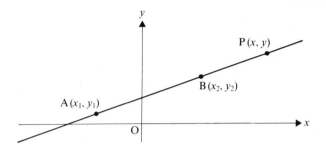

The gradient of AB is $\dfrac{y_2 - y_1}{x_2 - x_1}$

so the formula given in [1] becomes

$$y - y_1 = \left[\frac{y_2 - y_1}{x_2 - x_1}\right](x - x_1) \qquad [2]$$

Examples 9c

1. Find the equation of the line with gradient $-\frac{1}{3}$ and passing through $(2, -1)$

Using [1] with $m = -\frac{1}{3}$, $x_1 = 2$ and $y_1 = -1$ gives

$$y - (-1) = -\tfrac{1}{3}(x - 2)$$

$\Rightarrow \qquad\qquad x + 3y + 1 = 0$

Alternatively the equation of this line can be found from the standard form of the equation of a straight line, i.e. $y = mx + c$

Using $y = mx + c$ and $m = -\frac{1}{3}$ we have

$$y = -\tfrac{1}{3}x + c$$

The point $(2, -1)$ lies on this line so its coordinates satisfy the equation, i.e.

$$-1 = -\tfrac{1}{3}(2) + c \quad \Rightarrow \quad c = -\tfrac{1}{3}$$

Therefore $\qquad\qquad y = -\tfrac{1}{3}x - \tfrac{1}{3}$

$\Rightarrow \qquad\qquad x + 3y + 1 = 0$

2. Find the equation of the line through the points $(1, -2)$, $(3, 5)$

Using formula [2] with $x_1 = 1$, $y_1 = -2$, $x_2 = 3$ and $y_2 = 5$
gives

$$y - (-2) = \frac{-2 - 5}{1 - 3}(x - 1)$$

$\Rightarrow \qquad\qquad 7x - 2y - 11 = 0$

The worked examples in this book necessarily contain a lot of explanation but this should not mislead readers into thinking that their solutions must be equally long. The temptation to 'overwork' a problem should be avoided, particularly in the case of coordinate geometry problems which are basically simple. With practice, any of the methods illustrated above enable the equation of a straight line to be written down directly.

3. Find the equation of the line through $(1, 2)$ which is perpendicular to the line $3x - 7y + 2 = 0$

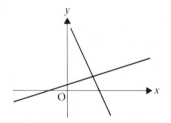

Expressing $3x - 7y + 2 = 0$ in standard form gives $y = \frac{3}{7}x + \frac{2}{7}$

Hence the given line has gradient $\frac{3}{7}$.

So the required line has a gradient of $-\frac{7}{3}$ and it passes through $(1, 2)$

Using $y - y_1 = m(x - x_1)$ gives its equation as

$$y - 2 = -\frac{7}{3}(x - 1) \qquad \Rightarrow \qquad 7x + 3y - 13 = 0$$

In the last example note that the line perpendicular to

$$3x - 7y + 2 = 0$$

has equation $\qquad\qquad 7x + 3y - 13 = 0$

i.e. the coefficients of x and y have been transposed and the sign between the x and y terms has changed. This is a particular example of the general fact that

given a line with equation $ax + by + c = 0$ then the equation of any perpendicular line is $bx - ay + k = 0$

This property of perpendicular lines can be used to shorten the working of problems,

e.g. to find the equation of the line passing through $(2, -6)$ which is perpendicular to the line $5x - y + 3 = 0$, we can say that the required line has an equation of the form $y + 5x + k = 0$ and then use the fact that the coordinates $(2, -6)$ satisfy this equation to find the value of k.

EXERCISE 9c

1. Find the equation of the line with the given gradient and passing through the given point.

   (a) $3, (4, 9)$ (b) $-5, (2, -4)$ (c) $\frac{1}{4}, (4, 0)$

   (d) $0, (-1, 5)$ (e) $-\frac{2}{5}, (\frac{1}{2}, 4)$ (f) $-\frac{3}{8}, (\frac{22}{5}, -\frac{5}{2})$

2. Find the equation of the line passing through the points

   (a) $(0, 1), (2, 4)$ (b) $(-1, 2), (1, 5)$ (c) $(3, -1), (3, 2)$

3. Determine which of the following pairs of lines are perpendicular.

   (a) $x - 2y + 4 = 0$ and $2x + y - 3 = 0$

   (b) $x + 3y - 6 = 0$ and $3x + y + 2 = 0$

   (c) $x + 3y - 2 = 0$ and $y = 3x + 2$

   (d) $y + 2x + 1 = 0$ and $x = 2y - 4$

4. Find the equation of the line through the point $(5, 2)$ and perpendicular to the line $x - y + 2 = 0$

5. Find the equation of the perpendicular bisector of the line joining

   (a) $(0, 0), (2, 4)$ (b) $(3, -1), (-5, 2)$ (c) $(5, -1), (0, 7)$

6. Find the equation of the line through the origin which is parallel to the line $4x + 2y - 5 = 0$

7. The line $4x - 5y + 20 = 0$ cuts the x-axis at A and the y-axis at B. Find the equation of the median through O of $\triangle$OAB.

8. Find the equation of the altitude through O of the triangle OAB defined in Question 7.

9. Find the equation of the perpendicular from $(5, 3)$ to the line $2x - y + 4 = 0$

10. The points $A(1, 4)$ and $B(5, 7)$ are two adjacent vertices of a parallelogram ABCD. The point $C(7, 10)$ is another vertex of the parallelogram. Find the equation of the side CD.

THE GRADIENT OF A LINE

When the equation of a line is written in standard form, i.e. $y = mx + c$, then m is the gradient of the line and c is its intercept on the y-axis.

When m is positive, the line makes an acute angle with the positive direction of the x-axis, and when m is negative, the line makes an obtuse angle with the positive direction of the x-axis.

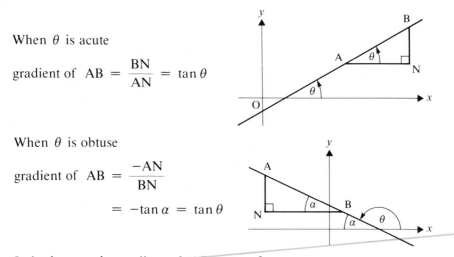

When θ is acute

gradient of AB $= \dfrac{BN}{AN} = \tan \theta$

When θ is obtuse

gradient of AB $= \dfrac{-AN}{BN}$

$= -\tan \alpha = \tan \theta$

In both cases the gradient of AB $= \tan \theta$

Therefore the gradient of a line is equal to the tangent of the angle between the line and the positive direction of the x-axis.

i.e. $m = \tan \theta$

Example 9d _____

Find the equation of the line through the point $(5, 1)$ which is inclined at $60°$ to the positive direction of the x-axis.

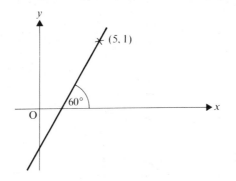

As the line is inclined at $60°$ to Ox, $m = \tan 60° = \sqrt{3}$

Therefore the required line has gradient $\sqrt{3}$ and passes through $(5, 1)$.

Using $y - y_1 = m(x - x_1)$ gives its equation as

$$y - 1 = \sqrt{3}(x - 5)$$

$\Rightarrow$ $$y = \sqrt{3}x + 1 - 5\sqrt{3}$$

EXERCISE 9d

1. Find the equation of the line with given inclination to Ox and passing through the given point.
   Use $\tan 60° = \sqrt{3}$ and $\tan 30° = \frac{1}{3}\sqrt{3}$
   (a) $45°, (1, -2)$ (b) $30°, (0, 0)$
   (c) $135°, (-1, 2)$ (d) $120°, (2, -1)$
   (e) $150°, (3, 1)$ (f) θ where $\tan \theta = \frac{1}{3}, (1, 5)$

2. Find the inclination to Ox of the line whose equation is given.
   Give answers in degrees correct to 1 decimal place.
   (a) $y = 2x - 1$ (b) $y = 5x - 3$
   (c) $x - 2y + 1 = 0$ (d) $3x + 2y - 1 = 0$

3. $A(2, 1)$, $B(7, 1)$ and $C(2, 8)$ are the vertices of triangle ABC. Find the equation of the line that bisects angle BAC.

THE COORDINATES OF A POINT DIVIDING A LINE IN A GIVEN RATIO

$A(x_1, y_1)$ and $B(x_2, y_2)$ are two fixed points and the point $P(X, Y)$ divides the line joining A to B in the ratio $\lambda : \mu$

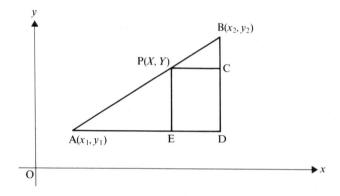

From the diagram, $\triangle APE$ and $\triangle PBC$ are similar,

$$\therefore \qquad \frac{AE}{PC} = \frac{AP}{PB} = \frac{\lambda}{\mu}$$

But $AE = X - x_1$, $PC = x_2 - X$ $\qquad \therefore \ X = \dfrac{\lambda x_2 + \mu x_1}{\lambda + \mu}$

A similar result can be found for Y, so

> if the point P divides the line joining $A(x_1, y_1)$ to $B(x_2, y_2)$ in the ratio $\lambda : \mu$, then the coordinates of P are
>
> $$\left(\frac{\lambda x_2 + \mu x_1}{\lambda + \mu}, \ \frac{\lambda y_2 + \mu y_1}{\lambda + \mu} \right)$$

These formulae, which are quotable, apply to both internal and external division. Their use is not always necessary however. When the coordinates of A and B are known numbers, a diagram, together with simple mental arithmetic, is often adequate.

Example 9e _____

Find the coordinates of the points P and Q which divide the line joining $A(-2, 5)$ and $B(4, 2)$ in the ratio $2 : 1$
(a) internally   (b) externally.

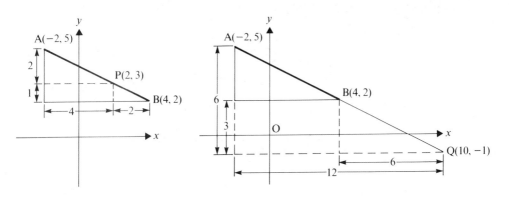

(a) Internal division   (b) External division

(a) From the diagram, P is the point $(2, 3)$

Alternatively, the formula can be used as follows.

P divides AB internally in the ratio $2 : 1$ so $\lambda = 2$ and $\mu = 1$

$$\Rightarrow \qquad x = \frac{2(4) + 1(-2)}{2 + 1} = 2 \quad \text{and} \quad y = \frac{2(2) + 1(5)}{2 + 1} = 3$$

(b) The diagram shows that Q is the point $(10, -1)$

Or, using the formula,

Q divides AB externally in the ratio $2 : 1$, so $\lambda = 2$ and $\mu = -1$

$$\Rightarrow \qquad x = \frac{2(4) - 1(-2)}{2 - 1} = 10 \quad \text{and} \quad y = \frac{2(2) - 1(5)}{2 - 1} = -1$$

Note that in the external division, the sign of μ is opposite to the sign of λ because the direction of the line segment QB is opposite that of the line segment AQ. For this reason, external division is sometimes denoted by a negative ratio, e.g. $2 : -1$.

_____

INTERSECTION

The point where two lines (or curves) cut is called a point of intersection.

> The coordinates of the point(s) of intersection can be found by solving the equations simultaneously.

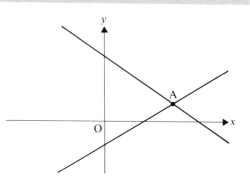

If A is the point of intersection of the lines $y - 3x + 1 = 0$ [1]

and $y + x - 2 = 0$ [2]

then the coordinates of A satisfy both of these equations. A can be found by solving [1] and [2] simultaneously, i.e.

$[2] - [1]$ $\Rightarrow$ $4x - 3 = 0$ $\Rightarrow$ $x = \frac{3}{4}$ and $y = \frac{5}{4}$

Therefore $(\frac{3}{4}, \frac{5}{4})$ is the point of intersection.

Note that the coordinates of A can also be found using a graphics calculator or a graph drawing package on a computer.

EXERCISE 9e

1. Find the coordinates of the point that divides AB in the given ratio in each of the following cases.
   (a) A(2, 4), B(−3, 9)   1 : 4 internally
   (b) A(−3, −4), B(3, 5)   3 : 1 externally
   (c) A(1, 5), B(8, −2)   4 : 3
   (d) A(−1, 6), B(3, −2)   3 : −2

2. Find the coordinates of the point of intersection of each pair of lines.
   (a) $y = 3x - 5$, $y = 4 - 5x$
   (b) $2x + y - 3 = 0$, $x - 3y + 4 = 0$
   (c) $y = 2x - 9$, $4x - 7y + 5 = 0$

PROBLEMS

Example 9f _____

A circle has radius 4 and its centre is the point $C(5, 3)$.

(a) Show that the points $A(5, -1)$ and $B(1, 3)$ are on the circumference of the circle.

(b) Prove that the perpendicular bisector of AB goes through the centre of the circle.

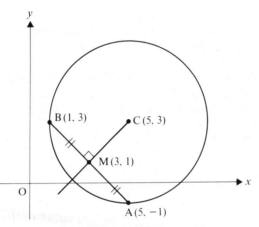

(a) From the diagram, $BC = 4$

$\therefore$ B is on the circumference.

Similarly $AC = 4$,

$\therefore$ A is on the circumference.

(b) The midpoint, M, of AB is $\left[\dfrac{5 + 1}{2}, \dfrac{-1 + 3}{2} \right]$ i.e. $(3, 1)$

The gradient of AB is $\dfrac{-1 - 3}{5 - 1} = -1$

If l is the perpendicular bisector of AB, its gradient is 1 and it goes through $(3, 1)$.

$\therefore$ the equation of l is

$$y - 1 = 1(x - 3) \quad \Rightarrow \quad y = x - 2 \qquad [1]$$

In equation [1], when $x = 5$, $y = 3$

$\therefore$ the point $(5, 3)$ is on l,

i.e. the perpendicular bisector of AB goes through C.

MIXED EXERCISE 9

1. Show that the triangle whose vertices are $(1, 1)$, $(3, 2)$ and $(2, -1)$ is isosceles.

2. Find the area of the triangular region enclosed by the x and y axes and the line $2x - y - 1 = 0$

3. Find the coordinates of the triangular region enclosed by the lines $y = 0$, $y = x + 5$ and $x + 2y - 6 = 0$

4. Write down the equation of the perpendicular bisector of the line joining the points $(2, -3)$ and $(-\frac{1}{2}, 3\frac{1}{2})$

5. Find the equation of the line through $A(5, 2)$ which is perpendicular to the line $y = 3x - 5$. Hence find the coordinates of the foot of the perpendicular from A to the line.

6. Find, in terms of a and b, the coordinates of the foot of the perpendicular from the point (a, b) to the line $x + 2y - 4 = 0$

7. The coordinates of a point P are $(t + 1, 2t - 1)$. Sketch the position of P when $t = -1, 0, 1$ and 2 Show that these points are collinear and write down the equation of the line on which they lie.

8. Write down the equation of the line which goes through $(7, 3)$ and which is inclined at $45°$ to the positive direction of the x-axis.

9. Find the equation of the perpendicular bisector of the line joining the points (a, b) and $(2a, -3b)$

10. The centre of a circle is at the point $C(3, 7)$ and the point $A(5, 3)$ is on the circumference of the circle. Find

    (a) the radius of the circle,

    (b) the equation of the line through A that is perpendicular to AC.

11. The equations of two sides of a square are $y = 3x - 1$ and $x + 3y - 6 = 0$. If $(0, -1)$ is one vertex of the square find the coordinates of the other vertices.

12. The lines $y = 2x$, $2x + y - 12 = 0$ and $y = 2$ enclose a triangular region of the xy-plane. Find

    (a) the coordinates of the vertices of this region,

    (b) the area of this region.

13. $A(5, 0)$ and $B(0, 8)$ are two vertices of triangle OAB.

   (a) What is the equation of the bisector of angle AOB.

   (b) If E is the point of intersection of this bisector and the line through A and B, find the coordinates of E.
   Hence show that $OA:OB = AE:EB$.

14. The points $A(6, 3)$ and $B(9, 0)$ are two vertices of a triangle OAB. The point P divides AB externally in the ratio $5:2$ and the point Q divides OB externally in the ratio $5:2$. Find the coordinates of P and Q and verify that PQ is parallel to OA.

CONSOLIDATION B

SUMMARY

PLANE GEOMETRY

Intercept Theorem

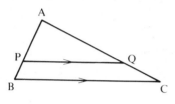

$$PQ \parallel BC \iff AP:BP = AQ:QC$$

Pythagoras' Theorem

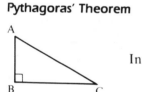

$$\text{In } \triangle ABC, \; \angle B = 90° \iff AC^2 = AB^2 + BC^2$$

Similar Triangles

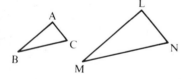

$$\angle A = \angle L, \; \angle B = \angle M, \; \angle C = \angle N$$
$$\iff AB:LM = BC:MN = AC:LN$$

Angle Bisector Theorem

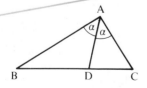

$$AD \text{ bisects } \angle A \iff BD:DC = AB:AC$$

142

TRIGONOMETRY

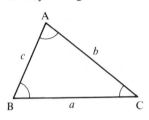

$$\sin \theta = \sin (180° - \theta)$$
$$\cos \theta = -\cos (180° - \theta)$$
$$\tan \theta = -\tan (180° - \theta)$$

In any triangle ABC,

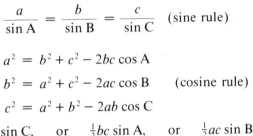

$$\frac{a}{\sin A} = \frac{b}{\sin B} = \frac{c}{\sin C} \quad \text{(sine rule)}$$

$$a^2 = b^2 + c^2 - 2bc \cos A$$
$$b^2 = a^2 + c^2 - 2ac \cos B \quad \text{(cosine rule)}$$
$$c^2 = a^2 + b^2 - 2ab \cos C$$

The area of $\triangle ABC$ is $\quad \frac{1}{2}ab \sin C, \quad$ or $\quad \frac{1}{2}bc \sin A, \quad$ or $\quad \frac{1}{2}ac \sin B$

COORDINATE GEOMETRY

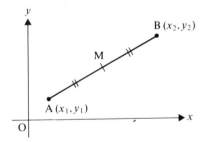

Length of AB is $\sqrt{[(x_2 - x_1)^2 + (y_2 - y_1)^2]}$

Midpoint, M, of AB is $[\frac{1}{2}(x_1 + x_2), \frac{1}{2}(y_1 + y_2)]$

Gradient of AB is $\dfrac{y_2 - y_1}{x_2 - x_1}$

Parallel lines have equal gradients.

When two lines are perpendicular, the product of their gradients is -1

The standard equation of a straight line is $y = mx + c$, where m is its gradient and c its intercept on the y-axis.

Any equation of the form $ax + by + c = 0$ is a straight line.

The equation of a line passing through (x_1, y_1) and with gradient m is

$$y - y_1 = m(x - x_1)$$

Given a line with equation $ax + by + c = 0$,
then any perpendicular line has equation $bx - ay + k = 0$

For two points, $A(x_1, y_1)$ and $B(x_2, y_2)$, the point which divides AB in the ratio $\lambda : \mu$ has coordinates

$$\left(\frac{\lambda x_2 + \mu x_1}{\lambda + \mu}, \frac{\lambda y_2 + \mu y_1}{\lambda + \mu} \right)$$

For internal division, λ and μ are both positive while, for external division, the smaller of λ and μ is negative, i.e. the ratio is negative.

When a line with gradient m
makes an angle θ with the
positive x-axis, then $m = \tan \theta$

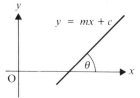

MULTIPLE CHOICE EXERCISE B

TYPE I

1. A line AB is 10 cm long. P divides AB externally in the ratio 9:4. The length of PB is

   A 18 cm   **C** 8 cm   **E** 4 cm
   B $\frac{40}{13}$ cm   **D** 40 cm

2. In $\triangle ABC$, $a = 3$ cm, $b = 4$ cm and $c = 5$ cm,

   A $A = 90°$ **C** $B = 45°$ **E** $C = 90°$
   B $C = 60°$ **D** $B = 90°$

3.

ABCDE is a right square-based pyramid (i.e. E is vertically above the centre of the base). The angle between the face BEC and the base ABCD is

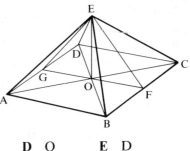

A F **B** B **C** C **D** O **E** D

4. The length of the line joining $(3, -4)$ to $(-7, 2)$ is

A $2\sqrt{13}$ **C** $2\sqrt{34}$ **E** 6
B 16 **D** $2\sqrt{5}$

5. The midpoint of the line joining $(-1, -3)$ to $(3, -5)$ is

A $(1, 1)$ **C** $(2, -8)$ **E** $(1, -1)$
B $(0, 0)$ **D** $(1, -4)$

6. The gradient of the line joining $(1, 4)$ and $(-2, 5)$ is

A $\frac{1}{3}$ **B** $-\frac{1}{3}$ **C** 3 **D** -3 **E** 1.3

7. The gradient of the line perpendicular to the join of $(-1, 5)$ and $(2, -3)$ is

A $\frac{3}{8}$ **B** $-2\frac{2}{3}$ **C** $\frac{1}{2}$ **D** 2 **E** $2\frac{2}{3}$

8. The equation of the line through the origin and perpendicular to $3x - 2y + 4 = 0$ is

A $3x + 2y = 0$ **C** $2x + 3y = 0$ **E** $3x - 2y = 0$
B $2x + 3y + 1 = 0$ **D** $2x - 3y - 1 = 0$

TYPE II

9. In triangle ABC

$$\textbf{A} \quad \frac{a}{\sin A} = \frac{b}{\sin a}$$

B $b^2 = a^2 + c^2 + 2ac \cos a$
C $a^2 = b^2 + c^2 + 2bc \cos A$

10. A and B are two points with coordinates $(3, 4)$, $(-1, 6)$
     A   Gradient of AB is $-\frac{1}{2}$
     B   Midpoint of AB is the point $(2, 5)$
     C   Length of AB is $2\sqrt{5}$

11. A, B and C are the points $(5, 0)$, $(-5, 0)$, $(2, 3)$
     A   AB and BC are perpendicular.
     B   Area of triangle ABC is 15 square units.
     C   A, B and C are collinear.

12. The equation of a line l is $y = 2x - 1$
     A   The line through the origin perpendicular to l is $y + 2x = 0$
     B   The line through $(1, 2)$ parallel to l is $y = 2x - 3$
     C l passes through $(1, 1)$.

13. The equation of a line l is $7x - 2y + 4 = 0$

     A l has a gradient of $3\frac{1}{2}$
     B l is parallel to $7x + 2y - 3 = 0$
     C l is perpendicular to $2x + 7y - 5 = 0$

TYPE III

14. The line joining $(0, 0)$ and $(1, 3)$ is equal in length to the line joining $(0, 1)$ and $(3, 0)$

15. If a line has gradient m and intercept d on the x-axis, its equation is $y = mx - md$

16. The line passing through $(3, 1)$ and $(-2, 5)$ is perpendicular to the line $4y = 5x - 3$

MISCELLANEOUS EXERCISE B

1. Find the equation of the straight line which passes through the point $(2, 5)$ and has gradient 3. Find also the coordinates of the point of intersection of this line and the line with equation $x + 2y = 0$
                                                                    (O/C, SU & C)

2. The points A and B have coordinates $(-5, -2)$ and $(7, 4)$ respectively. Find
    (a) the coordinates of the point C which divides AB internally in the ratio $2 : 1$
    (b) an equation of the line through C perpendicular to AB.
                                                                    (U of L)

3. The vertices of the quadrilateral $ABCD$ are the points $A(-4, 3)$, $B(2, -1)$, $C(4, 2)$ and $D(1, 4)$. Showing relevant calculations, prove that the quadrilateral has two parallel sides and that one of its other sides is perpendicular to them.

Calculate the area of the quadrilateral.

Find the equation of the line which passes through the midpoint of CD and is parallel to BC.

(JMB)

4. A line passes through the point of intersection of the lines with equations $x + 3y - 3 = 0$ and $2x - 3y - 6 = 0$. Find the equation of this line if

(a) its gradient is $\frac{3}{2}$

(b) it passes through the point $(0, 3)$

5. The curve $y = 1 + \dfrac{1}{2 + x}$ crosses the x-axis at the point A and the y-axis at the point B.

(a) Calculate the coordinates of A and of B.

(b) Find the equation of the line AB.

(c) Calculate the coordinates of the point of intersection of the line AB and the line with equation $3y = 4x$

(O/C, SU & C)

6. Coastguard stations are situated at Portland (P) and Swanage (S). The station at Swanage is $32\ \text{km}$ from the station at Portland on a bearing of $080°$. From P the bearing of a container ship (C) is $130°$ and the bearing of a car ferry (F) is $135°$. From S the bearing of C is $220°$ and the bearing of F is $200°$

(a) Write down the size of $\angle PCS$.

Calculate, in km to 2 decimal places, the distance,

(b) CS (c) FS (d) CF

Calculate to the nearest degree,

(e) the bearing of F from C. (U of L)

7. The equations of two adjacent sides of a rhombus are $y = 2x + 4$, $y = -\frac{1}{3}x + 4$. If $(12, 0)$ is one vertex and all vertices have positive coordinates, find the coordinates of the other three vertices.

8. Show that the volume V of a right circular cone of slant height l and semi-vertical angle θ, where $0 < \theta < \pi/2$, is given by

$$V = \tfrac{1}{3}\pi l^3 \sin^2\theta \cos\theta$$

(U of L)p

9. A vertical wall, 2.7 m high, runs parallel to the wall of a house and is at a horizontal distance of 6.4 m from the house. An extending ladder is placed to rest on the top B of the wall with one end C against the house and the other end A resting on horizontal ground, as shown in the figure.

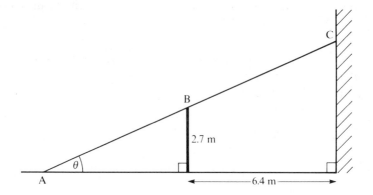

The points A, B and C are in a vertical plane at right angles to the wall and the ladder makes an angle θ, where $0 < \theta < \pi/2$, with the horizontal. Show that the length, y metres, of the ladder is given by

$$y = \frac{2.7}{\sin \theta} + \frac{6.4}{\cos \theta}$$ (U of L)p

10. In the triangle ABC, $AB = 8$, $BC = 7$, $\angle A = 60°$. Given that $AC < 4$, find AC.   (C)

11. In the triangle ABC, the point D is the foot of the perpendicular from A to BC. Show that

$$AD = \frac{BC \sin B \sin C}{\sin A}$$

The triangle ABC lies in a horizontal plane. A vertical pole FT stands with its foot F on AD and between A and D. The top T of the pole is at a height 3 m above the plane, and the angle of elevation of T from D is 65°. Given that $BC = 7$ m, $\angle B = 62°$ and $\angle C = 34°$, calculate AF correct to two decimal places.

Find, to the nearest degree, the angle between the planes ATC and ABC.   (JMB)

12. A pyramid *VABCD* has a horizontal square base *ABCD*, of side 6*a*. The vertex *V* of the pyramid is at a height 4*a*, vertically above the centre of the base. Calculate, in degrees, to one decimal place, the acute angle between

(a) the edge *VA* and the horizontal

(b) the plane *VAB* and the horizontal

(c) the planes *VBA* and *VBC*   (AEB)

13. The rectangle *ABCD* is the horizontal base of a pyramid, and the vertex *V* of the pyramid is vertically above the centre of the base. The length of *AB* is 6 cm, the length of *BC* is 12 cm and the length of *VA* is 20 cm. The points *X* on *VB* and *Y* on *BC* are such that *AX* and *YX* are both perpendicular to *VB*.

(a) Show that the length of *BX* is 0.9 cm, and calculate the length of *BY*.

(b) Calculate, to the nearest degree, the angle between the planes *VAB* and *VBC*.   (C)

14.

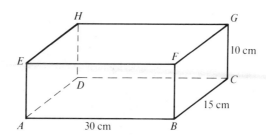

The diagram shows a rectangular solid *ABCDEFGH*, with *AB* = 30 cm, *BC* = 15 cm and *CG* = 10 cm. Calculate

(a) the angle between *DG* and *BG*, giving your answer correct to the nearest 0.1°

(b) the perpendicular distance of *C* from *BD*, giving your answer correct to three significant figures

(c) the angle between the planes *DGB* and *DCB*, giving your answer correct to the nearest 0.1°   (C)

15. Two of the sides of a parallelogram lie on the lines with equations $y = \frac{1}{2}x$ and $y = 3x$. The vertex that is not on these sides is at the point (4, 7). Find the equations of the lines containing the other two sides of the parallelogram. Hence find the coordinates of the remaining vertices.

16. The points $A(-7, 12)$, $B(2, 10)$ and $C(-1, 4)$ are the vertices of a triangle ABC. Write down the coordinates of D, the midpoint of AC. Find the equation of the line through D parallel to AB. This line meets BC at E. Determine the coordinates of E.

17. The coordinates of the midpoints of the sides of a triangle are $(3, 2)$, $(1, 5)$ and $(-3, -1)$. Find the equations of the lines on which the sides of the triangle lie.

CHAPTER 10

CIRCLE GEOMETRY

PARTS OF A CIRCLE

We start this chapter with a reminder of the language used to describe parts of a circle.

Part of the circumference is called an *arc*.
If the arc is less than half the circumference it is called a *minor arc*; if it is greater than half the circumference it is called a *major arc*.

A straight line which cuts a circle in two distinct points is called a *secant*. The part of the line inside the circle is called a *chord*.

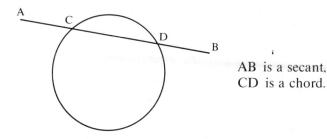

AB is a secant,
CD is a chord.

The area enclosed by two radii and an arc is called a *sector*.

The area enclosed by a chord and an arc is called a *segment*. If the segment is less than half a circle it is called a *minor segment*; if it is greater than half a circle it is called a *major segment*.

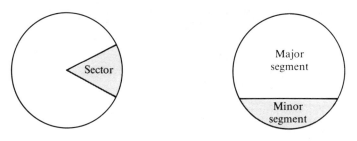

THE ANGLE SUBTENDED BY AN ARC

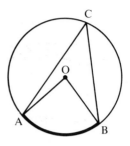

Consider the points A, B and C
on the circumference of a circle
whose centre is O.

We say that $\angle ACB$ stands on the minor arc AB.

The minor arc AB is said to *subtend* the angle ACB at the
circumference (and the angle is *subtended* by the arc).

In the same way, the arc AB is said to subtend the angle AOB at the
centre of the circle.

Example 10a _____

A circle of radius 2 units which has its centre at the origin, cuts the
x-axis at the points A and B and cuts the y-axis at the point C.
Prove that $\angle ACB = 90°$

All the information given in the
question, and gleaned from the known
properties of the figure, can be marked
in the diagram as shown. The diagram
can then be referred to as justification
for steps taken in the solution.

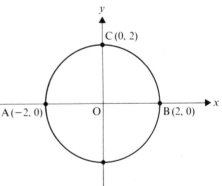

From the diagram, the gradient of AC is $\dfrac{2-0}{0-(-2)} = 1$

and the gradient of BC is $\dfrac{2-0}{0-2} = -1$

$\therefore$ (gradient of AC) $\times$ (gradient of BC) $= -1$

i.e. AC is perpendicular to BC $\quad\Rightarrow\quad$ $\angle ACB = 90°$

EXERCISE 10a

1.

Name the angles subtended

(a) at the circumference by the minor arc AE

(b) at the circumference by the major arc AE

(c) at the centre by the minor arc AC

(d) at the circumference by the major arc AC

(e) at the centre by the minor arc CE

(f) at the circumference by the minor arc CD

(g) at the circumference by the minor arc BC.

2. AB is a chord of a circle, centre O, and M is its midpoint. The radius from O is drawn through M. Prove that OM is perpendicular to AB.

3. $C(5, 3)$ is the centre of a circle of radius 5 units.

    (a) Show that this circle cuts the x-axis at $A(1, 0)$ and $B(9, 0)$

    (b) Prove that the radius that is perpendicular to AB goes through the midpoint of AB.

    (c) Find the angle subtended at C by the minor arc AB.

    (d) The point D is on the major arc AB and DC is perpendicular to AB. Find the coordinates of D and hence find the angle subtended at D by the minor arc AB.

4. A and B are two points on the circumference of a circle centre O. C is a point on the major arc AB. Draw the lines AC, BC, AO, BO and CO, extending the last line to a point D inside the sector AOB. Prove that ∠AOD is twice ∠ACO and that ∠BOD is twice angle ∠BCO. Hence show that the angle subtended by the minor arc AB at the centre of the circle is twice the angle that it subtends at the circumference of the circle.

ANGLES IN A CIRCLE

The solutions to questions in the last exercise illustrate two important results.

1) The perpendicular bisector of a chord of a circle goes through the centre of the circle.

2) The angle subtended by an arc at the centre of a circle is twice the angle subtended at the circumference by the same arc.

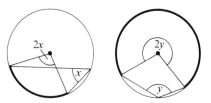

Further important results follow from the last fact.

3)

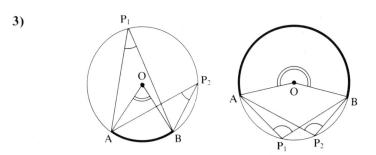

In both diagrams, $\angle AOB = 2\angle P_1 = 2\angle P_2$

So it follows that $\angle P_1 = \angle P_2$

i.e.

all angles subtended at the circumference by the same arc are equal.

4) A semicircle subtends an angle of 180° at the centre of the circle; therefore it subtends an angle half that size, i.e. 90°, at any point on the circumference. This angle is called the angle in a semicircle.

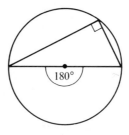

Hence   the angle in a semicircle is 90°

5) If all four vertices of a quadrilateral ABCD lie on the circumference of a circle, ABCD is called a *cyclic quadrilateral*.

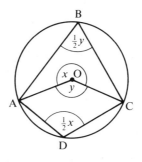

In the diagram, O is the centre of the circle.

∴ $\angle ADC = \tfrac{1}{2}x$ and $\angle ABC = \tfrac{1}{2}y$

But $x + y = 360°$, therefore $\angle ADC + \angle ABC = 180°$

i.e.   the opposite angles of a cyclic quadrilateral are supplementary

Example 10b

A circle circumscribes a triangle whose vertices are at the points A(0, 4), B(2, 3) and C(−2, −1). Find the centre of the circle.

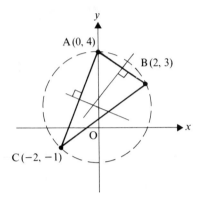

When a circle circumscribes a figure, the vertices of the figure lie on the circumference of the circle. The centre of the circle can be found by locating the point of intersection of the perpendicular bisectors of two chords.

The gradient of AC is $\dfrac{4 - (-1)}{0 - (-2)} = \dfrac{5}{2}$

and its midpoint is $\left[\dfrac{0 - 2}{2}, \dfrac{4 - 1}{2}\right] \quad \Rightarrow \quad (-1, \tfrac{3}{2}),$

∴ the gradient of the perpendicular bisector of AC is $-\tfrac{2}{5}$ and its equation is

$$y = -\tfrac{2}{5}x + \tfrac{11}{10} \quad \Rightarrow \quad 4x + 10y - 11 = 0 \qquad [1]$$

Similarly the gradient of AB is $-\tfrac{1}{2}$ and its midpoint is $(1, \tfrac{7}{2})$

∴ the gradient of the perpendicular bisector of AB is 2 and its equation is

$$y = 2x + \tfrac{3}{2} \quad \Rightarrow \quad 4x - 2y + 3 = 0 \qquad [2]$$

Solving equations [1] and [2] simultaneously gives

$$12y - 14 = 0 \quad \Rightarrow \quad y = \tfrac{7}{6} \text{ and } x = -\tfrac{1}{6}$$

Therefore the centre of the circle is the point $(-\tfrac{1}{6}, \tfrac{7}{6})$.

EXERCISE 10b

1. Find the size of each marked angle.

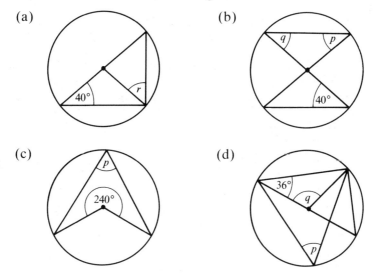

(a)　　　　　　　　　　(b)

(c)　　　　　　　　　　(d)

2. AB is a diameter of a circle centre O. C is a point on the circumference. D is a point on AC such that OD bisects ∠AOC. Prove that OD is parallel to BC.

3. A triangle has its vertices at the points A(1, 3), B(5, 1) and C(7, 5). Prove that △ABC is right-angled and hence find the coordinates of the centre of the circumcircle of △ABC.

4. Find the size of the angle marked
   e in the diagram.

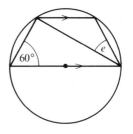

5. AB and CD are two chords of a circle that cut at E. (E is not the centre of the circle.) Show that △s ACE and BDE are similar.

6. A circle with centre O circumscribes an equilateral triangle ABC. The radius drawn through O and the midpoint of AB meets the circumference at D. Prove that △ADO is equilateral.

7. The line joining $A(5, 3)$ and $B(4, -2)$ is a diameter of a circle. If $P(a, b)$ is a point on the circumference find a relationship between a and b

8. ABCD is a cyclic quadrilateral. The side CD is produced to a point E outside the circle. Show that $\angle ABC = \angle ADE$.

9. A triangle has its vertices at the points $A(1, 3)$, $B(-2, 5)$ and $C(4, -2)$. Find the coordinates of the centre, and correct to 3 s.f., the radius of the circle that circumscribes $\triangle ABC$.

10. In the diagram, O is the centre of the circle and CD is perpendicular to AB. If $\angle CAB = 30°$ find the size of each marked angle.

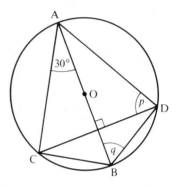

TANGENTS TO CIRCLES

If a line and a circle are drawn on a plane then there are three possibilities for the position of the line in relation to the circle. The line can miss the circle, or it can cut the circle in two distinct points, or it can touch the circle at one point. In the last case the line is called a *tangent* and the point at which it touches the circle is called the *point of contact*.

The length of a tangent drawn from a point to a circle is the distance from that point to the point of contact.

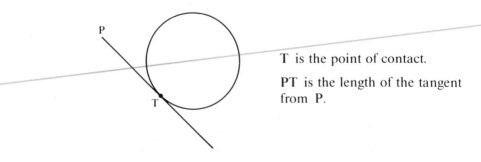

T is the point of contact.

PT is the length of the tangent from P.

Properties of Tangents to Circles

There are two important and useful properties of tangents to circles.

A tangent to a circle is
perpendicular to the radius drawn
through the point of contact,
i.e. AB is perpendicular to OT.

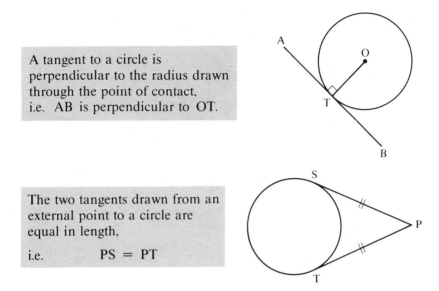

The two tangents drawn from an
external point to a circle are
equal in length,

i.e.   PS = PT

The second property is proved in the following worked example.

Examples 10c

1. PS and PT are two tangents drawn from a point P to a circle
   whose centre is O. Prove that PT = PS.

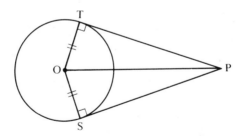

In △s OTP, OSP   ∠T = ∠S = 90°

                      OS = OT (radii)

                      and OP is common

∴ △s OTP, OSP are congruent.

Hence PT = PS.

Another useful property follows from the last example, namely

> when two tangents are drawn from a point to a circle, the line
> joining that point to the centre of the circle bisects the angle
> between the tangents.

---

2. A circle of radius 10 units is circumscribed by a right-angled
    isosceles triangle. Find the lengths of the sides of the triangle.

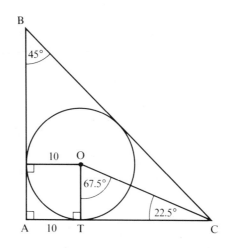

A circle is *circumscribed* by a figure when all the sides of the figure
touch the circle. Note also that the circle is inscribed in the figure.

From the diagram

in △OTC, $TC = 10 \tan 67.5°$

                                     $= 24.14$

                                $AT = 10$

∴ $AC = 34.14 = AB$

in △ABC, $BC = \sqrt{(34.14^2 + 34.14^2)}$ (Pythagoras)

                                     $= 48.28$

∴ correct to 3 s.f. the lengths of the sides of the triangle are
                34.1 units, 34.1 units and 48.3 units.

---

3. The centre of a circle of radius 3 units is the point $C(2, 5)$. The equation of a line, l, is $x + y - 2 = 0$

   (a) Find the equation of the line through C, perpendicular to l

   (b) Find the distance of C from l and hence determine whether l is a tangent to the circle.

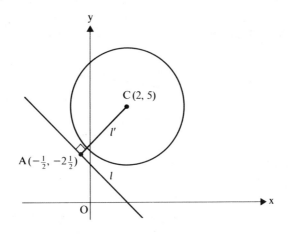

(a) The line l' is perpendicular to $x + y - 2 = 0$

   so its equation is $x - y + k = 0$

The point $(2, 5)$ lies on l'

$\therefore$ $2 - 5 + k = 0$ $\Rightarrow$ $k = 3$

$\therefore$ the equation of l' is $x - y + 3 = 0$

(b) To find the distance of C from l we need the coordinates of A, the point of intersection of l and l'.

Adding the equations of l and l' gives $2x + 1 = 0$

$\Rightarrow$ $x = -\frac{1}{2}$ and $y = \frac{5}{2}$

so A is the point $(-\frac{1}{2}, \frac{5}{2})$

$\therefore$ $CA = \sqrt{[\{2 - (-\frac{1}{2})\}^2 + \{5 - \frac{5}{2}\}^2]} = 3.54$ to 3 s.f.

For the line to be a tangent, CA would have to be 3 units exactly (i.e. equal to the radius).

CA > 3, therefore l is not a tangent.

EXERCISE 10c

1. The two tangents from a point to a circle of radius 12 units are each of length 20 units. Find the angle between the tangents.

2. Two circles with centres C and O have radii 6 units and 3 units respectively and the distance between O and C is less than 9 units. AB is a tangent to both circles, touching the larger circle at A and the smaller circle at B, where AB is of length 4 units. Find the length of OC.

3. The two tangents from a point A to a circle touch the circle at S and at T. Find the angle between one of the tangents and the chord ST given that the radius of the circle is 5 units and that A is 13 units from the centre of the circle.

4. An equilateral triangle of side 25 cm circumscribes a circle. Find the radius of the circle.

5. AB is a diameter of the circle and C is a point on the circumference. The tangent to the circle at A makes an angle of 30° with the chord AC. Find the angles in △ABC.

6. The centre of a circle is at the point C(4, 8) and its radius is 3 units. Find the length of the tangents from the origin to the circle.

7. A circle touches the y-axis at the origin and goes through the point A(8, 0). The point C is on the circumference. Find the greatest possible area of △OAC.

8. A triangular frame is made to enclose six identical spheres as shown. Each sphere has a radius of 2 cm. Find the lengths of the sides of the frame.

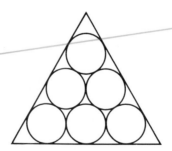

9. The line $y = 3x - 4$ is a tangent to the circle whose centre is the point $(5, 2)$. Find the radius of the circle.

10. A circle of radius 6 units has its centre at the point $(9, 0)$. If the two tangents from the origin to the circle are inclined to the x-axis at angles α and β, find $\tan \alpha$ and $\tan \beta$

11. A, B and C are three points on the circumference of a circle. The tangent to the circle at A makes an angle α with the chord AB. The diameter through A cuts the circle again at D and D is joined to B. Prove that $\angle ACB = \alpha$

12. The equations of the sides of a triangle are $y = 3x$, $y + 3x = 0$ and $3y - x + 12 = 0$ Find the coordinates of the circumcircle of this triangle.

13. The line $x - 2y + 4 = 0$ is a tangent to the circle whose centre is the point $C(-1, 2)$.
    (a) Find the equation of the line through C that is perpendicular to the line $x - 2y + 4 = 0$
    (b) Hence find the coordinates of the point of contact of the tangent and the circle.

14. The point $A(6, 8)$ is on the circumference of a circle whose centre is the point $C(3, 5)$. Find the equation of the tangent that touches the circle at A.

CHAPTER 11

RADIANS, ARCS AND SECTORS

ANGLE UNITS

An angle is a measure of rotation and the units which have been used up to now are the revolution and the degree. It is interesting to note why the number of degrees in a revolution was taken as 360. The ancient Babylonian mathematicians, in the belief that the length of the solar year was 360 days, divided a complete revolution into 360 parts, one for each day as they thought. We now know they did not have the length of the year quite right but the number they used, 360, remains as the number of degrees in one revolution.

Part of an angle smaller than a degree is usually given as a decimal part but until recently the common practice was to divide a degree into 60 minutes (60′) and each minute into 60 seconds (60″). Limited use is still made of this system.

Now we consider a different unit of rotation which is of great importance in much of the mathematics that follows.

THE RADIAN

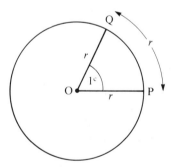

If O is the centre of a circle and an arc PQ is drawn so that its length is equal to the radius of the circle then the angle POQ is called a *radian* (one radian is written 1 rad or 1^c), i.e.

> An arc equal in length to the radius of a circle subtends an angle of 1 radian at the centre.

164

It follows that the number of radians in a complete revolution is the number of times the radius divides into the circumference.

Now the circumference of a circle is of length $2\pi r$

Therefore the number of radians in a revolution is

$$\frac{2\pi r}{r} = 2\pi$$

i.e. 2π radians $= 360°$

Further π radians $= 180°$

and $\tfrac{1}{2}\pi$ radians $= 90°$

When an angle is given in terms of π it is usual to omit the radian symbol, i.e. we would write $180° = \pi$ (not $180° = \pi^{c}$)

If an angle is a simple fraction of $180°$, it is easily expressed in terms of π

e.g. $60° = \tfrac{1}{3}$ of $180° = \tfrac{1}{3}\pi$

and $135° = \tfrac{3}{4}$ of $180° = \tfrac{3}{4}\pi$

Conversely, $\tfrac{7}{6}\pi = \tfrac{7}{6}$ of $180° = 210°$

and $\tfrac{2}{3}\pi = \tfrac{2}{3}$ of $180° = 120°$

Angles that are not simple fractions of $180°$, or of π, can be converted by using the relationship $\pi = 180°$ and the value of π from a calculator,

e.g. $73° = \dfrac{73}{180} \times \pi = 1.27^{c}$ (correct to 3 s.f.)

and $2.36^{c} = \dfrac{2.36}{\pi} \times 180° = 135°$ (correct to the nearest degree)

It helps in visualising the size of a radian to remember that 1 radian is just a little less than $60°$.
($180° = \pi$ rad $= 3.142$ rad $\Rightarrow$ 1 rad $= 180°/3.142 \approx 57°$)

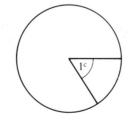

EXERCISE 11a

1. Express each of the following angles in radians as a fraction of π.
   $45°$, $150°$, $30°$, $270°$, $225°$, $22.5°$, $240°$, $300°$, $315°$

2. Without using a calculator express each of the following angles in degrees.
   $\frac{1}{6}\pi$, π, $\frac{1}{10}\pi$, $\frac{1}{4}\pi$, $\frac{5}{6}\pi$, $\frac{1}{12}\pi$, $\frac{1}{8}\pi$, $\frac{4}{3}\pi$, $\frac{1}{9}\pi$, $\frac{3}{2}\pi$, $\frac{4}{9}\pi$

3. Use a calculator to express each of the following angles in radians.
   $35°$, $47.2°$, $93°$, $233°$, $14.1°$, $117°$, $370°$

4. Use a calculator to express each of the following angles in degrees.
   1.7 rad, 3.32 rad, 1 rad, 2.09 rad, 5 rad, 6.283 19 rad

MENSURATION OF A CIRCLE

The reader will already be familiar with the formulae for the circumference and the area of a circle, i.e.

$$\text{Circumference} = 2\pi r \quad \text{and} \quad \text{Area} = \pi r^2$$

Now that we have defined a radian, these formulae can be used to derive other results.

The Length of an Arc

Consider an arc which subtends an angle θ at the centre of a circle, *where θ is measured in radians*

From the definition of a radian, the arc which subtends an angle of 1 radian at the centre of the circle is of length r. Therefore an arc which subtends an angle of θ radians at the centre is of length $r\theta$

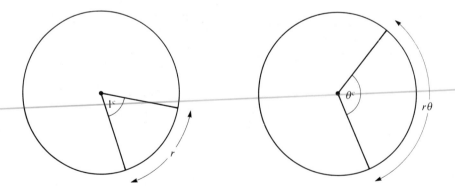

The Area of a Sector

The area of a sector can be thought of as a fraction of the area of the whole circle.

Consider a sector containing an angle of θ radians at the centre of the circle.

The complete angle at the centre of the circle is 2π, hence

$$\frac{\text{Area of sector}}{\text{Area of circle}} = \frac{\theta}{2\pi}$$

$\Rightarrow \qquad \text{Area of sector} = \dfrac{\theta}{2\pi} \times \pi r^2$

$$= \tfrac{1}{2}r^2\theta$$

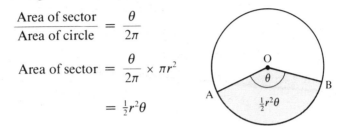

We now have two important facts about a circle in which an arc AB subtends an angle θ at the centre of the circle

> The length of arc AB $= r\theta$
>
> The area of sector AOB $= \tfrac{1}{2}r^2\theta$

When solving problems involving arcs and sectors, answers are usually given in terms of π. If a numerical answer is required it will be asked for specifically.

Examples 11b

1. An elastic belt is placed round the rim of a pulley of radius 5 cm. One point on the belt is pulled directly away from the centre, P, of the pulley, until it is at A, 10 cm from P. Find the length of the belt that is in contact with the rim of the pulley.

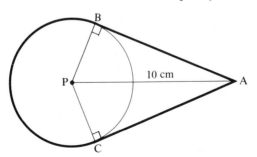

The belt leaves the pulley at B and at C. At these two points the belt is a tangent to the rim, so AB is perpendicular to the radius BP. Similarly, AC and PC are perpendicular.

In $\triangle ABP$ $BP = 5$ cm, $AP = 10$ cm and $\angle ABP = 90°$

Therefore $\cos APB = \frac{5}{10} = \frac{1}{2}$

$\Rightarrow$ $\angle APB = 60° = \frac{1}{3}\pi$

Similarly $\angle APC = \frac{1}{3}\pi$

The angle subtended at P by the major arc BC is given by

$$2\pi - \angle BPA - \angle CPA = 2\pi - \tfrac{1}{3}\pi - \tfrac{1}{3}\pi = \tfrac{4}{3}\pi$$

The length of an arc is given by $r\theta$, therefore

the length of the major arc BC is $5 \times \tfrac{4}{3}\pi = \tfrac{20}{3}\pi$

i.e. the length of belt in contact with the pulley is $\tfrac{20}{3}\pi$ cm.

---

2. AB is a chord of a circle with centre O and radius 4 cm. AB is of
    length 4 cm and divides the circle into two segments. Find, correct
    to two decimal places, the area of the minor segment.

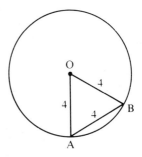

Each side of $\triangle ABC$ is 4 cm.
$\therefore$ ABC is an equilateral triangle
$\therefore$ each angle is $60°$.

Area of sector $AOB = \tfrac{1}{2}r^2\theta$

$\qquad\qquad\qquad = \tfrac{1}{2}(4^2)(\tfrac{1}{3}\pi)$

$\qquad\qquad\qquad = \tfrac{8}{3}\pi$

Area of minor segment = area of sector AOB − area of $\triangle AOB$

$\qquad\qquad\qquad\qquad = \tfrac{8}{3}\pi - \tfrac{1}{2}(4)(4)(\sin 60°)$

$\qquad\qquad\qquad\qquad = 8.378 - 6.928$

$\qquad\qquad\qquad\qquad = 1.450$

The area of the minor segment is 1.45 cm^2 correct to 2 d.p.

EXERCISE 11b

In questions 1 to 10, s is the length of an arc subtending an angle θ at the centre of a circle of radius r, and a is the area of the corresponding sector. Complete the table.

	s (cm)	θ	r (cm)	a (cm^2)
1.		30°	4	
2.		$\frac{5}{6}\pi$	10	
3.	15	π		
4.	20	$\frac{4}{5}\pi$		
5.		135°	8	
6.			2	π
7.			5	12
8.		$\frac{1}{6}\pi$		3π
9.	5π			20π

10. Calculate, in degrees, the angle subtended at the centre of a circle of radius 2.7 cm by an arc of length 6.9 cm.

11. Calculate, in radians, the angle at the centre of a circle of radius 83 mm contained in a sector of area 974 mm^2.

12. The diameter of the moon is about 3445 km and the distance between the moon and earth is about 382 100 km. Find the angle subtended at a point on the earth's surface by the moon (give your answer as a decimal part of a degree to 2 d.p.).

13. In a circle with centre O and radius 5 cm, AB is a chord of length 8 cm. Find
    (a) the area of triangle AOB
    (b) the area of the sector AOB (in square centimetres, correct to 3 s.f.).

14. A chord of length 10 mm divides a circle of radius 7 mm into two segments. Find the area of each segment.

15. A chord PQ, of length 12.6 cm, subtends an angle of $\frac{2}{3}\pi$ at the centre of a circle. Find

(a) the length of the arc PQ

(b) the area of the minor segment cut off by the chord PQ.

16. A curve in the track of a railway line is a circular arc of length 400 m and radius 1200 m. Through what angle does the direction of the track turn?

17. Two discs, of radii 5 cm and 12 cm, are placed, partly overlapping, on a table. If their centres are 13 cm apart find the perimeter of the 'figure-eight' shape.

18. Two circles, each of radius 14 cm, are drawn with their centres 20 cm apart. Find the length of their common chord. Find also the area common to the two circles.

The next three questions are a little more demanding.

19. A chord of a circle subtends an angle of θ radians at the centre of the circle. The area of the minor segment cut off by the chord is one eighth of the area of the circle.
Prove that $4\theta = \pi + 4\sin\theta$

20. A chord PQ of length $6a$ is drawn in a circle of radius $10a$. The tangents to the circle at P and Q meet at R. Find the area enclosed by PR, QR and the minor arc PQ.

21. Two discs are placed, in contact with each other, on a table. Their radii are 4 cm and 9 cm. An elastic band is stretched round the pair of discs. Calculate

(a) the angle subtended at the centre of the smaller disc by the arc that is in contact with the elastic band.

(b) the length of the part of the band that is in contact with the smaller disc.

(c) the length of the part of the band that is in contact with the larger disc.

(d) the total length of the stretched band.

(*Hint.* The straight parts of the stretched band are common tangents to the two circles.)

FUNCTIONS

MAPPINGS

If, on a calculator, the number 2 is entered and then the x^2 button is pressed, the display shows the number 4

We say that 2 is mapped to 2^2 or $2 \rightarrow 2^2$

Under the same rule, i.e. squaring the input number,
$3 \rightarrow 9$, $25 \rightarrow 625$, $0.5 \rightarrow 0.25$, $-4 \rightarrow 16$ and in fact,

(any real number) $\rightarrow$ (the square of that number)

The last statement can be expressed more briefly as

$$x \rightarrow x^2$$

where x is any real number.

This mapping can be represented graphically by plotting values of x^2 against values of x

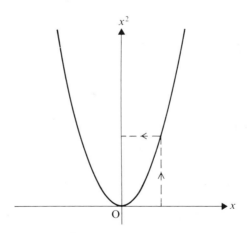

The graph, and knowledge of what happens when we square a number, show that one input number gives just one output number.

171

Now consider the mapping which maps a number to its square roots; the rule by which, for example, $4 \rightarrow +2$ and -2

This rule gives a real output only if the input number is greater than zero (negative numbers do not have real square roots). This mapping can now be written in general terms as

$$x \rightarrow \pm\sqrt{x} \quad \text{for} \quad x \geqslant 0$$

The graphical representation of this mapping is shown below.

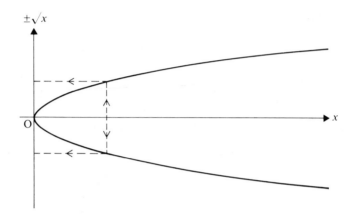

This time we notice that one input value gives two output values.

From these two examples, we can see that a mapping is a rule for changing a number to another number or numbers.

FUNCTIONS

Under the first mapping, $x \rightarrow x^2$, one input number gives one output number. However, for the second mapping, $x \rightarrow \pm\sqrt{x}$, one input number gives two output numbers.

We use the word *function* for any rule that gives the same kind of result as the first mapping, i.e. one input value gives one output value.

A function is a rule which maps a single number to another single number.

The second mapping does not satisfy this condition so we cannot call it a function.

Consider again what we can now call the function for which $x \rightarrow x^2$
Using f for 'function' and the symbol : to mean 'such that',
we can write $f : x \rightarrow x^2$

We use the notation $f(x)$ to represent the output values of the
function.

e.g.

$$\text{for } f : x \rightarrow x^2, \text{ we have } f(x) = x^2$$

Examples 12a

1. Determine whether these mappings are functions,

   (a) $x \rightarrow \dfrac{1}{x}$ (b) $x \rightarrow y$ where $y^2 - x = 0$

(a) For any value of x, except $x = 0$, $\dfrac{1}{x}$ has a single value,

therefore $x \rightarrow \dfrac{1}{x}$ is a function provided that $x = 0$ is excluded.

Note that $\dfrac{1}{0}$ is meaningless, so to make this mapping a function we have to

exclude 0 as an input value. The function can be described by $f(x) = \dfrac{1}{x}, \ x \neq 0$

(b) If, for example, we input $x = 4$, then the output is the value
of y when

$$y^2 - 4 = 0 \ \Rightarrow \ y = 2 \text{ and } y = -2$$

therefore an input gives more than one value for the output,
so $x \rightarrow y$ where $y^2 - x = 0$ is not a function.

_____

2. If $f(x) = 2x^2 - 5$, find f(3) and f(-1)

As $f(x)$ is the output of the mapping, f(3) is the output when 3 is the input,
i.e. f(3) is the value of $2x^2 - 5$ when $x = 3$

$$f(3) = 2(3)^2 - 5 = 13$$
$$f(-1) = 2(-1)^2 - 5 = -3$$

EXERCISE 12a

1. Determine which of these mappings are functions.

(a) $x \rightarrow 2x - 1$ (b) $x \rightarrow x^3 + 3$ (c) $x \rightarrow \dfrac{1}{x - 1}$

(d) $x \rightarrow k$ where $k^2 = x$ (e) $x \rightarrow \sqrt{x}$

(f) $x \rightarrow$ the length of the line from the origin to $(0, x)$

(g) $x \rightarrow$ the greatest integer less than or equal to x

(h) $x \rightarrow$ the height of a triangle whose area is x

2. If $f(x) = 5x - 4$ find $f(0)$, $f(-4)$

3. If $f(x) = 3x^2 + 25$ find $f(0)$, $f(8)$

4. If $f(x) = $ the value of x correct to the nearest integer, find $f(1.25)$, $f(-3.5)$, $f(12.49)$

5. If $f(x) = \sin x$, find $f(\tfrac{1}{2}\pi)$, $f(\tfrac{2}{3}\pi)$

DOMAIN AND RANGE

We have assumed that we can use any real number as an input for a function unless some particular numbers have to be excluded because they do not give real numbers as output.

> The set of inputs for a function is called the *domain* of the function.

The domain does not have to contain all possible inputs; it can be as wide, or as restricted, as we choose to make it. Hence to define a function fully, the domain must be stated.

If the domain is not stated, we assume that it is the set of all real numbers ($\mathbb{R}$).

Consider the mapping $x \rightarrow x^2 + 3$

We can define a function f for this mapping over any domain we choose. Some examples, together with their graphs are given overleaf.

1) $f(x) = x^2 + 3$ for $x \in \mathbb{R}$

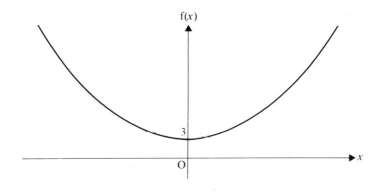

2) $f(x) = x^2 + 3$ for $x \geqslant 0$

Note that the point on the curve where $x = 0$ is included and we denote this on the curve by a solid circle.

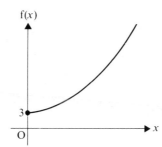

If the domain were $x > 0$, then the point would not be part of the curve and we indicate this fact by using an open circle.

3) $f(x) = x^2 + 3$ for $x \in \{1, 2, 3, 4, 5\}$

This time the graphical representation consists of just five discrete (i.e. separate) points.

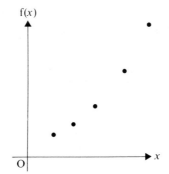

For each domain, there is a corresponding set of output numbers.

> The set of output numbers is called the *range* or *image-set* of the function.

Thus for the function defined in **(1)** above, the range is $f(x) \geqslant 3$ and for the function given in **(2)**, the range is also $f(x) \geqslant 3$ For the function defined in **(3)**, the range is the set $\{4, 7, 12, 19, 28\}$.

Sometimes a function can be made up from more than one mapping, where each mapping is defined for a different domain. This is illustrated in the next worked example.

Example 12b _____

The function, f, is defined by $f(x) = x^2$ for $x \leqslant 0$
and $f(x) = x$ for $x > 0$

(a) Find $f(4)$ and $f(-4)$ (b) Sketch the graph of f

(c) Give the range of f

(a) For $x > 0$, $f(x) = x$, $\therefore$ $f(4) = 4$

   For $x \leqslant 0$, $f(x) = x^2$, $\therefore$ $f(-4) = (-4)^2 = 16$

(b) To sketch the graph of a function, we can apply our knowledge of lines and curves in the xy-plane to the equation $y = f(x)$. In this way we can interpret $f(x) = x$ for $x > 0$, as that part of the line $y = x$ which corresponds to positive values of x

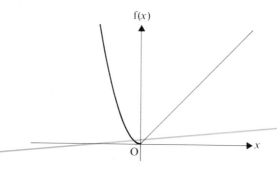

(c) The range of f is $f(x) \geqslant 0$

EXERCISE 12b

1. Find the range of each of the following functions.
   (a) $f(x) = 2x - 3$ for $x \geqslant 0$
   (b) $f(x) = x^2 - 5$ for $x \leqslant 0$
   (c) $f(x) = 1 - x$ for $x \leqslant 1$
   (d) $f(x) = 1/x$ for $x \geqslant 2$

2. Draw a sketch graph of each function given in Question 1.

3. The function f is such that $f(x) = -x$ for $x < 0$
                          and $f(x) = x$ for $x \geqslant 0$
   (a) Find the value of $f(5)$, $f(-4)$, $f(-2)$ and $f(0)$
   (b) Sketch the graph of the function.

4. The function f is such that $f(x) = x$ for $0 \leqslant x \leqslant 5$
                          and $f(x) = 5$ for $x > 5$
   (a) Find the value of $f(0)$, $f(2)$, $f(4)$, $f(5)$ and $f(7)$
   (b) Sketch the graph of the function.
   (c) Give the range of the function.

5. In Utopia, the tax on earned income is calculated as follows. The first £20 000 is tax free and remaining income is taxed at 20%.
   (a) Find the tax payable on an earned income of £15 000 and of £45 000
   (b) Taking x as the number of pounds of earned income and y as the number of pounds of tax payable, define a function f such that $y = f(x)$. Draw a sketch of the function and state the domain and range.

CURVE SKETCHING

When functions have similar definitions they usually have common properties and graphs of the same form. If the common characteristics of a group of functions are known, the graph of any one particular member of the group can be sketched without having to plot points.

Quadratic Functions

The general form of a quadratic function is

$$f(x) = ax^2 + bx + c \quad \text{for} \quad x \in \mathbb{R}$$

where a, b and c are constants and $a \neq 0$

When a graphics calculator, or a computer, is used to draw the graphs of quadratic functions for a variety of values of a, b and c, the basic shape of the curve is always the same. This shape is called a *parabola*.

Every parabola has an axis of symmetry which goes through the vertex, i.e. the point where the curve turns back upon itself.

If the coefficient of x^2 is positive, i.e. $a > 0$, then $f(x)$ has a least value, and the parabola looks like this.

If the coefficient of x^2 is negative, i.e. $a < 0$, then $f(x)$ has a greatest value and the curve is this way up.

These properties of the graph of a quadratic function can be proved algebraically.

For $f(x) = ax^2 + bx + c$, 'completing the square' on the RHS and simplifying, gives

$$f(x) = \left[\frac{4ac - b^2}{4a} \right] + a \left[x + \frac{b}{2a} \right]^2 \qquad [1]$$

Now the first bracket is constant and, as the second bracket is squared, its value is zero when $x = -\dfrac{b}{2a}$ and greater than zero for all other values of x.

Hence

when a is positive,

$f(x) = ax^2 + bx + c$ has a least value when $x = -\dfrac{b}{2a}$

and when a is negative,

$f(x) = ax^2 + bx + c$ has a greatest value when $x = -\dfrac{b}{2a}$

Further, taking values of x that are symmetrical about $x = -\dfrac{b}{2a}$,
e.g. $x = \pm k - \dfrac{b}{2a}$, we see from [1] that

$$f\left(k - \frac{b}{2a}\right) = f\left(-k - \frac{b}{2a}\right) = \left[\frac{4ac - b^2}{4a}\right] + ak^2$$

i.e.   the value of $f(x)$ is symmetrical about $x = -\dfrac{b}{2a}$

These properties can now be used to draw *sketches* of the graphs of
quadratic functions.

Examples 12c _____

1. Find the greatest or least value of the function given by
    $f(x) = 2x^2 - 7x - 4$ and hence sketch the graph of $f(x)$.

$f(x) = 2x^2 - 7x - 4$ $\Rightarrow$ $a = 2,\ b = -7$ and $c = -4$

As $a > 0,$ $f(x)$ has a least value

and this occurs when $x = -\dfrac{b}{2a} = \dfrac{7}{4}$

$\therefore$ the least value of $f(x)$ is $f(\tfrac{7}{4}) = 2(\tfrac{7}{4})^2 - 7(\tfrac{7}{4}) - 4$

$$= -\tfrac{81}{8}$$

We now have one point on the graph of $f(x)$ and we know that the curve is
symmetrical about this value of x. However, to locate the curve more accurately
we need another point; we use $f(0)$ as it is easy to find.

$f(0) = -4$

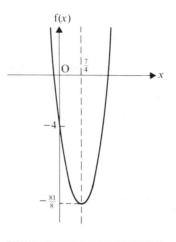

---

2. Draw a quick sketch of the graph of $f(x) = (1 - 2x)(x + 3)$

The coefficient of x^2 is negative, so $f(x)$ has a greatest value.
The curve cuts the x-axis when $f(x) = 0$

When $f(x) = 0$, $(1 - 2x)(x + 3) = 0$

$\Rightarrow$ $x = \frac{1}{2}$ or -3

The average of these values is $-\frac{5}{4}$, so the curve is symmetrical about $x = -\frac{5}{4}$

We now have enough information to draw a quick sketch, but note that this method is suitable only when the quadratic function factorises.

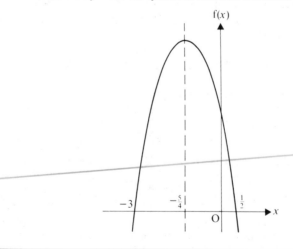

EXERCISE 12c

1. Find the greatest or least value of $f(x)$ where $f(x)$ is
   (a) $x^2 - 3x + 5$ (b) $2x^2 - 4x + 5$ (c) $3 - 2x - x^2$

2. Find the range of f where $f(x)$ is
   (a) $7 + x - x^2$ (b) $x^2 - 2$ (c) $2x - x^2$

3. Sketch the graph of each of the following quadratic functions, showing the greatest or least value and the value of x at which it occurs.
   (a) $x^2 - 2x + 5$ (b) $x^2 + 4x - 8$ (c) $2x^2 - 6x + 3$
   (d) $4 - 7x - x^2$ (e) $x^2 - 10$ (f) $2 - 5x - 3x^2$

4. Draw a quick sketch of each of the following functions.
   (a) $(x - 1)(x - 3)$ (b) $(x + 2)(x - 4)$ (c) $(2x - 1)(x - 3)$
   (d) $(1 + x)(2 - x)$ (e) $x^2 - 9$ (f) $3x^2$

CUBIC FUNCTIONS

> The general form of a cubic function is
> $$f(x) = ax^3 + bx^2 + cx + d$$
> where a, b, c and d, are constants and $a \neq 0$

Investigating the curve $y = ax^3 + bx^2 + cx + d$ for a variety of values of a, b, c and d shows that the shape of the curve is

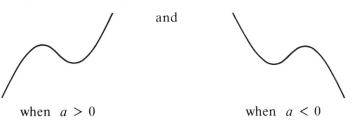

and

when $a > 0$ when $a < 0$

Sometimes there are no turning points and the curve looks like this

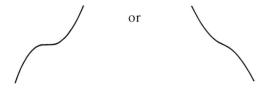

or

THE FUNCTION $f(x) = \dfrac{1}{x}$

Now consider the familiar function $f(x) = 1/x$ and its graph. From its form we can infer various properties of $f(x)$

1) As the value of x increases, the value of $f(x)$ gets closer to zero, e.g. when $x = 100$, $f(x) = 1/100$
and when $x = 1000$, $f(x) = 1/1000$
We write this as $x \to \infty$, $f(x) \to 0$

Similarly as the value of x decreases, i.e. as $x \to -\infty$, the value of $f(x)$ again gets closer to zero, i.e. $f(x) \to 0$

2) $f(x)$ does not exist when $x = 0$, so this value of x must be excluded from the domain of f
x can get as close as we like to zero however, and can approach zero in two ways.

If $x \to 0$ from above (i.e. from positive values, $--\!|\!\longleftarrow--$)
then $f(x) \to \infty$

If $x \to 0$ from below (i.e. from negative values, $--\longrightarrow\!|--$)
then $f(x) \to -\infty$

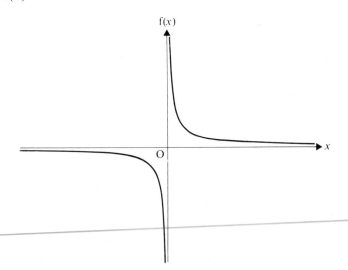

Notice that, as $x \to \pm\infty$, the curve gets closer to the x-axis but does not cross it. Also, as $x \to 0$, the curve approaches the y-axis but again does not cross it.
We say that the x-axis and the y-axis are *asymptotes* to the curve.

EXERCISE 12d

1. Find the values of x where the curve $y = f(x)$ cuts the x-axis and sketch the curve when
   (a) $f(x) = x(x - 1)(x + 1)$ (b) $f(x) = (x^2 - 1)(2 - x)$

2. The basic cubic function is given by $f : x \rightarrow x^3$, $x \in \mathbb{R}$. Draw an accurate graph of this function for values of x from -3 to 3 by plotting points at intervals of 0.25

SIMPLE TRANSFORMATION OF CURVES

Transformations of curves are best appreciated if they can be 'seen', so this section starts with an investigative approach using a graphics calculator or a computer with a graph-drawing package. This exercise is not essential and all the necessary conclusions are drawn analytically in the next part of the text.

EXERCISE 12e

You will need a graphics calculator or computer for this exercise.

1. (a) On the screen, draw the graph of $y = x^3$. Superimpose the graphs of $y = x^3 + 2$ and $y = x^3 - 1$. Clear the screen and again draw the graph of $y = x^3$. This time superimpose the graph of $y = x^3 + c$ for a variety of values of c.
   (b) Describe the transformation that maps the graph of $f(x) = x^3$ to the graph of $g(x) = x^3 + c$
   (c) Repeat (a) and (b) for other simple functions, e.g., x^2, $3x$, $1/x$

2. Use a procedure similar to that described in Question 1 to investigate the relationship between the graphs of $f(x)$ and $f(x + c)$

3. Investigate the relationship between the graphs of
   (a) $f(x)$ and $-f(x)$ (b) $f(x)$ and $f(-x)$

TRANSLATIONS

Consider the function $f(x) = x^3$ and the function $g(x) = x^3 + 2$

For any given value of x, $g(x) = f(x) + 2$. Therefore for equal values of x, points on the curve $y = x^3 + 2$ are 2 units above points on the curve $y = x^3$,

i.e. the curve $y = f(x) + 2$ is a translation of the curve $y = f(x)$ by 2 units in the direction Oy

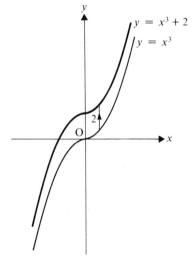

In general, for any function f, the curve $y = f(x) + c$ is the translation of the curve $y = f(x)$ by c units parallel to the y-axis.

Now consider the curves $y = x^3$ and $y = (x - 2)^3$

The values of y are equal when the value of x in $y = (x - 2)^3$ is 2 units *greater* than the value of x in $y = x^3$, therefore for equal values of y, points on the curve $y = (x - 2)^3$ are 2 units to the *right* of the points on $y = x^3$.

i.e. the curve $y = f(x - 2)$ is a translation of the curve $y = f(x)$ by 2 units in the direction Ox.

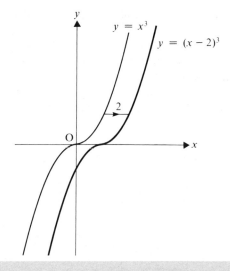

In general, the curve $y = f(x + c)$ is a translation of the curve $y = f(x)$ by c units parallel to the x-axis.

If c is negative the translation is in the direction Ox
If c is positive, the translation is in the direction xO

REFLECTIONS

Consider the function $f(x) = x^2$ and the function $g(x) = -x^2$
For any given value of x, $g(x)$ is equal to $-f(x)$. Therefore for equal values of x, points on the curve $y = -x^2$ are the reflection in the x-axis of points on the curve $y = x^2$

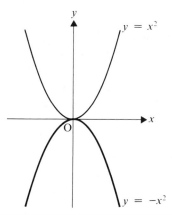

In general, the curve $y = -f(x)$ is the reflection of the curve $y = f(x)$ in the x-axis.

Now consider the function $f(x) = x(x - 1)$ and the function
$g(x) = f(-x)$

Comparing $f(x) = x(x - 1)$ with $g(x) = -x(-x - 1)$ we see that
when the inputs to $f(x)$ and to $g(x)$ are equal in value but
opposite in sign, $f(x)$ and $g(x)$ have the same value.

Therefore for equal values of y, points on the curve $y = -x(-x - 1)$
are the reflection in the y-axis of points on the curve $y = x(x - 1)$

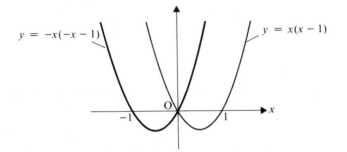

In general, the curve $y = f(-x)$ is the reflection of the curve
$y = f(x)$ in the y-axis.

EVEN FUNCTIONS

A function is even if $f(x) = f(-x)$

Since the curve $y = f(-x)$ is the reflection of the curve $y = f(x)$
in the y-axis, it follows that

when $f(x)$ is an even function,
the curve $y = f(x)$ is symmetrical about Oy

The function $f:x \rightarrow x^2$ is a familiar even function.

$$f(x) = x^2$$

ODD FUNCTIONS

A function is odd if $f(x) = -f(-x)$

As the curve $y = -f(-x)$ is a reflection of the curve $y = f(x)$ in Oy followed by a reflection in Ox, it follows that

when $f(x) = -f(-x)$ the curve $y = f(x)$ has rotational symmetry of order 2 about the origin.

Some familiar odd functions and their graphs are shown below.

$$f(x = x) \qquad\qquad f(x) = x^3 \qquad\qquad f(x) = \tfrac{1}{x}$$

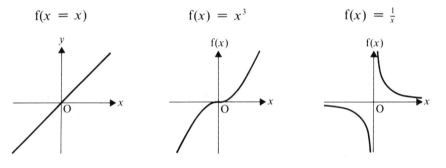

Example 12f

Sketch the curve $y = (x - 1)^3$.

The shape and position of the curve $y = x^3$ is known. If $f(x) = x^3$, then $(x - 1)^3 = f(x - 1)$, so the curve $y = (x - 1)^3$ is a translation of the first curve by one unit in the positive direction of the x-axis.

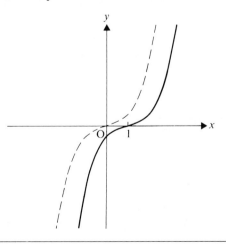

EXERCISE 12f

Sketch each of the following curves, and state whether y is an even or an odd function of x or neither of these.

1. $y = -x^2$

2. $y = -\dfrac{1}{x}$

3. $y = -x^3$

4. $y = 1 + \dfrac{1}{x}$

5. $y = x^2 + 3$

6. $y = \dfrac{1}{x} - 2$

7. $y = (x - 4)^3$

8. $y = x^2 - 9$

9. $y = \dfrac{1}{x - 2}$

10. On the same set of axes sketch the graphs of $f(x) = x^3$, $f(x) = (x + 1)^3$, $f(x) = -(x + 1)^3$ and $f(x) = 2 - (x + 1)^3$

11. On the same set of axes sketch the lines $y = 2x - 1$ and $y = \frac{1}{2}(x + 1)$. Describe a transformation which maps the first line to the second line.

12. Repeat Question 11 for the curves $y = 1 + \dfrac{1}{x}$ and $y = \dfrac{1}{x - 1}$

.13. Find the coordinates of the reflection of the point $(2, 5)$ in the line $y = x$

14. P' is the reflection of the point $P(a, b)$ in the line $y = x$. Find the coordinates of P' in terms of a and b

INVERSE FUNCTIONS

Consider the function f where $f(x) = 2x$ for $x \in \{2, 3, 4\}$

Under this function, the domain $\{2, 3, 4\}$ maps to the image-set $\{4, 6, 8\}$ and this is illustrated by the arrow diagram.

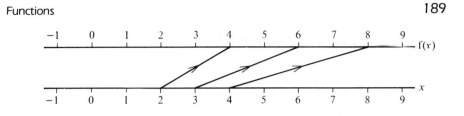

It is possible to reverse this mapping, i.e. we can map each member of the image-set back to the corresponding member of the domain by halving each member of the image-set.

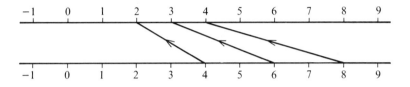

This procedure can be expressed algebraically, i.e.
for $x \in \{4, 6, 8\}$, $x \rightarrow \frac{1}{2}x$ maps 4 to 2, 6 to 3 and 8 to 4

This reverse mapping is a function in its own right and it is called the *inverse* function of f where $f(x) = 2x$

Denoting this inverse function by f^{-1} we can write $f^{-1}(x) = \frac{1}{2}x$
In fact, $f(x) = 2x$ can be reversed for all real values of x and the procedure for doing this is a function.

Therefore, for $f(x) = 2x$, $f^{-1}(x) = \frac{1}{2}x$ is such that f^{-1} reverses f for all real values of x, i.e. f^{-1} maps the output of f to the input of f.

In general, for any function f,

if there exists a function, g, that maps the output of f back to its input, i.e. $g: f(x) \rightarrow x$, then this function is called the inverse of f and it is denoted by f^{-1}.

THE GRAPH OF A FUNCTION AND ITS INVERSE

Consider the curve that is obtained by reflecting $y = f(x)$ in the line $y = x$. The reflection of a point $A(a,b)$ on the curve $y = f(x)$, is the point A' whose coordinates are (b,a), i.e. interchanging the x and y coordinates of A gives the coordinates of A'.

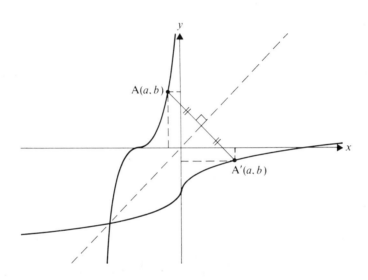

We can therefore obtain the equation of the reflected curve by interchanging x and y in the equation $y = f(x)$

Now the coordinates of A on $y = f(x)$ can be written as $[a, f(a)]$. Therefore the coordinates of A' on the reflected curve are $[f(a), a]$, i.e. the equation of the reflected curve is such that the output of f is mapped to the input of f.

Hence if the equation of the reflected curve can be written in the form $y = g(x)$, then g is the inverse of f, i.e. $g = f^{-1}$.

To illustrate these properties, consider the curve $y = x^3 + 1$ and its reflection in the line $y = x$

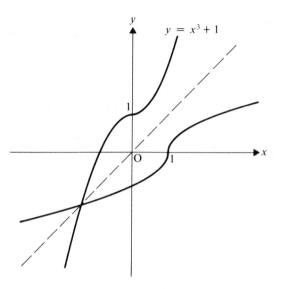

The equation of the reflected curve is given by $x = y^3 + 1$

We can write this equation in the form $y = \sqrt[3]{(x - 1)}$

Therefore for the function $f(x) = x^3 + 1$ the inverse function is given by $f^{-1}(x) = \sqrt[3]{(x - 1)}$

Any curve whose equation can be written in the form $y = f(x)$ can be reflected in the line $y = x$. However this reflected curve may not have an equation that can be written in the form $y = f^{-1}(x)$

Consider the curve $y = x^2$ and its reflection in the line $y = x$

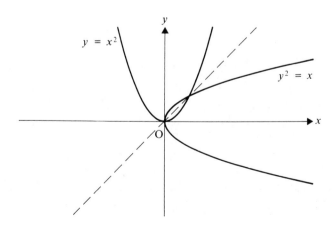

The equation of the image curve is $x = y^2 \Rightarrow y = \pm\sqrt{x}$ and $x \rightarrow \pm\sqrt{x}$ is not a function.

(We can see this from the diagram as, on the reflected curve, one value of x maps to two values of y. So in this case y cannot be written as a function of x.)

Therefore the function $f: x \rightarrow x^2$ does not have an inverse, i.e.

> not every function has an inverse.

If we change the definition of f to $f: x \rightarrow x^2$ for $x \in \mathbb{R}^+$ then the inverse mapping

is $x \rightarrow \sqrt{x}$ for $x \in \mathbb{R}^+$ and this *is* a function, i.e.

$$f^{-1}(x) = \sqrt{x} \text{ for } x \in \mathbb{R}^+$$

(graph with y-axis and x-axis, showing curve $y = x^2, x > 0$, the line $y = x$ dashed, and curve $y = \sqrt{x}, x > 0$, meeting at origin O)

To summarise:

> The inverse of a function undoes the function, i.e. it maps the output of a function back to its input.
>
> The inverse of the function f is written f^{-1}
>
> Not all functions have an inverse.
>
> When the curve whose equation is $y = f(x)$ is reflected in the line $y = x$, the equation of the reflected curve is $x = f(y)$
> If this equation can be written in the form $y = g(x)$ then g is the inverse of f, i.e. $g(x) = f^{-1}(x)$

Examples 12g _____ _____

1. Determine whether there is an inverse of the function f given by

   $$f(x) = 2 + \frac{1}{x}$$

   If f^{-1} exists, express it as a function of x

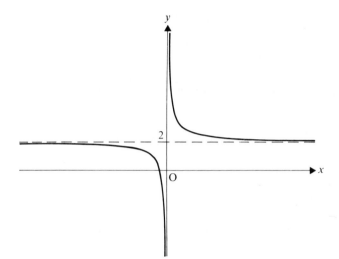

From the sketch of $f(x) = 2 + \dfrac{1}{x}$, we see that one value of $f(x)$
maps to one value of x, therefore the reverse mapping is a function.
The equation of the reflection of $y = 2 + \dfrac{1}{x}$ can be written as

$$x = 2 + \frac{1}{y} \quad \Rightarrow \quad y = \frac{1}{x - 2}$$

$\therefore$ when $f(x) = 2 + \dfrac{1}{x}$, $\qquad f^{-1}(x) = \dfrac{1}{x - 2}$

_____

2. Find $f^{-1}(4)$ when $f(x) = 5x - 1$

If $y = f(x)$, i.e. $y = 5x - 1$,

then $\qquad\qquad\qquad x = 5y - 1 \quad \Rightarrow \quad y = \tfrac{1}{5}(x + 1)$

i.e. $\qquad\qquad\qquad\qquad f^{-1}(x) = \tfrac{1}{5}(x + 1)$

$\therefore \qquad\qquad\qquad\qquad f^{-1}(4) = \tfrac{1}{5}(4 + 1) = 1$

_____

EXERCISE 12g

1. Sketch the graphs of $f(x)$ and $f^{-1}(x)$ on the same axes.
   (a) $f(x) = 3x - 1$ (b) $f(x) = x^3 - 1$ (c) $f(x) = (x - 1)^3$
   (d) $f(x) = 2 - x$ (e) $f(x) = \dfrac{1}{x - 3}$ (f) $f(x) = \dfrac{1}{x}$

2. Which of the functions given in Question 1 are their own inverses?

3. Determine whether f has an inverse function, and if it does, find it when
   (a) $f(x) = x + 1$ (b) $f(x) = x^2 + 1$ (c) $f(x) = x^3 + 1$
   (d) $f(x) = x^2 - 4,\ x \geqslant 0$ (e) $f(x) = (x + 1)^4,\ x \geqslant -1$

4. The function f is given by $f(x) = 1 - \dfrac{1}{x}$. Find
   (a) $f^{-1}(4)$ (b) the value of x for which $f^{-1}(x) = 2$
   (c) any values of x for which $f^{-1}(x) = x$

5. If $f(x) = x^3$, find
   (a) $f(2)$ (b) $f^{-1}(27)$ (c) $f^{-1}(1)$

FUNCTION OF A FUNCTION

Consider the two functions f and g where

$$f:x \rightarrow x^2 \quad \text{and} \quad g:x \rightarrow \frac{1}{x}$$

These functions can be combined in several ways; they can be added, subtracted, multiplied or divided.

As well as adding, subtracting, multiplying and dividing them, the output of f can be made the input of g,

i.e. $x \xrightarrow{\ f\ } x^2 \xrightarrow{\ g\ } \dfrac{1}{x^2}$ or $g[f(x)] = g(x^2) = \dfrac{1}{x^2}$

Therefore the function $x \rightarrow 1/x^2$ is obtained by taking the function g of the function f.

A compound function formed in this way is known as a *function of a function* and it is denoted by gf,

For example, if $f(x) = x^2$ and $g(x) = 1 - x$ then $gf(x)$ means the function g of $f(x)$,

i.e. $\qquad\qquad\qquad\qquad gf(x) = g(x^2) = 1 - x^2$

Similarly $\qquad\qquad\qquad fg(x) = f(1 - x) = (1 - x)^2$

Note that $gf(x)$ is *not* the same as $fg(x)$

EXERCISE 12h

1. If f, g and h are functions defined by $f(x) = x^2$, $g(x) = 1/x$, $h(x) = 1 - x$ find as a function of x

    (a) fg   (b) fh   (c) hg   (d) hf   (e) gf

2. If $f(x) = 2x - 1$ and $g(x) = x^3$ find the value of

    (a) gf(3)   (b) fg(2)   (c) fg(0)   (d) gf(0)

3. Given that $f(x) = 2x$, $g(x) = 1 + x$ and $h(x) = x^2$, find as a function of x

    (a) hg   (b) fhg   (c) ghf

4. The function $f(x) = (2 - x)^2$ can be expressed as a function of a function. Find g and h as functions of x such that $gh(x) = f(x)$

5. Repeat Question 4 when $f(x) = (x + 1)^4$

6. Express the function $f(x)$ as a function g of a function h and define $g(x)$ and $h(x)$, if $f(x)$ is

    (a) $\dfrac{1}{2x + 1}$ (b) $(5x - 6)^4$ (c) $\sqrt{(x^2 - 2)}$

MIXED EXERCISE 12

1. A function f is defined by $f(x) = 1/(1 - x)$, $x \neq 1$
    (a) Why is 1 excluded from the domain of f?
    (b) Find the value of $f(-3)$
    (c) Sketch the curve $y = f(x)$
    (d) Find $f^{-1}(x)$ in terms of x and give the domain of f^{-1}

2. Find the greatest or least value of each of the following functions, stating the value of x at which they occur.

(a) $f(x) = x^2 - 3x + 5$ (b) $f(x) = 2x^2 - 7x + 1$

(c) $f(x) = (x - 1)(x + 5)$

3. If $f(x) = x^3$, sketch the following curves on the same set of axes.

(a) $y = f(x)$ (b) $y = f(x + 3)$ (c) $y = f^{-1}(x)$

4. Given that $f(x) = 2x + 1$, $g(x) = x^2$ and $h(x) = 1/x$,

(a) state whether f, g and h are even or odd functions or neither

(b) find $fg(2)$, $hg(3)$ and $gf(-1)$

(c) find, in terms of x, $hfg(x)$ and $gfh(x)$

(d) if they exist, find $f^{-1}(x)$, $g^{-1}(x)$ and $h^{-1}(x)$

(e) find the value(s) of x for which $gh(x) = 9$

(f) does the function $(gh)^{-1}$ exist?

5. Draw sketches of the following curves, showing any asymptotes.

(a) $(x - 2)(x - 3)(x - 4)$ (b) $y = \dfrac{1}{3 - x}$ (c) $y = x^2 - 4$

6. The function f is given by $f(x) = \dfrac{1}{2x - 1}$

(a) find $g(x)$ and $h(x)$ such that $f = gh$

(b) evaluate $ff(2)$ and $f^{-1}(2)$

7. The functions f, g and h are defined by $f(x) = 2x$, $g(x) = 3x^2$ and $h(x) = x - 1$

(a) Sketch the curves $y = g(x - 3)$ and $y = gf^{-1}(x)$

(b) Find the value(s) of x for which $f^{-1}(x) = g(x)$

CHAPTER 13

FURTHER ALGEBRA

RELATIONSHIPS BETWEEN ROOTS AND COEFFICIENTS OF A QUADRATIC EQUATION

The general quadratic equation $ax^2 + bx + c = 0$ can be written

$$x^2 + \frac{b}{a}x + \frac{c}{a} = 0 \qquad [1]$$

If the roots of the equation are α and β then the equation can also be written in the form

$$(x - \alpha)(x - \beta) = 0$$

$\Rightarrow \qquad x^2 - (\alpha + \beta)x + \alpha\beta = 0 \qquad [2]$

Comparing the terms in equations [1] and [2] shows that

$$\alpha + \beta = -\frac{b}{a} \text{ and } \alpha\beta = \frac{c}{a}$$

e.g. if the equation $2x^2 - 3x + 6 = 0$ has roots α and β, then

the sum of its roots, i.e. $\alpha + \beta$, is $-(-\frac{3}{2}) = \frac{3}{2}$

and   the product of its roots, i.e. $\alpha\beta$, is $\frac{6}{2} = 3$

It also follows from the relationships above that any quadratic equation can be expressed in the form

$$x^2 - (\text{sum of roots})x + (\text{product of roots}) = 0$$

e.g. the quadratic equation with roots whose sum is 7 and whose product is 10, is $x^2 - 7x + 10 = 0$

Examples 13a _____

1. The roots of the equation $2x^2 - 7x + 4 = 0$ are α and β

   Find the values of $\dfrac{1}{\alpha} + \dfrac{1}{\beta}$ and $\dfrac{1}{\alpha\beta}$ and hence write down the

   equation whose roots are $\dfrac{1}{\alpha}$ and $\dfrac{1}{\beta}$

From $2x^2 - 7x + 4 = 0$ we have

$$\alpha + \beta = -(-\tfrac{7}{2}) = \tfrac{7}{2} \quad \text{and} \quad \alpha\beta = \tfrac{4}{2} = 2$$

To evaluate $\dfrac{1}{\alpha} + \dfrac{1}{\beta}$ we must first express it in terms of $\alpha + \beta$ and $\alpha\beta$ as these have known values.

Expressing $\dfrac{1}{\alpha} + \dfrac{1}{\beta}$ as a single fraction gives

$$\frac{1}{\alpha} + \frac{1}{\beta} = \frac{\alpha + \beta}{\alpha\beta} = \frac{\tfrac{7}{2}}{2} = \frac{7}{4}$$

and

$$\frac{1}{\alpha\beta} = \frac{1}{2}$$

The required equation has roots $\dfrac{1}{\alpha}$ and $\dfrac{1}{\beta}$ therefore

the sum of its roots is $\left(\dfrac{1}{\alpha} + \dfrac{1}{\beta}\right) = \dfrac{7}{4}$

and the product of its roots is $\alpha\beta = \tfrac{1}{2}$

Hence the required equation is $x^2 - \tfrac{7}{4}x + \tfrac{1}{2} = 0$

i.e. $\qquad\qquad\qquad\qquad\qquad 4x^2 - 7x + 2 = 0$

Alternatively the following method can be used.

For the given equation, $2x^2 - 7x + 4 = 0$, $x = \alpha$ and β

and for the required equation, $aX^2 + bX + c = 0$, $X = \dfrac{1}{\alpha}$ and $\dfrac{1}{\beta}$

Therefore $\qquad\qquad X = \dfrac{1}{x} \quad \Rightarrow \quad x = \dfrac{1}{X}$

Substituting $\dfrac{1}{X}$ for x in the given equation we get

$$2\left(\frac{1}{X}\right)^2 - 7\left(\frac{1}{X}\right) + 4 = 0$$

i.e. $\qquad\qquad\qquad\qquad 4X^2 - 7X + 2 = 0$

and this is the required equation.

The alternative method can be used only if each new root depends in the same way on each original root. For example, if the given equation has roots α, β and the required equation has roots α^2, β^2 it can be used, but if the required equation has roots $\alpha + \beta$, $\alpha - \beta$ it cannot.

------

2. If α and β are the roots of $x^2 + 3x - 2 = 0$ find the values of $\alpha^3 + \beta^3$ and $\alpha^3\beta^3$. Write down the equation whose roots are α^3 and β^3

From $x^2 + 3x - 2 = 0$ we see that $\alpha + \beta = -3$

$$\text{and}\qquad \alpha\beta = -2$$

To express $\alpha^3 + \beta^3$ in terms of $\alpha + \beta$ and $\alpha\beta$ we can use

$$(\alpha + \beta)^3 \equiv \alpha^3 + 3\alpha^2\beta + 3\alpha\beta^2 + \beta^3$$

$$\equiv \alpha^3 + \beta^3 + 3\alpha\beta(\alpha + \beta)$$

therefore $\qquad \alpha^3 + \beta^3 \equiv (\alpha + \beta)^3 - 3\alpha\beta(\alpha + \beta)$

$$= (-3)^3 - 3(-2)(-3)$$

$$= -45$$

$$\alpha^3\beta^3 \equiv (\alpha\beta)^3 = (-2)^3 = -8$$

As the required equation has roots α^3 and β^3, the sum of its roots is $\alpha^3 + \beta^3 = -45$ and the product of its roots is $\alpha^3\beta^3 = -8$

Therefore the required equation is $x^2 - (-45)x + (-8) = 0$

i.e. $x^2 + 45x - 8 = 0$

Note that, although the alternative method could be used in this example, as $X = x^3$, it is not recommended because the resulting equation $(\sqrt[3]{X})^2 + 3(\sqrt[3]{X}) + 4 = 0$ is not easy to simplify.

------

3. Find the range of values of k for which the roots of the equation $x^2 - 2x - k = 0$ are real. If the roots of this equation differ by 1, find the value of k

If $x^2 - 2x - k = 0$ has real roots then $b^2 - 4ac \geqslant 0$

i.e. $\qquad\qquad (-2)^2 - 4(1)(-k) \geqslant 0$

$\Rightarrow \qquad\qquad 4 + 4k \geqslant 0$

$\therefore \qquad\qquad k \geqslant -1$

If one root of the equation is α, then the other is $\alpha + 1$

The sum of the roots is $\qquad\qquad 2\alpha + 1 = -(-2)$

$\Rightarrow \qquad\qquad 2\alpha = 1$

$\therefore \qquad\qquad \alpha = \frac{1}{2}$

The product of the roots is $\qquad \alpha(\alpha + 1) = -k$

Therefore $\qquad\qquad k = -\frac{3}{4}$

EXERCISE 13a

1. Write down the sums and products of the roots of the following equations.

   (a) $x^2 - 3x + 2 = 0$

   (b) $4x^2 + 7x - 3 = 0$

   (c) $x(x - 3) = x + 4$

   (d) $\dfrac{x - 1}{2} = \dfrac{3}{x + 2}$

   (e) $x^2 - kx + k^2 = 0$

   (f) $ax^2 - x(a + 2) - a = 0$

2. Write down the equation, the sum and product of whose roots are

   (a) 3, 4
   (b) $-2, \frac{1}{2}$
   (c) $\frac{1}{3}, -\frac{2}{5}$
   (d) $-\frac{1}{4}, 0$

   (e) a, a^2
   (f) $-(k + 1), k^2 - 3$
   (g) $\dfrac{b}{a}, \dfrac{c^2}{b}$

3. The roots of the equation $2x^2 - 4x + 5 = 0$ are α and β. Find the value of

   (a) $\dfrac{1}{\alpha} + \dfrac{1}{\beta}$

   (b) $(\alpha + 1)(\beta + 1)$

   (c) $\alpha^2 + \beta^2$

   (d) $\alpha^2\beta + \alpha\beta^2$

   (e) $(\alpha - \beta)^2$

   (f) $\dfrac{\alpha}{\beta} + \dfrac{\beta}{\alpha}$

4. The roots of $x^2 - 2x + 3 = 0$ are α and β. Find the equation whose roots are

(a) $\alpha + 2, \beta + 2$ (b) $\dfrac{1}{\alpha}, \dfrac{1}{\beta}$ (c) α^2, β^2 (d) $\dfrac{\alpha}{\beta}, \dfrac{\beta}{\alpha}$

DIVISION OF ONE POLYNOMIAL BY ANOTHER POLYNOMIAL

Long division can be used to divide say $x^3 + 4x^2 - 7$ by $x^2 - 3$ ($x^3 + 4x^2 - 7$ is called the *dividend* and $x^2 - 3$ is called the *divisor*.)

$$
\begin{array}{r}
x + 4 \\
x^2 - 3 \overline{)\, x^3 + 4x^2 \qquad\; - 7} \\
x^3 \qquad\quad - 3x \\
\hline
4x^2 + 3x - 7 \\
4x^2 \qquad\; - 12 \\
\hline
+ 3x + 5
\end{array}
$$

Divide x^3 by x^2: it goes in x times. Multiply the divisor by x and then subtract it from the dividend. The result is the new dividend; repeat the process until the dividend is not divisible by x^2

The number over the division line is the *quotient*, and what is left is called the *remainder*.

The relationship between the divisor, the dividend, the quotient and the remainder can be expressed as

$$x^3 + 4x^2 - 7 \equiv (x + 4)(x^2 - 3) + 3x + 5$$

IMPROPER FRACTIONS

When the highest power of x in the numerator of a fraction is greater than or *equal* to the highest power of x in the denominator, the fraction is called *improper*.

For example, $\dfrac{x^3 + 4x^2 - 7}{x^2 - 3}$ and $\dfrac{x^2 + 7}{x^2 - 2}$ are improper fractions.

Improper fractions can be expressed in a form in which any fractions are proper, by dividing the numerator by the denominator,

i.e. $\dfrac{x^3 + 4x^2 - 7}{x^2 - 3}$ can be written as $x + 4 + \dfrac{3x + 5}{x^2 - 3}$

Long division is not always necessary; it is often simpler to rearrange the numerator by adding and subtracting appropriate terms. This is illustrated in the following worked example.

Example 13b _____

Express $\dfrac{x^3 + 5x^2 - 3x}{x^3 + 1}$ in a form without an improper fraction.

$$\frac{x^3 + 5x^2 - 3x}{x^3 + 1} \equiv \frac{x^3 + 1 + 5x^2 - 3x - 1}{x^3 + 1} \equiv \frac{x^3 + 1}{x^3 + 1} + \frac{5x^2 - 3x - 1}{x^3 + 1}$$

$$\equiv 1 + \frac{5x^2 - 3x - 1}{x^3 + 1}$$

Notice that we add 1 to x^3 so that the expression can be split into two fractions, one of which divides out exactly. It is important to realise that, having added 1 to the numerator we also have to subtract 1 so that the value of the numerator is not altered.

After some practice, it is possible to do the intermediate steps mentally.

_____

EXERCISE 13b

1. Find the quotient and the remainder for each of the following divisions.

   (a) $(x^3 + x^2 - 3x + 6) \div (x^2 + 3)$

   (b) $(x^4 - 5x^2 + 2) \div (x + 1)$

   (c) $(2x^3 - 4x^2 + 3x - 1) \div (x^2 - 1)$

   (d) $(3x^3 - 5) \div (x - 2)$

   (e) $(x^5 - 5x^2 + 1) \div (x^3 + 1)$

   (f) $(2x^3 - 5x^2 + 6x + 2) \div (x - 3)$

   (g) $(x^2 - 7x + 2) \div (x + 3)$

   (h) $(5x^3 - x^2 + 1) \div (x^2 - 1)$

   (i) $(3x^2 - 7) \div (x^2 + 1)$

   (j) $(4x^3 - 9x + 1) \div (2x - 1)$

2. Express the following fractions as the sum of a polynomial and a proper fraction.

   (a) $\dfrac{x + 4}{x + 1}$ (b) $\dfrac{2x}{x - 2}$ (c) $\dfrac{x^2 + 3}{x^2 - 1}$

   (d) $\dfrac{x^2}{x - 2}$ (e) $\dfrac{x^2 + 3x}{x - 4}$ (f) $\dfrac{x^2 - 4}{x(x + 1)}$

THE REMAINDER THEOREM

When $f(x) = x^3 - 7x^2 + 6x - 2$ is divided by $x - 2$, we get a quotient and a remainder. The relationship between these quantities can be written as

$$f(x) = x^3 - 7x^2 + 6x - 2 \equiv (\text{quotient})(x - 2) + \text{remainder}$$

Now substituting 2 for x eliminates the term containing the quotient, giving

$$f(2) = \text{remainder}$$

This is a particular illustration of the more general case, namely if a polynomial $f(x)$, is divided by $(x - a)$ then

$$f(x) \equiv (\text{quotient})(x - a) + \text{remainder}$$
$$\Rightarrow \qquad f(a) = \text{remainder}$$

This result is called the *remainder theorem* and can be summarised as

when a polynomial $f(x)$ is divided by $(x - a)$, the remainder is $f(a)$

Examples 13c _____

1. Find the remainder when
    (a) $x^3 - 2x^2 + 6$ is divided by $x + 3$
    (b) $6x^2 - 7x + 2$ is divided by $2x - 1$

(a) When $f(x) = x^3 - 2x^2 + 6$ is divided by $x + 3$, the remainder is
$$f(-3) = (-3)^3 - 2(-3)^2 + 6 = -39$$

(b) If $f(x) = 6x^2 - 7x + 2$, then
$$f(x) = (2x - 1)(\text{quotient}) + \text{remainder}$$
$$\Rightarrow \qquad \text{remainder} = f(\tfrac{1}{2}) = 0$$

Note that as the remainder is zero, $2x - 1$ is a factor of $f(x)$

_____

The Factor Theorem

This is a special case of the remainder theorem because if $x - a$ is a factor of a polynomial $f(x)$ then there is no remainder when $f(x)$ is divided by $x - a$,

i.e. $$f(a) = 0$$

This result, which is called the factor theorem, states that

> if, for a polynomial $f(x)$, $f(a) = 0$
> then $x - a$ is a factor of $f(x)$

The factor theorem is very helpful when factorising cubics or higher degree polynomials.

Examples 13c (continued) _____

2. Factorise $x^3 - x^2 + 2x - 8$

$$f(x) \equiv x^3 - x^2 + 2x - 8$$

We will test for factors of the form $x - a$ by finding $f(a)$ for various values of a. Note that, as the factors of 8 are 1, 2, 4 and 8, the values we choose for a must belong to the set $\{\pm 1, \pm 2, \pm 4, \pm 8\}$

$f(1) = 1 - 1 + 2 - 8 \neq 0,$ so $x - 1$ is not a factor of $f(x)$

$f(-1) = -1 - 1 - 2 - 8 \neq 0,$ so $x + 1$ is not a factor of $f(x)$

$f(2) = 8 - 4 + 4 - 8 = 0,$ therefore $(x - 2)$ is a factor of $f(x)$

Now that a factor has been found, it should be taken out; this can be done by inspection or by long division.

$$x^3 - x^2 + 2x - 8 = (x - 2)(x^2 + x + 4)$$

and $x^2 + x + 4$ does not factorise.

Therefore $x^3 - x^2 + 2x - 8 = (x - 2)(x^2 + x + 4)$

The Factors of $a^3 - b^3$ and $a^3 + b^3$

$a^3 - b^3 = 0$ when $a = b$, hence $a - b$ is a factor of $a^3 - b^3$

Therefore $\qquad\qquad\qquad a^3 - b^3 \equiv (a - b)(a^2 + ab + b^2)$

in particular $\qquad\qquad\quad x^3 - 1 \equiv (x - 1)(x^2 + x + 1)$

Also, $a^3 + b^3 = 0$ when $a = -b$, so $a + b$ is a factor of $a^3 + b^3$

Therefore $\qquad\qquad\qquad a^3 + b^3 \equiv (a + b)(a^2 - ab + b^2)$

in particular $\qquad\qquad\quad x^3 + 1 \equiv (x + 1)(x^2 - x + 1)$

EXERCISE 13c

1. Find the remainder when the following functions are divided by the linear factors indicated.

   (a) $x^3 - 2x + 4$, $\quad x - 1$ $\qquad$ (b) $x^3 + 3x^2 - 6x + 2$, $\quad x + 2$

   (c) $2x^3 - x^2 + 2$, $\quad x - 3$ $\qquad$ (d) $x^4 - 3x^3 + 5x$, $\quad 2x - 1$

   (e) $9x^5 - 5x^2$, $\quad 3x + 1$ $\qquad$ (f) $x^3 - 2x^2 + 6$, $\quad x - a$

   (g) $x^2 + ax + b$, $\quad x + c$ $\qquad$ (h) $x^4 - 2x + 1$, $\quad ax - 1$

2. Determine whether the following linear functions are factors of the given polynomials.

   (a) $x^3 - 7x + 6$, $\quad x - 1$ $\qquad$ (b) $2x^2 + 3x - 4$, $\quad x + 1$

   (c) $x^3 - 6x^2 + 6x - 2$, $\quad x - 2$ $\qquad$ (d) $x^3 - 27$, $\quad x - 3$

   (e) $2x^4 - x^3 - 1$, $\quad 2x - 1$ $\qquad$ (f) $x^3 + ax^2 - a^2x - a^3$, $\quad x + a$

3. Factorise the following functions as far as possible.

   (a) $x^3 + 2x^2 - x - 2$ $\qquad\qquad$ (b) $x^3 - x^2 - x - 2$

   (c) $x^4 - 1$ $\qquad\qquad\qquad\qquad$ (d) $x^3 + 3x^2 + 3x + 2$

   (e) $2x^3 - x^2 + 2x - 1$ $\qquad\qquad$ (f) $27x^3 - 1$

   (g) $x^3 + a^3$ $\qquad\qquad\qquad\qquad$ (h) $x^3 - y^3$

4. If $x^2 - 7x + a$ has a remainder 1 when divided by $x + 1$, find a

5. If $x - 2$ is a factor of $ax^2 - 12x + 4$ find a

6. One solution of the equation $x^2 + ax + 2 = 0$ is $x = 1$, find a

7. One root of the equation $x^2 - 3x + a = 0$ is 2. Find the other root.

SOLUTION OF CUBIC EQUATIONS

Consider the general cubic function $f: x \rightarrow ax^3 + bx^2 + cx + d$
From the shape of the graph we know that $f(x)$ takes *all* real values from $-\infty$ to ∞,
so there is at least one value of x for which $f(x) = 0$,

i.e.   a cubic equation has at least one real root.

If this root is rational, i.e. of the form $\frac{a}{b}$ where a and b are integers, then we can find it by using the factor theorem.

Examples 13d _____

1. Solve the equation $x^3 - 2x^2 - x + 2 = 0$

Possible factors of $x^3 - 2x^2 - x + 2$ are $(x \pm 1)$ and $(x \pm 2)$

Using the factor theorem with $f(x) = x^3 - 2x^2 - x + 2$ gives

$f(1) = 1 - 2 - 1 + 2 = 0$ so $(x - 1)$ is a factor of $f(x)$

Hence $x^3 - 2x^2 - x + 2 = (x - 1)(x^2 - x - 2)$

$$= (x - 1)(x + 1)(x - 2)$$

$\therefore x^3 - 2x^2 - x + 2 = 0$ when $x = 1, -1$ or 2

_____

The factor and remainder theorems can be useful when dealing with some problems involving quadratic equations.

_____

2. The equation $f(x) = 0$ has a repeated root, where
$f(x) = 4x^2 + px + q$ When $f(x)$ is divided by $x + 1$ the
remainder is 1. Find the values of p and q.

$$f(-1) = 4 - p + q = 1 \quad \Rightarrow \quad p = q + 3 \qquad [1]$$

If $4x^2 + px + q = 0$ has a repeated root then '$b^2 - 4ac$' $= 0$

i.e. $\qquad\qquad\qquad p^2 - 16q = 0 \qquad [2]$

Solving equations [1] and [2] simultaneously gives

$$(q + 3)^2 - 16q = 0 \quad \Rightarrow \quad q^2 - 10q + 9 = 0$$
$$\Rightarrow \quad (q - 9)(q - 1) = 0$$

$\therefore$ either $q = 9$ and $p = 12$ or $q = 1$ and $p = 4$

EXERCISE 13d

1. Factorise $2x^3 - x^2 - 2x + 1$. Hence find the values of x for which
$2x^3 - x^2 - 2x + 1 = 0$

2. Given that $f(x) = x^3 - x^2 - x - 2$ show that $f(x) = 0$ has only
one root.

3. Find the value of p for which $x = \frac{1}{2}$ is a solution of the equation
$4x^2 - px + 3 = 0$

4. Show that the x coordinates of the points of intersection of the
curves $\qquad\qquad y = \dfrac{1}{x}$ and $x^2 + 4y = 5$
satisfy the equation $x^3 - 5x + 4 = 0$
Solve this equation.

5. Factorise $x^3 - 4x^2 + x + 6$.
Hence sketch the curve $y = x^3 - 4x^2 + x + 6$

6. A function f is defined by
$$f(x) = 5x^3 - px^2 + x - q$$
When $f(x)$ is divided by $x - 2$, the remainder is 3. Given that
$(x - 1)$ is a factor of $f(x)$
(a) find p and q
(b) find the number of real roots of the equation
$$5x^3 - px^2 + x - q = 0$$

CHAPTER 14

INEQUALITIES

MANIPULATING INEQUALITIES

An inequality compares two unequal quantities.

Consider, for example, the two real numbers 3 and 8 for which

$$8 > 3$$

The inequality remains true, i.e. the inequality sign is unchanged, when the same term is added or subtracted on both sides, e.g.

$$8 + 2 > 3 + 2 \quad \Rightarrow \quad 10 > 5$$

and
$$8 - 1 > 3 - 1 \quad \Rightarrow \quad 7 > 2$$

The inequality sign is unchanged also when both sides are multiplied or divided by a positive quantity, e.g.

$$8 \times 4 > 3 \times 4 \quad \Rightarrow \quad 32 > 12$$

and
$$8 \div 2 > 3 \div 2 \quad \Rightarrow \quad 4 > 1\tfrac{1}{2}$$

If, however, both sides are multiplied or divided by a *negative* quantity the inequality is no longer true. For example, if we multiply by -1, the LHS becomes -8 and the RHS becomes -3 so the correct inequality is now LHS < RHS, i.e.

$$8 \times -1 < 3 \times -1 \quad \Rightarrow \quad -8 < -3$$

Similarly, dividing by -2 gives $-4 < -1\tfrac{1}{2}$

These examples are illustrations of the following general rules.

> Adding or subtracting a term, or multiplying or dividing both sides by a positive number, does not alter the inequality sign.

> Multiplying or dividing both sides by a *negative* number reverses the inequality sign.

i.e.   if a, b and k are real numbers, and $a > b$ then,

$$a + k > b + k \quad \text{for } \textit{all} \text{ values of } k$$

$$ak > bk \qquad\quad \text{for } \textit{positive} \text{ values of } k$$

$$ak < bk \qquad\quad \text{for } \textit{negative} \text{ values of } k$$

SOLVING LINEAR INEQUALITIES

SG

When an inequality contains an unknown quantity, the rules given above can be used to 'solve' it. Whereas the solution of an equation is a value, or values, of the variable, the solution of an inequality is a range, or ranges of values, of the variable.

If the unknown quantity appears only in linear form, we have a *linear inequality* and the solution range has only *one boundary*.

Example 14a _____

Find the set of values of x that satisfy the inequality $x - 5 < 2x + 1$

$$x - 5 < 2x + 1$$

$\Rightarrow$ $\qquad\qquad x < 2x + 6$ $\quad$ adding 5 to each side

$\Rightarrow$ $\qquad\qquad -x < 6$ $\qquad$ subtracting $2x$ from each side

$\Rightarrow$ $\qquad\qquad x > -6$ $\qquad$ multiplying both sides by -1

Therefore the set of values of x satisfying the given inequality is

$$x > -6$$

EXERCISE 14a

Solve the following inequalities.

1. $x - 4 < 3 - x$ $\qquad$ **2.** $x + 3 < 3x - 5$ $\qquad$ **3.** $x < 4x + 9$

4. $7 - 3x < 13$ $\qquad$ **5.** $x > 5x - 2$ $\qquad$ **6.** $2x - 1 < x - 4$

7. $1 - 7x > x + 3$ $\qquad$ **8.** $2(3x - 5) > 6$ $\qquad$ **9.** $3(3 - 2x) < 2(3 + x)$

Double Inequalities

Example 14b

Find the range of values of x for which $4 - x < x + 8 < 5 - 2x$

$4 - x < x + 8 < 5 - 2x$ is called a *double inequality* because it contains two inequalities.

i.e. $4 - x < x + 8$ and $x + 8 < 5 - 2x$

We are looking for the set of values of x for which *both* inequalities are satisfied so first we solve them separately.

$$4 - x < x + 8 \qquad\qquad x + 8 < 5 - 2x$$

$\Rightarrow \qquad\qquad -4 < 2x \qquad\qquad\qquad\qquad 3x < -3$

$\Rightarrow \qquad\qquad x > -2 \qquad\qquad\qquad\qquad x < -1$

The required set of values must satisfy both of these conditions

i.e.

The solution set is $-2 < x < -1$

EXERCISE 14b

For what range(s) of values of x are the following inequalities valid?

1. $x + 2 < 4x - 1 < 2x + 5$

2. $2 - x > 2x + 4 > x$

3. $x - 1 < 3x + 1 < x + 5$

4. $2x - 1 < 3x + 2 < x + 6$

5. $4x - 5 > 5x + 8 > x$

6. $x + 11 > 3x + 1 > 2x + 3$

7. $4 + 5x < 8 + 4x < 6x + 10$

8. $x - 6 > 2 - x > 3x - 18$

SOLVING QUADRATIC INEQUALITIES

A quadratic inequality is one in which the variable appears to the power 2, e.g. $x^2 - 3 > 2x$
The solution is a range or ranges of values of the variable with *two boundaries*.

If the terms in the inequality can be collected and factorised, a graphical solution is easy to find.

Example 14c _____

Find the range(s) of values of x that satisfy the inequality
$x^2 - 3 > 2x$

$$x^2 - 3 > 2x$$

$\Rightarrow$ $$x^2 - 2x - 3 > 0$$

$\Rightarrow$ $$(x - 3)(x + 1) > 0$$

or $$f(x) > 0 \quad \text{where} \quad f(x) = (x - 3)(x + 1)$$

If we sketch the graph of $f(x)$ then $f(x) > 0$ where the graph is above the x-axis.
The values of x corresponding to these portions of the graph satisfy $f(x) > 0$
The points where $f(x) = 0$, i.e. where $x = 3$ and -1 are not part of this
solution and this is indicated on the sketch by open circles.

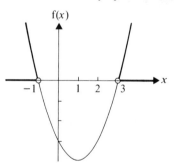

From the graph we see that the ranges of values of x which satisfy the given inequality are

$$x < -1 \quad \text{and} \quad x > 3$$

_____

Note that the solution in the above example was two separate ranges each with its own boundary. If however we consider the inequality $(x - 3)(x + 1) < 0$ the part of the graph of $f(x)$ for which $f(x) < 0$ is below the x-axis and the corresponding values of x are $-3 < x < 1$
This time there is only one range, but it still has two boundaries.

EXERCISE 14c

Find the ranges of values of x that satisfy the following inequalities.

1. $(x - 2)(x - 1) > 0$

2. $(x + 3)(x - 5) \geqslant 0$

3. $(x - 2)(x + 4) < 0$

4. $(2x - 1)(x + 1) \geqslant 0$

5. $x^2 - 4x > 3$

6. $4x^2 < 1$

7. $(2 - x)(x + 4) \geqslant 0$

8. $5x^2 > 3x + 2$

9. $(3 - 2x)(x + 5) \leqslant 0$

10. $(x - 1)^2 > 9$

11. $(x + 1)(x + 2) \leqslant 4$

12. $(1 - x)(4 - x) > x + 11$

Problems

The types of problems which involve inequalities are very varied. Their solutions depend not only on all the methods used so far in this chapter but also on other facts known to the reader, for example:

1)   a perfect square can never be negative.

2)   the nature of the roots of a quadratic equation $ax^2 + bx + c = 0$ depends upon whether $b^2 - 4ac = 0$ or $b^2 - 4ac > 0$ or $b^2 - 4ac < 0$. As two of the above conditions are inequalities, many problems about the roots of a quadratic equation require the solution or interpretation, of inequalities.

The worked examples that follow are intended to give the reader some ideas to use in problem solving and make no claim to cover every situation.

Examples 14d _____

1. Find the range(s) of values of k for which the roots of the equation
    $kx^2 + kx - 2 = 0$ are real.

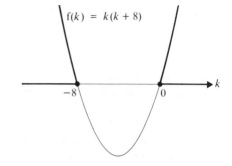

$$kx^2 + kx - 2 = 0$$

For real roots '$b^2 - 4ac$' $\geqslant 0$

i.e. $k^2 - 4(k)(-2) \geqslant 0$

$\Rightarrow$ $k(k + 8) \geqslant 0$

From the sketch we see that
$k(k + 8) \geqslant 0$ for
$k \leqslant -8$ and $k \geqslant 0$

Therefore the equation $kx^2 + kx - 2 = 0$ has real roots if the value
of k lies in either of the ranges $k \leqslant -8$ or $k \geqslant 0$

_____

Note. This type of question is sometimes expressed in another, less
obvious, way, i.e. 'If x is real and $kx^2 + kx - 2 = 0$, find the values
that k can take'. Once the reader appreciates that, because x is real
the roots of the equation are real, the solution is identical to that
above.

_____

2. Prove that $x^2 + 2xy + 2y^2$ cannot be negative.

Knowing that a perfect square cannot be negative, we rearrange the given
expression in the form of perfect squares.

$$x^2 + 2xy + 2y^2 = x^2 + 2xy + y^2 + y^2$$
$$= (x + y)^2 + y^2$$

Each of the two terms on the RHS is a square and so cannot be
negative.

Therefore $x^2 + 2xy + 2y^2$ cannot be negative.

_____

EXERCISE 14d

1. Find the range(s) of values of k for which the given equation has real different roots

   (a) $kx^2 - 2x + k = 0$ (b) $x^2 + 3kx + k = 0$

2. Find the ranges of values of p for which the given equation has real roots.

   (a) $x^2 + (p + 3)x + 4p = 0$ (b) $x^2 + 3x + 1 = px$

3. Find the range of values of a for which the equation
   $x^2 - ax + (a + 3) = 0$ has no real roots.

4. Given that $px^2 + (p - 3)x + 1 = 0$

   (a) find the range(s) of values of p for which the equation has real different roots

   (b) write down the values of p for which the equation has equal roots

   (c) write down the range(s) of values of p for which the roots of the equation are not real.

5. Show that $x^2 - 4xy + 5y^2 \geq 0$ for all real values of x and y

6. Prove that $(a + b)^2 \geq 4ab$ for all real values of a and b

INEQUALITIES IN TWO VARIABLES

Consider the equation $y = x + 3$ which can be represented by a straight line in the xy-plane. This line divides the plane into two areas.

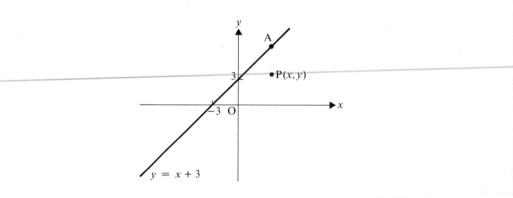

The area below the line contains the set of points $P(x,y)$ for which

$$y \text{ at } P < y \text{ at } A$$

At A, $y = x + 3$

So at P $\hspace{4cm} y < x + 3$

This is true for all points P in the area below the line therefore this area represents the inequality $y < x + 3$

It can be shown similarly that the area above the line $y = x + 3$ represents the inequality $y > x + 3$

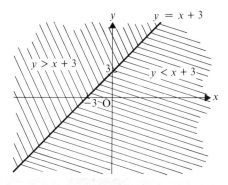

In general, the line with equation $y = mx + c$ divides the xy-plane into two areas which represent the inequalities

$$y < mx + c \quad \text{and} \quad y > mx + c$$

Any linear equation in x and y can be expressed in the form $y = mx + c$. In the same way any linear inequality in x and y can be expressed in the form $y < mx + c$ or $y > mx + c$ and hence can be represented by an area in the xy-plane.

Example 14e _____

Shade the area in the xy-plane which represents
(a) $2y - x > 4$ (b) $3x + 4y \leqslant 2$

(a) $2y - x > 4$ can be written $y > \frac{1}{2}x + 2$

First we draw $y = \frac{1}{2}x + 2$, using a broken line because points on the line are not included in the inequality.

Points for which $y > \frac{1}{2}x + 2$ lie in the area above the line therefore this area represents the inequality $2y - x > 4$

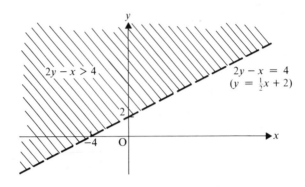

(b) $3x + 4y \leqslant 2$ can be written $y \leqslant \frac{1}{2} - \frac{3}{4}x$

i.e. $y \leqslant -\frac{3}{4}x + \frac{1}{2}$

This relationship includes both the inequality

$$y < -\frac{3}{4}x + \frac{1}{2}$$

and the equation $y = -\frac{3}{4}x + \frac{1}{2}$

Therefore it is represented by points both in the area below the line $y = -\frac{3}{4}x + \frac{1}{2}$ *and* on the line.

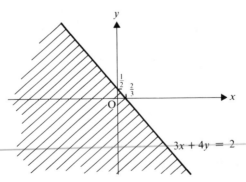

Note that a solid line is drawn to indicate that points on the line are included.

EXERCISE 14e

Shade the area in the xy-plane which represents each of the following inequalities.

1. $y < 2x - 1$ 2. $y \geqslant x + 1$

3. $2y - x \geqslant 6$ 4. $3y + 6x < 4$

5. $2x + y > 3$ 6. $y - x + 1 < 0$

7. $3x + 2y - 4 \leqslant 0$ 8. $2x + y + 5 > 0$

9. $x < 7 - 2y$ 10. $4 \geqslant 3x - y$

SIMULTANEOUS LINEAR INEQUALITIES

Suppose that we require the set of points for which two linear inequalities, $y < m_1 x + c_1$ and $y > m_2 x + c_2$ are *both* satisfied.

Each inequality can be represented separately by an area in the xy-plane.

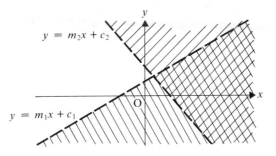

Points in the shaded area \\\\\\ represent $y < m_1 x + c_1$ and points in the area shaded ////// represent $y > m_2 x + c_2$

Therefore points with coordinates which satisfy both inequalities lie in the area where the hatching overlaps, i.e.

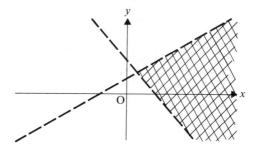

Examples 14f _____

   1. Shade the area which represents both of the inequalities
$$y > x \quad \text{and} \quad y \geqslant 3 - 2x$$

$y > x$ is represented by the area above the line $y = x$ but not by points on that line.

$y \geqslant 3 - 2x$ is represented by points both on and above the line $y = 3 - 2x$

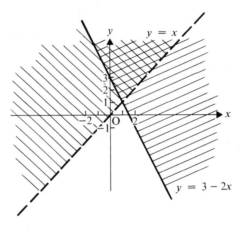

Therefore the area representing both inequalities is

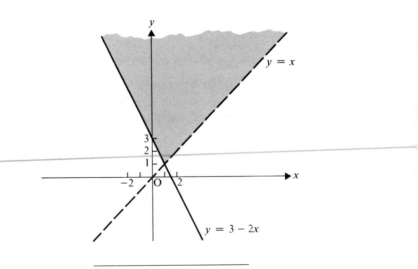

2. Shade the set of points in the xy-plane for which $2x - y \leqslant 3$ and $4x + 3y \leqslant 6$ are both satisfied.

$$2x - y \leqslant 3 \quad \Rightarrow \quad y \geqslant 2x - 3$$

and $\qquad 4x + 3y \leqslant 6 \quad \Rightarrow \quad y \leqslant 2 - \tfrac{4}{3}x$

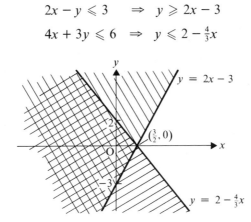

Therefore the set of points which satisfy both inequalities lie in the area shaded below.

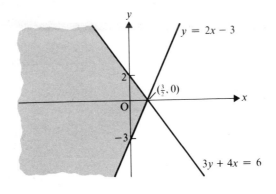

EXERCISE 14f

Shade the area in the xy-plane containing the set of points which satisfy each pair of inequalities.

1. $y \geqslant x - 2$ and $2y \leqslant x + 2$

2. $y < 2x + 3$ and $y > 3 - 2x$

3. $y \leqslant 3x$ and $y + x > 0$

4. $x - 2y + 6 \geqslant 0$ and $y < 2x$

THREE INEQUALITIES

The method shown above for illustrating the set of points which satisfy two inequalities, can be extended to three inequalities.

Example 14g

Find the area in the xy-plane which represents points with coordinates that satisfy simultaneously, $y < x + 1$, $y + 2 \geqslant 0$ and $x + y - 3 < 0$

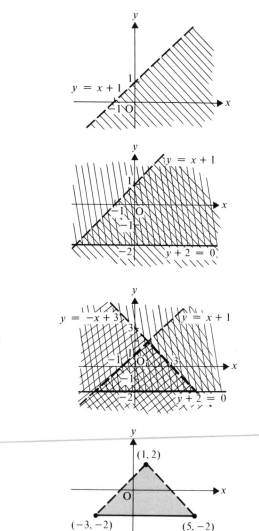

$y < x + 1 \implies$

Then adding $y + 2 \geqslant 0 \implies$

Finally adding $x + y - 3 < 0$

i.e. $\qquad y < -x + 3 \implies$

Therefore the required area is

EXERCISE 14g

In Questions 1 to 5 represent, by an area in the xy-plane, the solution of each set of simultaneous inequalities.

1. $x \leqslant 4$, $y \geqslant 1$ and $y \leqslant x + 5$ 2. $x > 4$, $y > 1$ and $y < x + 5$
3. $2y \geqslant x$, $y - 2x \leqslant 4$ and $x + y - 4 < 0$
4. $y \leqslant 3$, $y < x$ and $x + y > 0$
5. $x + 2 > 0$, $2y - x - 2 \leqslant 0$ and $x < 2$

In Questions 6 to 8,
(a) find the area of the region of the xy-plane that represents the given set of inequalities.
(b) write down the coordinates of all points $P(x, y)$ which satisfy all the given inequalities and whose coordinates are integers.

6. $x > 0$, $y > 0$ and $x + y - 2 \leqslant 0$
7. $x < 2$, $y + 2 > 0$ and $x - 2y > 0$
8. $y \geqslant 1$, $x + y \leqslant 3$ and $x > y - 3$

MIXED EXERCISE 14

Solve each of the inequalities given in Questions 1 to 8.

1. $2x + 1 < 4 - x$ 2. $x - 5 > 1 - 3x$
3. $6x - 5 > 1 + 2x$ 4. $(x - 3)(x + 2) > 0$
5. $(2x - 3)(3x + 2) < 0$ 6. $x^2 - 3 < 10$
7. $(x - 3)^2 > 2$ 8. $(3 - x)(2 - x) < 20$

In each question from 9 to 12, find the set of values of x that satisfy the given inequalities.

9. $-3 < 5 - 2x < 3$ 10. $1 + 4x > 3x - 5 > 5x + 1$
11. $1 - 3x < 5 + x < 9 - x$ 12. $2x + 4 > 3x + 7 > 2x - 1$

13. Prove that $x^2 + y^2 - 10y + 25 \geqslant 0$ for all real values of x and y.

14. For what values of k does the equation $4x^2 + 8x - 8 = k(4x - 3)$ have real roots?

15. Shade the region of the xy-plane representing the given inequality. Find the coordinates of the vertices of this region. If the region is finite, find its area.
    (a) $y \leqslant 0$, $x \geqslant 1$, $y \geqslant x - 2$ (b) $x \geqslant 0$, $y < 3$, $x - y > 1$
    (c) $x - 1 > 0$, $y + 1 > 0$, $x - y + 1 \geqslant 0$

CONSOLIDATION C

SUMMARY

CIRCLE THEOREMS

The perpendicular bisector of any chord goes through O, and conversely.

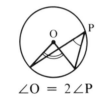

$\angle O = 2\angle P$

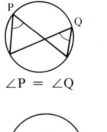

$\angle P = \angle Q$

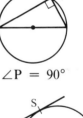

$\angle P = 90°$

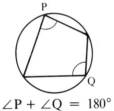

$\angle P + \angle Q = 180°$

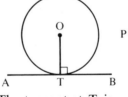

The tangent at T is
perpendicular to OT

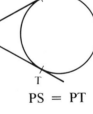

$PS = PT$

CIRCULAR MEASURE

One radian (1ᶜ) is the size of the angle
subtended at the centre of a circle by an arc
equal in length to the radius of the circle.

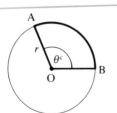

The length of arc AB is $r\theta$

The area of sector AOB is $\frac{1}{2}r^2\theta$

ALGEBRA

Quadratic Equations

If the roots of the quadratic equation $ax^2 + bx + c = 0$ are α and β
then $\alpha + \beta = -\dfrac{b}{a}$ and $\alpha\beta = \dfrac{c}{a}$

Any quadratic equation is of the form

$$x^2 - (\text{sum of roots})x + (\text{product of roots}) = 0$$

The Remainder Theorem

When a polynomial, $f(x)$, is divided by $(x - a)$ the remainder is equal to $f(a)$

The Factor Theorem

If $f(a) = 0$ then $(x - a)$ is a factor of $f(x)$

INEQUALITIES

If $a > b$ then $a + k > b + k$ for all values of k

                                $ak > bk$ for all positive values of k

                                $ak < bk$ for all negative values of k

FUNCTIONS

A function f is a rule that maps a number x to another single number $f(x)$. The domain of a function is the set of input numbers, i.e. the set of values of x.

The range of a function is the set of output values, i.e. the set of values of $f(x)$.

The general form of a quadratic function is $f(x) = ax^2 + bx + c$ where $a \neq 0$
If $a > 0$, $f(x)$ has a minimum value where $x = -b/2a$
If $a < 0$, $f(x)$ has a maximum value where $x = -b/2a$

A function is *even* if $f(x) = f(-x)$
Even functions are symmetrical about the y-axis

A function is *odd* if $f(x) = -f(-x)$
Odd functions have rotational symmetry about the origin.

The function that maps the output of f to its input is called the inverse function of f, and is denoted by f^{-1}, i.e. $f^{-1}: f(x) \rightarrow x$

Note that while it is always possible to reverse a mapping, the rule that does this may not be a function, so not all functions have an inverse.

When a function f operates on a function g we have a function of a function, or a composite function, which is denoted by fg, or $f \circ g$

TRANSFORMATIONS OF CURVES

$y = f(x) + c$ is a translation of $y = f(x)$ by c units in the direction Oy

$y = f(x + c)$ is a translation of $y = f(x)$ by c units in the direction xO

$y = -f(x)$ is the reflection of $y = f(x)$ in the x-axis

$y = f(-x)$ is the reflection of $y = f(x)$ in the y-axis

MULTIPLE CHOICE EXERCISE C

TYPE I

1. The least value of $(x - 1)(x - 3)$ is when x equals

    A 1   **B** 3   **C** 2   **D** 0   **E** −2

2. If $f(x) = 2x - 1$ then $f^{-1}(x)$ is

    A $1 - 2x$ **C** $2y - 1$ **E** $2x + 1$

    B $\frac{1}{2}(x + 1)$ **D** $\frac{1}{2}x - 1$

3. The values of x for which $(x - 1)(x - 5) < 0$ are

    A $x < 3$ **C** $x < 0$ **E** $1 \leqslant x \leqslant 5$

    B $x < 1, x > 5$ **D** $1 < x < 5$

4. An angle of 1 radian is equivalent to:

    A 90°   **B** 60°   **C** 67.3°   **D** 57.3°   **E** 45°

5. An arc PQ subtends an angle of 60° at the centre of a circle of radius 1 cm. The length of the arc PQ is

    A 60 cm   **B** 30 cm   **C** $\frac{1}{6}\pi$ cm   **D** $\frac{1}{3}\pi$ cm   **E** $\frac{1}{18}\pi^2$ cm

6. $x^3 - 3x^2 + 6x - 2$ has remainder 2 when divided by

A $x - 1$ C x E $2x - 1$
B $x + 1$ D $x + 2$

7. The value of k for which $x - 1$ a factor of $4x^3 - 3x^2 - kx + 2$ is

A -1 B 0   C 1   D 2   E 3

8. The curve $y = f(x)$ is

The curve $y = f(-x)$ could be

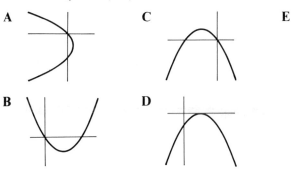

A   C   E

B   D

9. The equation of a curve C is $y = ax^2 + 1$, where a is a constant. Which of these curves cannot be C?

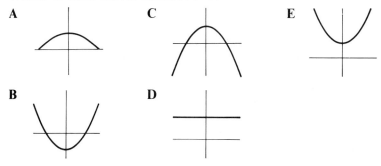

A   C   E

B   D

10. The sum of the roots of the equation $2x^2 - 5x + 3 = 0$ is

A $-2\frac{1}{2}$ B $1\frac{1}{2}$ C 5   D $-1\frac{1}{2}$ E $2\frac{1}{2}$

TYPE II

11. π can represent

    A   the ratio of the circumference to the radius of a circle

    B   half a revolution

    C   half the circumference of a circle.

12. $f(x) \equiv 2x^2 + 3x - 2$

    A $f(x)$ has a remainder -2 when divided by $x - 1$

    B   the equation $f(x) = 0$ has two real distinct roots

    C $x + 2$ is a factor of $f(x)$.

13. A region of the xy-plane is defined by $y > 0$, $y < x$ and $y < 2 - x$

    A   The area of the region is 1 square unit.

    B   The origin is in the region.

    C   The region contains no points with coordinates that are integers.

14. $f(x) = x^2$, $x \in \mathbb{R}$

    A   f is an even function.

    B   The domain of f is the set of positive real numbers.

    C $f(x + a)$ has a least value when $x = 0$

15. If α and β are the roots of the equation $x^2 - 5x - k^2 = 0$

    A $\alpha^2 + \beta^2 = 25$

    B α and β are real and distinct.

    C   either α or β, but not both, is negative.

TYPE III

16. If an arc of a circle of radius 0.5 cm subtends an angle of $60°$ at the centre of the circle, then the length of the arc is 30 cm.

17. If $\dfrac{1}{x} < 2$ then $\dfrac{1}{2} < x$

18. If $x - a$ is a factor of $x^2 + px + q$, the equation $x^2 + px + q = 0$ has a root equal to a

19. If $f : x \rightarrow 2x - 1$ and $g : x \rightarrow x^2$ then $fg : x \rightarrow (2x - 1)^2$

20. The point $(-1, 1)$ lies in the region defined by the inequalities $y > 0$, $x < 0$, $y > x + 1$

MISCELLANEOUS EXERCISE C

1. Find the set of values of x for which $x^2 - 3 < \frac{1}{2}x$ (O/C, SU & C)

2. On a sheet of graph paper shade the region for which $y + x \leqslant 2$ and $2y - x \geqslant 4$ (U of L)

3. If α and β are the roots of the equation $x^2 + px + q = 0$, find the equation whose roots are 2α and 2β

4. $f(x) \equiv (x + 1)(3 - 2x)$
   (a) Sketch the graph of $y = f(x)$. Write down the coordinates of the points where the curve crosses the coordinate axes.
   (b) Write down the set of values of x for which $f(x) \leqslant 0$ (U of L)

5. Shade the region for which $x \leqslant 4$, $y \leqslant 3x - 2$ and $y \geqslant x - 5$

6. The equation $2x^2 - 5x + 1 = 0$ has roots α and β. Find the value of   (a) $\alpha^2 + \beta^2$ (b) $\alpha^2 - \beta^2$

7. The polynomial $g(x) \equiv 2x^3 + Ax^2 + Bx + 6$ is exactly divisible by $(2x - 3)$ and has remainder -28 when divided by $(x + 2)$.
   (a) Find the values of A and B
   (b) Using your values of A and B solve the equation $g(x) = 0$
                                                                           (U of L)

8. Find all the points $P(a, b)$, where a and b are integers, that lie in the region defined by $x \geqslant 0$, $y > 0$, $y \leqslant 4 - 2x$

9. The function f is defined by
$$f : x \rightarrow \frac{3x + 1}{x - 2}, \quad x \in \mathbb{R}, \quad x \neq 2$$
   Find, in a similar form, the functions
   (a) ff   (b) f^{-1} (U of L)

10. Given that $f(x) = x^4$ and $g(x) = x + 2$, simplify
$$fg(x) - gf(x)$$ (JMB)

11. Given that $f(x) \equiv x^3 + 2x^2 - 5x - 6$, find
   (a) $f(2)$
   (b) the complete set of values of x for which $f(x) < 0$
                                                                           (U of L)

12. Given that $(x + 2)$ is a factor of $x^4 + kx^2 + 4x + 1$, find the value of k.

13. Given that $f(x)$, where $f(x) \equiv x^2 + ax + 3$ and a is a constant, is such that the remainder on dividing $f(x)$ by $x - 1$ is three times the remainder on dividing $f(x)$ by $x + 1$, find the value of a (AEB)p

14. (a) Write down the coordinates of the mid-point M of the line joining $A(0, 1)$ and $B(6, 5)$.

(b) Show that the line $3x + 2y - 15 = 0$ passes through M and is perpendicular to AB.

(c) Calculate the coordinates of the centre of the circle which passes through A, B and the origin O.   (U of L)

15. The function

$$f : x \rightarrow 2x^3 + ax^2 + bx + 36 \quad x \in \mathbb{R}$$

is such that $f(3) = 0$ and the remainder when $f(x)$ is divided by $(x + 2)$ is -30. Find the values of a and b and express $f(x)$ as the product of three linear factors.   (AEB)

16. A function f is defined by

$$f : x \rightarrow 1 - \frac{1}{x}, \quad x \subset \mathbb{R}, \quad x \neq 0, \quad x \neq 1$$

Find (a) $ff(x)$ (b) $fff(x)$ (c) $f^{-1}(x)$ (U of L)

17. The functions f and g are defined by
$$f : x \rightarrow x^2 - 3, \quad x \subset \mathbb{R}$$
$$g : x \rightarrow 2x + 5, \quad x \subset \mathbb{R}$$
Find in a similar form the composite function $f \circ g$

Sketch on separate axes the graphs of f and $f \circ g$

Hence, or otherwise, show that the range of f corresponding to the domain $-4 \leqslant x \leqslant 4$ is $-3 \leqslant f(x) \leqslant 13$, and find the range of $f \circ g$ corresponding to this domain.   (AEB)

18. A circle, centre O and radius a, has AB as a diameter and C is a point on AB produced such that $BC = a$. Points P and Q lie on the circle and $PC = QC$. Given that angle $POC = \theta$, show that L, the perimeter of the area enclosed by the lines CP, CQ and the arc PAQ, is given by

$$L = 2a\pi - 2a\theta + 2a(5 - 4\cos\theta)^{1/2} \qquad \text{(U of L)p}$$

19. A chord divides a circle, centre O, into two regions whose areas are in the ratio $2:1$. Prove that the angle θ, subtended by this chord at O, satisfies the equation $f(\theta) = 0$, where

$$f(\theta) = \theta - \sin\theta - 2\pi/3 \qquad\qquad \text{(U of L)p}$$

20. A is the point $(0,6)$ and B is the point $(4,0)$. Calculate the coordinates of the centre of the circle which passes through A, B and the origin O. Hence find the radius of this circle.

21. A circular sector, of area A cm^2, has bounding radii, each of length x cm, and the angle between these radii is θ radians.
Given that the perimeter of the sector is 12 cm,

(a) express θ in terms of x

(b) show that $A = 6x - x^2$ \qquad\qquad (AEB)p

CHAPTER 15

DIFFERENTIATION

CHORDS, TANGENTS AND NORMALS

Consider any two points, A and B, on any curve.

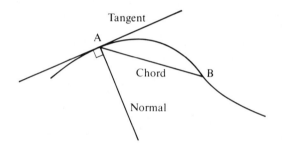

The line joining A and B is called a chord.

The line that touches the curve at A is called the tangent at A.

Note that the word *touch* has a precise mathematical meaning, i.e. a line that meets a curve at a point and carries on without crossing to the other side of the curve at that point, is said to *touch* the curve at the *point of contact*.

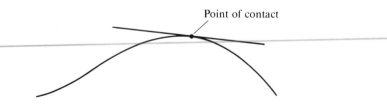

The line perpendicular to the tangent at A is called the normal at A.

THE GRADIENT OF A CURVE

107

108

Gradient, or slope, defines the direction of a line (lines can be straight or curved).

When walking along a straight line we walk in the same direction all the time, i.e. the gradient of a straight line is constant.

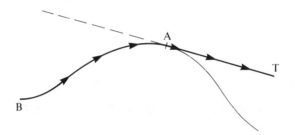

If, however, we move from B to A along a curve, our direction is changing all the time, i.e. the gradient of a curve is not constant but has different values at different points on the curve.

Now if at A we continue to move, but without any further change in direction, we go along the straight line AT which is the tangent at A, so

> the gradient of a curve at a point A is the same as the gradient of the tangent at A

For a straight line the numerical value of the gradient is found by taking the coordinates of any two points on the line and working out

$$\frac{\text{increase in } y}{\text{increase in } x}$$

This can be used also to find the gradient of a tangent to a curve but, if the tangent is just drawn by eye, the value obtained can only be an approximation.

A more precise method is needed for determining the gradient of a
curve whose equation is known, so that further analysis can be made
of the properties of such a curve.

Let us consider the problem of finding the gradient of the tangent at a
point A on a curve.

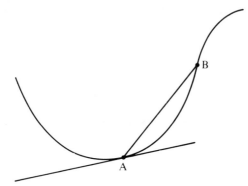

If B is another point on the curve, fairly close to A, then the gradient
of the chord AB gives an *approximate* value for the gradient of the
tangent at A. As B gets nearer to A, the chord AB gets closer to the
tangent at A, so the approximation becomes more accurate.

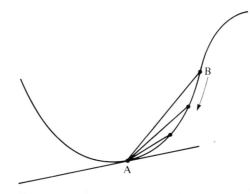

So, as B gets closer and closer to A, we can say,

as B → A

the gradient of chord AB → the gradient of the tangent at A

This fact can also be expressed in the form

limit (gradient of chord AB) = gradient of tangent at A
as B → A

This definition can be applied to a particular point on a particular curve. Suppose, for instance, that we want the gradient of the curve $y = x^2$ at the point where $x = 1$

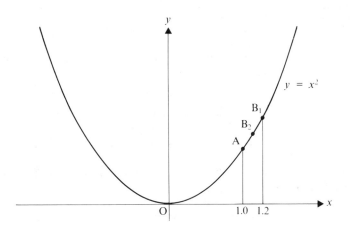

A is the point $(1, 1)$ and we need a succession of points, B_1, B_2,.... getting closer and closer to A. Let us take the points where $x = 1.2, 1.1, 1.05, 1.01, 1.001$, then calculate the corresponding y coordinates, find the increases in x and y between A and B and hence find the gradient of AB.

x	1.2	1.1	1.05	1.01	1.001
y $(= x^2)$	1.44	1.21	1.1025	1.0201	1.002 001
Increase in y	0.44	0.21	0.1025	0.0201	0.002 001
Increase in x	0.2	0.1	0.05	0.01	0.001
Gradient of chord AB	2.2	2.1	2.05	2.01	2.001

From the numbers in the last row of the table it is clear that, as B gets nearer to A, the gradient of the chord gets nearer to 2, i.e.

$$\underset{\text{as } B \to A}{\text{limit}} \text{ (gradient of chord AB)} = 2$$

It is equally clear that it is much too tedious to go through this process each time we want the gradient at just one point on just one curve and that we need a more general method. For this we use a general point $A(x, y)$ and a variable small change in the value of x between A and B.

A new symbol, δ, is used to denote this small change.

When δ appears as a prefix to any letter representing a variable quantity, it denotes a small increase in that quantity,

e.g. δx means a small increase in x

δy means a small increase in y

δt means a small increase in t

Note that δ is only a prefix. It does not have an independent value and cannot be treated as a factor.

Now consider again the gradient of the curve with equation $y = x^2$

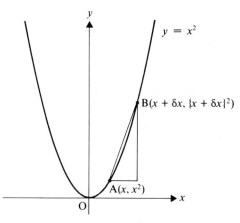

This time we will look for the gradient at *any* point $A(x, y)$ on the curve and use a point B where the x-coordinate of B is $x + \delta x$.

For any point on the curve, $y = x^2$

So, at B, the y-coordinate is $(x + \delta x)^2 = x^2 + 2x\delta x + (\delta x)^2$

Therefore the gradient of chord AB, which is given by $\dfrac{\text{increase in } y}{\text{increase in } x}$,

is $\dfrac{(x + \delta x)^2 - x^2}{(x + \delta x) - x} = \dfrac{2x\delta x + (\delta x)^2}{\delta x}$

$= 2x + \delta x$

Now, as $B \rightarrow A$, $\delta x \rightarrow 0$, therefore

$$\text{gradient of curve at } A = \underset{\text{as } B \to A}{\text{limit}} \; (\text{gradient of chord AB})$$

$$= \underset{\text{as } \delta x \to 0}{\text{limit}} \; (\text{gradient of chord AB})$$

$$= \underset{\text{as } \delta x \to 0}{\text{limit}} \; (2x + \delta x)$$

$$= 2x$$

This result can now be used to give the gradient at any point on the curve with equation $y = x^2$, where the x-coordinate is given, e.g.

at the point where $x = 3$, the gradient is $2(3) = 6$
and at the point $(4, 16)$, the gradient is $2(4) = 8$

Looking back at the longer method we used on page 233 to find the gradient at the point where $x = 1$, we see that the value obtained there is confirmed by using the general result,
i.e. gradient $= 2x = 2(1) = 2$

DIFFERENTIATION

109 The process of finding a general expression for the gradient of a curve at any point is known as differentiation.

110 The general gradient expression for a curve $y = f(x)$ is itself a function so it is called the *gradient function*. For the curve $y = x^2$ for example, the gradient function is $2x$.

Because the gradient function is derived from the given function, it is more often called the *derived function* or the *derivative*.

The method used above, in which the limit of the gradient of a chord was used to find the derived function, is known as *differentiating from first principles*. It is the fundamental way in which the gradient of each new type of function is found and, although many short cuts can be developed, it is important to understand this basic method, which can be applied to any function of x.

The General Gradient Function

Consider any curve with equation $y = f(x)$.

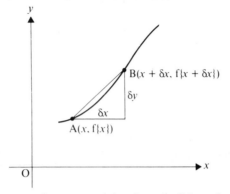

Taking two points on the curve, $A(x, y)$ and $B(x + \delta x, y + \delta y)$ we have,

at A, $y = f(x)$ and at B, $y + \delta y = f(x + \delta x)$

The gradient of AB is given by

$$\frac{\delta y}{\delta x} = \frac{f(x + \delta x) - f(x)}{\delta x}$$

Therefore   the gradient of the curve at A is given by

$$\lim_{\delta x \to 0} \frac{\delta y}{\delta x} = \lim_{\delta x \to 0} \frac{f(x + \delta x) - f(x)}{\delta x}$$

NOTATION

The gradient function of a curve $y = f(x)$ can be denoted by $\dfrac{dy}{dx}$ (we say dy by dx)

i.e. $$\frac{dy}{dx} = \lim_{\delta x \to 0} \frac{\delta y}{\delta x} = \lim_{\delta x \to 0} \frac{f(x + \delta x) - f(x)}{\delta x}$$

Note that d has no independent meaning and must never be regarded as a factor.

The complete symbol $\dfrac{d}{dx}$ means 'the derivative with respect to x of'

So $\dfrac{dy}{dx}$ means 'the derivative with respect to x of y'

(Because the phrase 'with respect to' is used frequently, it is often abbreviated to w.r.t.)

We have seen that the derivative of x^2 is $2x$ so we can write

$$\text{for } y = x^2, \quad \frac{dy}{dx} = 2x$$

An alternative notation concentrates on the function of x rather than the equation of the curve. Using this form we can write

$$\text{for } f(x) = x^2, \quad f'(x) = 2x$$

In this form, f' means 'the gradient function' or 'the derived function'.

Examples 15a

1. Given that $f : x \rightarrow x^3$, find the gradient function at any point on the curve $y = f(x)$. Find also the gradient at the point $(2, 13)$

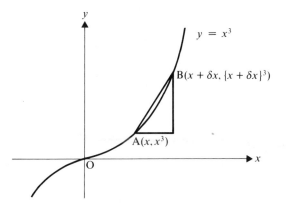

If A is a point (x, y) and B is a neighbouring point whose x-coordinate is $x + \delta x$, then

$\quad$ at A the y-coordinate is x^3

and $\quad$ at B the y-coordinate is $(x + \delta x)^3$

The gradient of the chord AB is $\dfrac{(x + \delta x)^3 - x^3}{(x + \delta x) - x}$

which simplifies to $3x^2 + 3x(\delta x) + (\delta x)^2$

Therefore $\qquad \dfrac{dy}{dx} = \underset{\delta x \rightarrow 0}{\text{limit}} \ \{3x^2 + 3x(\delta x) + (\delta x)^2\}$

$$= 3x^2$$

i.e. the gradient function of x^3 is $3x^2$

At the point $(2, 13)$ $x = 2$ so the gradient is $3(2^2)$, i.e. 12

2. Find the derivative of the function $1/x$

$$f(x) = \frac{1}{x} \quad \Rightarrow \quad f(x + \delta x) = \frac{1}{x + \delta x}$$

$$f(x + \delta x) - f(x) = \frac{1}{x + \delta x} - \frac{1}{x} = \frac{x - (x + \delta x)}{x(x + \delta x)}$$

$$= \frac{-\delta x}{x(x + \delta x)}$$

$$\frac{f(x + \delta x) - f(x)}{\delta x} = \frac{-\delta x}{x(x + \delta x)(\delta x)} = \frac{-1}{x(x + \delta x)}$$

Hence
$$f'(x) = \lim_{\text{as } \delta x \to 0} \frac{f(x + \delta x) - f(x)}{\delta x}$$

$$= \lim_{\text{as } \delta x \to 0} \frac{-1}{x(x + \delta x)}$$

$$= \frac{-1}{x^2}$$

i.e. the derivative of $1/x$ is $-1/x^2$

The reader may like to try using this method to find $\dfrac{dy}{dx}$ for $y = x^4$;
(Pascal's Triangle will help).

DIFFERENTIATING x^n WITH RESPECT TO x

Some of the results that have been produced so far can now be collected and tabulated.

y	x^2	x^3	x^4	x^{-1}
$\dfrac{dy}{dx}$	$2x$	$3x^2$	$4x^3$	$-x^{-2}$ (or $-1/x^2$)

From this table it *appears* that when we differentiate a power of x we multiply by the power and then reduce the power by 1, i.e. it looks as though

$$\text{if } y = x^n, \text{ then } \frac{dy}{dx} = nx^{n-1}$$

This result, although deduced from just a few examples, is in fact valid for all powers, including those that are fractional or negative. It is not possible to give a proof at this stage and this is one example of a 'rule' which, for the moment, we must just take on trust. It is easy to apply and makes the task of differentiating a power of x very much simpler, e.g.

$$\frac{d}{dx}(x^7) = 7x^6 \qquad \frac{d}{dx}(x^3) = 3x^2$$

$$\frac{d}{dx}(x^{-2}) = -2x^{-3} \qquad \frac{d}{dx}(x^{3/2}) = (3/2)x^{1/2}$$

Example 15b _____

Differentiate with respect to x

(a) $\sqrt{x^3}$ (b) $x^{-1/3}$

(a) Using $\dfrac{d}{dx}(x^n) = nx^{n-1}$ where $n = \frac{3}{2}$ gives

$$\frac{d}{dx}(x^{3/2}) = \tfrac{3}{2}x^{(3/2-1)} = \tfrac{3}{2}x^{1/2} \text{ or } \tfrac{3}{2}\sqrt{x}$$

(b) This time $n = -\frac{1}{3}$ so

$$\frac{d}{dx}(x^{-1/3}) = -\tfrac{1}{3}x^{(-1/3-1)} = -\tfrac{1}{3}x^{-4/3}$$

EXERCISE 15b

Differentiate with respect to x

1. x^5 2. x^{11} 3. x^{20} 4. x^{10}

5. x^{-3} 6. x^{-7} 7. x^{-5} 8. $x^{4/3}$

9. x 10. x^{-1} 11. $x^{1/3}$ 12. $\sqrt{x}$

13. $\dfrac{1}{x^2}$ 14. $\sqrt{\dfrac{1}{x}}$ 15. $\dfrac{1}{x^4}$ 16. $\dfrac{1}{x^{10}}$

17. $x^{-1/4}$ 18. $\sqrt{x^5}$ 19. $x^{1/7}$ 20. x^p

Differentiating a Constant

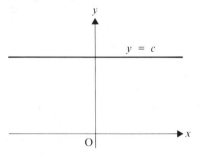

Consider the equation $y = c$. Graphically this is a horizontal straight line and its gradient is zero, i.e. $\dfrac{dy}{dx} = 0$

i.e.   if $y = c$ then $\dfrac{dy}{dx} = 0$

Differentiating a Linear Function of x

The graph of the equation $y = kx$, where k is a constant, is a straight line with gradient k.

Hence $\dfrac{d}{dx}(kx) = k$

Now if we apply the general rule for differentiating x^n to $y = x$,

i.e. to $y = x^1$ we get $\dfrac{d}{dx}(x^1) = 1x^0 = 1$

Combining these two facts shows that $\dfrac{d}{dx}(kx) = k \times \dfrac{d}{dx}(x)$

This conclusion applies, in fact, to a constant multiple of *any* function of x,

e.g.   if $y = 3x^5$, $\dfrac{dy}{dx} = 3 \times \dfrac{d}{dx}(x^5) = 3 \times 5x^4 = 15x^4$

and if $y = 4x^{-2}$, $\dfrac{dy}{dx} = 4 \times -2x^{-3} = -8x^{-3}$

In general then, if u is a constant,

$$\frac{d}{dx}(ax^n) = anx^{n-1}$$

This rule, although not proved for the general case, can be used freely.

Another very useful property is

a function of x which contains a number of different terms can be differentiated term by term, applying the basic rule to each in turn.

For example, if $y = x^4 + \dfrac{1}{x} - 6x$

then $\dfrac{dy}{dx} = \dfrac{d}{dx}(x^4) + \dfrac{d}{dx}(x^{-1}) - \dfrac{d}{dx}(6x) = 4x^3 - \dfrac{1}{x^2} - 6$

EXERCISE 15c

Differentiate each of the following functions w.r.t. x

1. $5x^3$

2. $7x$

3. $\dfrac{8}{x}$

4. $5\sqrt{x}$

5. $\dfrac{-1}{2x^3}$

6. $3x^3 - 4x^2$

7. $4x - \dfrac{4}{x}$

8. $\dfrac{3}{x} + \dfrac{x}{3}$

9. $x^3 - x^2 + 5x - 6$

10. $3x^2 + 7 - 4/x$

11. $x^3 - 2x^2 - 8x$

12. $2x^4 - 4x^2$

13. $x^2 + 5\sqrt{x}$

14. $3x^3 - 4x^2 + 9x - 10$

15. $x^{3/2} - x^{1/2} + x^{-1/2}$

16. $\sqrt{x} + \sqrt{x^3}$

17. $\dfrac{1}{x^2} - \dfrac{1}{x^3}$

18. $\dfrac{1}{\sqrt{x}} - \dfrac{2}{x}$

19. $x^{-1/2} + 3x^{3/2}$

20. $x^{1/4} - x^{1/5}$

21. $\dfrac{4}{x^3} + \dfrac{x^3}{4}$

22. $\dfrac{4}{x} + \dfrac{5}{x^2} - \dfrac{6}{x^3}$

23. $3\sqrt{x} - 3x$

24. $x - 2x^{-1} - 3x^{-3}$

Differentiating Products and Fractions

All the rules given above can be applied to the differentiation of expressions containing products or quotients provided that they are multiplied out or divided into separate terms.

Examples 15d _____

1. If $y = (x - 3)(x^2 + 7x - 1)$, find $\dfrac{dy}{dx}$

$$y = (x - 3)(x^2 + 7x - 1) = x^3 + 4x^2 - 22x + 3$$

$\Rightarrow$ $\qquad\qquad\qquad \dfrac{dy}{dx} = 3x^2 + 8x - 22$

_____

2. Find $\dfrac{dt}{dz}$ given that $t = \dfrac{6z^2 + z - 4}{2z}$

$$t = \frac{6z^2 + z - 4}{2z} = \frac{6z^2}{2z} + \frac{z}{2z} - \frac{4}{2z}$$

$$= 3z + \tfrac{1}{2} - 2/z$$

$\Rightarrow$ $\qquad\qquad\qquad \dfrac{dt}{dz} = 3 + 0 - 2(-z^{-2})$

$$= 3 + \frac{2}{z^2}$$

_____

EXERCISE 15d

In each question, differentiate with respect to the variable concerned.

1. $y = (x + 1)^2$

2. $z = x^{-2}(2 - x)$

3. $y = (3x - 4)(x + 5)$

4. $y = (4 - z)^2$

5. $s = \dfrac{t^{-1} + 3t^2}{2t^2}$

6. $s = \dfrac{t^2 + t}{2t}$

7. $y = \left(\dfrac{1}{x}\right)(x^2 + 1)$

8. $y = \dfrac{z^3 - z}{\sqrt{z}}$

9. $y = 2x(3x^2 - 4)$

10. $s = (t + 2)(t - 2)$

11. $s = \dfrac{t^3 - 2t^2 + 7t}{t^2}$

12. $y = \dfrac{\sqrt{x + 7}}{x^2}$

GRADIENTS OF TANGENTS AND NORMALS

If the equation of a curve is known, and the gradient function can be found, then the gradient, m say, at a particular point A on that curve can be calculated. This is also the gradient of the tangent to the curve at A.

The normal at A is perpendicular to the tangent at A, therefore its gradient is $-1/m$

Examples 15e

1. The equation of a curve is $s = 6 - 3t - 4t^2 - t^3$. Find the gradient of the tangent and of the normal to the curve at the point $(-2, 4)$.

$$s = 6 - 3t - 4t^2 - t^3 \quad \Rightarrow \quad \frac{ds}{dt} = 0 - 3 - 8t - 3t^2$$

At the point $(-2, 4)$, $\dfrac{ds}{dt} = -3 - 8(-2) - 3(-2)^2 = 1$

Therefore the gradient of the tangent at $(-2, 4)$ is 1 and the gradient of the normal is $-1/1$, i.e. -1

2. Find the coordinates of the points on the curve
$y = 2x^3 - 3x^2 - 8x + 7$ where the gradient is 4

$$y = 2x^3 - 3x^2 - 8x + 7 \quad \Rightarrow \quad \frac{dy}{dx} = 6x^2 - 6x - 8$$

If the gradient is 4 then $\dfrac{dy}{dx} = 4$

i.e. $6x^2 - 6x - 8 = 4 \quad \Rightarrow \quad 6x^2 - 6x - 12 = 0$

$\Rightarrow \quad x^2 - x - 2 = 0$

$\therefore \qquad (x - 2)(x + 1) = 0 \quad \Rightarrow \quad x = 2 \text{ or } -1$

When $x = 2$, $y = 16 - 12 - 16 + 7 = -5$

when $x = -1$, $y = -2 - 3 + 8 + 7 = 10$

Therefore the gradient is 4 at the points $(2, -5)$ and $(-1, 10)$

EXERCISE 15e

Find the gradient of the tangent and the gradient of the normal at the given point on the given curve.

1. $y = x^2 + 4$ where $x = 1$

2. $y = 3/x$ where $x = -3$

3. $y = \sqrt{z}$ where $z = 4$

4. $s = 2t^3$ where $t = -1$

5. $v = 2 - 1/u$ where $u = 1$

6. $y = (x + 3)(x - 4)$ where $x = 3$

7. $y = z^3 - z$ where $z = 2$

8. $s = t + 3t^2$ where $t = -2$

9. $z = x^2 - 2/x$ where $x = 1$

10. $y = \sqrt{x} + 1/\sqrt{x}$ where $x = 9$

11. $s = \sqrt{t}(1 + \sqrt{t})$ where $t = 4$

12. $y = \dfrac{x^2 - 4}{x}$ where $x = -2$

Find the coordinates of the point(s) on the given curve where the gradient has the value specified.

13. $y = 3 - 2/x; \frac{1}{2}$ 14. $z = x^2 - x^3; -1$

15. $s = t^3 - 12t + 9; 15$ 16. $v = u + 1/u; 0$

17. $s = (t + 3)(t - 5); 0$ 18. $y = 1/x^2; \frac{1}{4}$

19. $y = (2x - 5)(x + 1); -3$ 20. $y = z^3 - 3z; 0$

DIFFERENTIATING A FUNCTION OF A FUNCTION

Suppose that we want to differentiate $(2x - 1)^3$. We could expand the bracket and differentiate term by term, but this is tedious and, for powers higher than three, very long and not easy. We obviously need a more direct method for differentiating an expression of this kind.

Now $(2x - 1)^3$ is a cubic function of the linear function $(2x - 1)$, i.e. it is a *function of a function*.

A function of this type is of the form $gf(x)$.

Consider any equation of the form $y = gf(x)$

If we make the substitution $u = f(x)$ then $y = gf(x)$ can be expressed in two simple parts, i.e.

$$y = g(u) \text{ where } u = f(x)$$

A small increase of δx in the value of x causes a corresponding small increase of δu in the value of u.
Then if $\delta x \to 0$, it follows that $\delta u \to 0$

Hence
$$\frac{dy}{dx} = \lim_{\delta x \to 0} \left(\frac{\delta y}{\delta x} \right) = \lim_{\delta x \to 0} \left(\frac{\delta y}{\delta u} \right)\left(\frac{\delta u}{\delta x} \right)$$

$\Rightarrow$
$$\frac{dy}{dx} = \left(\lim_{\delta u \to 0} \frac{\delta y}{\delta u} \right) \times \left(\lim_{\delta x \to 0} \frac{\delta u}{\delta x} \right)$$

i.e.
$$\frac{dy}{dx} = \frac{dy}{du} \times \frac{du}{dx}$$

This is known as *the chain rule*.

Example 15f

Find $\dfrac{dy}{dx}$ if $y = (2x - 4)^4$

If $u = 2x - 4$ then $y = u^4$

Then $\dfrac{dy}{dx} = \dfrac{dy}{du} \times \dfrac{du}{dx}$ gives

$$\frac{dy}{dx} = (4u^3)(2) = 8u^3$$

But $u = 2x - 4$

$\therefore$
$$\frac{dy}{dx} = 8(2x - 4)^3$$

This example is a particular case of the equation $y = (ax + b)^n$.
Similar working shows that, in general,

$$\text{if } y = (ax + b)^n \text{ then } \frac{dy}{dx} = an(ax + b)^{n-1}$$

This fact is needed very often and is quotable.

EXERCISE 15f

Use the quotable result above to differentiate each function with respect to x.

1. $(3x + 1)^2$
2. $(x - 3)^4$
3. $(4x + 5)^5$

4. $(2 + 3x)^7$
5. $(6x - 2)^3$
6. $(4 - 2x)^5$

7. $(1 - 5x)^2$
8. $(3 - 2x)^3$
9. $(4 - 3x)^4$

10. $(3x + 1)^{-1}$
11. $(2x - 5)^{-4}$
12. $(1 - 2x)^{-5}$

13. $(2x + 3)^{1/2}$
14. $(8 - 3x)^{1/3}$
15. $\sqrt{(1 - 4x)}$

16. $\dfrac{1}{2 - 7x}$
17. $\dfrac{1}{\sqrt{(3 - x)}}$
18. $\dfrac{1}{(1 - 5x)^2}$

MIXED EXERCISE 15

1. Differentiate $3x^2 + x$ with respect to x from first principles.

2. Find the derivative of
   (a) $x^{-3} - x^3 + 7$ (b) $x^{1/2} - x^{-1/2}$ (c) $1/x^2 + 2/x^3$

3. Differentiate w.r.t. x.
   (a) $y = x^{3/2} - x^{2/3} + x^{-1/3}$ (b) $y = \sqrt{x} - 1/x + 1/x^3$
   (c) $(1 - 8x)^{1/3}$ (d) $\dfrac{1}{(3x - 2)^4}$

4. Find the gradient of the curve $y = 2x^3 - 3x^2 + 5x - 1$ at the point
   (a) $(0, -1)$ (b) $(1, 3)$ (c) $(-1, -11)$

5. Find the gradient of the given curve at the given point.
   (a) $y = x^2 + x - 9$; $x = 2$
   (b) $y = x(x - 4)$; $x = 5$
   (c) $y = (2x + 3)^2$; $x = 0$
   (d) $y = \sqrt{(10 - 3x)}$; $x = 2$

6. The equation of a curve is $y = (x - 3)(x + 4)$. Find the gradient of the curve
   (a) at the point where the curve crosses the y-axis
   (b) at each of the points where the curve crosses the x-axis

7. If the equation of a curve is $y = 2x^2 - 3x - 2$ find
   (a) the gradient at the point where $x = 0$
   (b) the coordinates of the points where the curve crosses the x-axis
   (c) the gradient at each of the points found in (b).

8. Find the coordinates of the point(s) on the curve $y = 3x^3 - x + 8$ at which the gradient is   (a) 8   (b) 0

9. Find $\dfrac{dy}{dx}$ if

   (a) $y = x^4 - x^2$ (b) $y = (3x + 4)^2$ (c) $y = \dfrac{x - 3}{\sqrt{x}}$

10. Find the gradient of the tangent at the point where $x = 1$ on the curve $y = \sqrt{(2 - x)}$.

11. Find the coordinates of the point on the curve $y = x^2$ where the gradient of the normal is $\frac{1}{4}$.

12. The equation of a curve is $s = 4t^2 + 5t$. Find the gradient of the normal at each of the points where the curve crosses the t-axis.

13. Find the coordinates of the points on the curve $y = x^3 - 6x^2 + 12x + 2$ at which the tangent is parallel to the line $y = 3x$

14. The curve $y = (x - 2)(x - 3)(x - 4)$ cuts the x-axis at the points $P(2, 0)$, $Q(3, 0)$ and $R(4, 0)$. Prove that the tangents at P and R are parallel and find the gradient of the normal at Q.

15. For a certain equation, $\dfrac{dy}{dx} = 2x + 1$ Which of the following could be the given equation?
    (a) $y = 2x^2 + x$ (b) $y = x^2 + x - 1$ (c) $y = x^2 + 1$
    (d) $y = x^2 + x$

CHAPTER 16

TANGENTS, NORMALS AND STATIONARY POINTS

THE EQUATIONS OF TANGENTS AND NORMALS

We have seen how to find the gradient of a tangent at a particular point, A, on a curve. We also know that the tangent passes through the point A. Therefore the tangent is a line passing through a known point and having a known gradient and its equation can be found using

$$y - y_1 = m(x - x_1)$$

The equation of a normal can be found in the same way.

Examples 16a _____

1. Find the equation of the normal to the curve $y = \dfrac{4}{x}$ at the point where $x = 1$

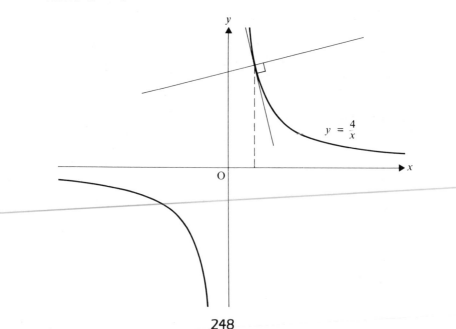

$$y = \frac{4}{x}$$

$$y = \frac{4}{x} \quad \Rightarrow \quad \frac{dy}{dx} = -\frac{4}{x^2}$$

When $x = 1$, $y = 4$ and $\dfrac{dy}{dx} = -4$

The gradient of the tangent at $(1, 4)$ is -4, therefore the gradient of the normal at $(1, 4)$ is $-\frac{1}{-4}$ i.e. $\frac{1}{4}$

The equation of the normal is given by $y - y_1 = m(x - x_1)$

i.e. $\qquad\qquad\qquad y - 4 = \frac{1}{4}(x - 1)$

$\Rightarrow \qquad\qquad\qquad 4y = x + 15$

---

2. Find the equation of the tangent to the curve $y = x^2 - 6x + 5$ at each of the points where the curve crosses the x-axis. Find also the coordinates of the point where these tangents meet.

The curve crosses the x-axis where $y = 0$,

i.e. where $\quad x^2 - 6x + 5 = 0 \quad\quad \Rightarrow \quad\quad (x - 5)(x - 1) = 0$

$\Rightarrow \qquad\qquad x = 5 \text{ and } x = 1$

Therefore the curve crosses the x-axis at $(5, 0)$ and $(1, 0)$

$$y = x^2 - 6x + 5 \quad \Rightarrow \quad \frac{dy}{dx} = 2x - 6$$

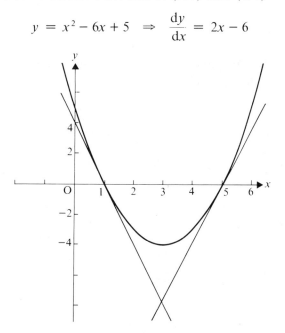

At $(5, 0)$, the gradient of the tangent is given by

$$\frac{dy}{dx} = 10 - 6 = 4$$

therefore the equation of this tangent is

$$y - 0 = 4(x - 5) \quad \Rightarrow \quad y = 4x - 20$$

At $(1, 0)$ the gradient of the tangent is given by

$$\frac{dy}{dx} = 2 - 6 = -4$$

Therefore the equation of the tangent is

$$y - 0 = -4(x - 1) \quad \Rightarrow \quad y + 4x = 4$$

If the two tangents meet at **P** then, at **P**,

$$y + 4x = 4 \tag{1}$$

and

$$y - 4x = -20 \tag{2}$$

[1] + [2] gives $\quad 2y = -16 \quad \Rightarrow \quad y = -8$

Using $y = -8$ in [1] gives $\quad -8 + 4x = 4 \quad \Rightarrow \quad x = 3$

Therefore the tangents meet at $(3, -8)$

EXERCISE 16a

In each question from 1 to 7 find, at the given point,

(a) the equation of the tangent

(b) the equation of the normal.

1. $y = x^2 - 4$ where $x = 1$

2. $y = x^2 + 4x - 2$ where $x = 0$

3. $y = 1/x$ where $x = -1$

4. $y = x^2 + 5$ where $x = 0$

5. $y = x^2 - 5x + 7$ where $x = 2$

6. $y = (x - 2)(x^2 - 1)$ where $x = -2$

7. $y = (1 - 2x)^4$ where $x = 1$

8. Find the equation of the normal to the curve $y = x^2 + 4x - 3$ at the point where the curve cuts the y-axis.

9. Find the equation of the tangent to the curve $y = x^2 - 3x - 4$ at the point where this curve cuts the line $x = 5$

10. Find the equation of the tangent to the curve $y = (2x - 3)(x - 1)$ at each of the points where this curve cuts the x-axis. Find the point of intersection of these tangents.

11. Find the equation of the normal to the curve $y = x^2 - 6x + 5$ at each of the points where the curve cuts the x-axis.

12. Find the equation of the tangent to the curve $y = 3x^2 + 5x - 1$ at each of the points of intersection of the curve and the line $y = x - 1$

13. Find the equations of the tangent to the curve $y = x^2 + 5x - 3$ at the points where the line $y = x + 2$ crosses the curve.

14. Find the coordinates of the point on the curve $y = 2x^2$ at which the gradient is 8 Hence find the equation of the tangent to $y = 2x^2$ whose gradient is 8

15. Find the coordinates of the point on the curve $y = 3x^2 - 1$ at which the gradient is 3

16. Find the equation of the tangent to the curve $y = 4x^2 + 3x$ which has a gradient of -1

17. Find the equation of the normal to the curve $y = 2x^2 - 2x + 1$ which has a gradient of $\frac{1}{2}$

18. Find the value of k for which $y = 2x + k$ is a tangent to the curve $y = 2x^2 - 3$

19. Find the equation of the tangent to the curve $y = (x - 5)(2x + 1)$ which is parallel to the x-axis.

20. Find the coordinates of the point(s) on the curve $y = x^2 - 5x + 3$ where the gradient of the normal is $\frac{1}{3}$

21. A curve has the equation $y = x^3 - px + q$. The tangent to this curve at the point $(2, -8)$ is parallel to the x-axis. Find the values of p and q.
Find also the coordinates of the other point where the tangent is parallel to the x-axis.

STATIONARY VALUES

110

Consider a function $f(x)$. The derived function, $f'(x)$, expresses the rate at which $f(x)$ increases with respect to x

At points where $f'(x)$ is positive, $f(x)$ is increasing as x increases, whereas if $f'(x)$ is negative then $f(x)$ is decreasing as x increases.

Now there may be points where $f'(x)$ is zero, i.e. $f(x)$ is momentarily neither increasing nor decreasing with respect to x

The value of $f(x)$ at such a point is called a *stationary value* of $f(x)$

i.e. $\qquad f'(x) = 0 \qquad \Rightarrow \qquad f(x)$ has a stationary value.

To look at this situation graphically we consider the curve with equation $y = f(x)$.

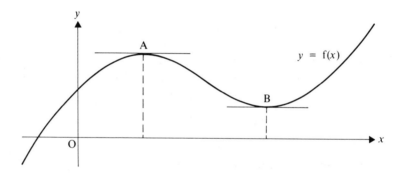

At A and B, $f(x)$, and therefore y, is neither increasing nor decreasing with respect to x. So the values of y at A and B are stationary values.

i.e. $\qquad \dfrac{dy}{dx} = 0 \qquad \Rightarrow \qquad y$ has a stationary value.

The point on a curve where y has a stationary value is called a *stationary point* and we see that, at any stationary point, the gradient of the tangent to the curve is zero, i.e. the tangent is parallel to the x-axis.

To sum up:

$$\text{at a stationary point}\begin{cases} y \quad \text{[or } f(x)\text{]} \quad \text{has a stationary value} \\ dy/dx \quad \text{[or } f'(x)\text{]} \quad \text{is zero} \\ \text{the tangent is parallel to the } x\text{-axis.} \end{cases}$$

Example 16b _____

Find the stationary values of the function $x^3 - 4x^2 + 7$.

If $\qquad\qquad$ $f(x) = x^3 - 4x^2 + 7$

then $\qquad\quad$ $f'(x) = 3x^2 - 8x$

At stationary points, $f'(x) = 0$, i.e. $3x^2 - 8x = 0$

$\Rightarrow \qquad\qquad x(3x - 8) = 0 \qquad \Rightarrow \qquad x = 0 \ \text{ and } \ x = \frac{8}{3}$

Therefore there are stationary points where $x = 0$ and $x = \frac{8}{3}$

When $x = 0$, $\qquad\quad f(x) = 0 - 0 + 7 = 7$

When $x = \frac{8}{3}$, $\qquad f(x) = (\frac{8}{3})^3 - 4(\frac{8}{3})^2 + 7 = -2\frac{13}{27}$

Therefore the stationary values of $x^3 - 4x^2 - 5$ are 7 and $-2\frac{13}{27}$

_____

EXERCISE 16b

Find the value(s) of x at which the following functions have stationary values.

1. $x^2 + 7$ $\qquad\qquad\qquad\qquad$ **2.** $2x^2 - 3x - 2$

3. $x^3 - 4x^2 + 6$ $\qquad\qquad\qquad$ **4.** $(2x - 3)^2$

5. $x^3 - 2x^2 + 11$ $\qquad\qquad\quad$ **6.** $x^3 - 3x - 5$

Find the value(s) of x for which y has a stationary value.

7. $y = x^2 - 8x + 1$ $\qquad\qquad\quad$ **8.** $y = x + 9/x$

9. $y = 2x^3 + x^2 - 8x + 1$ $\qquad\quad$ **10.** $y = 9x^3 - 25x$

11. $y = (2 - x)^3$ $\qquad\qquad\qquad$ **12.** $y = 3x^3 - 12x + 19$

Find the coordinates of the stationary points on the following curves.

13. $y = \dfrac{x^2 + 9}{2x}$ $\qquad\qquad\qquad$ **14.** $y = x^3 - 2x^2 + x - 7$

15. $y = (x - 3)(x + 2)$ $\qquad\qquad$ **16.** $y = \left(2 - \dfrac{x}{2}\right)^2$

17. $y = \sqrt{x} + \dfrac{1}{\sqrt{x}}$ $\qquad\qquad\quad$ **18.** $y = 8 + \dfrac{x}{4} + \dfrac{4}{x}$

TURNING POINTS

In the immediate neighbourhood of a stationary point a curve can have any one of the shapes shown in the following diagram.

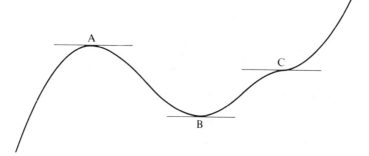

Moving through A from left to right we see that the curve is rising, then turns at A and begins to fall, i.e. the gradient changes from positive to zero at A and then becomes negative.

At A there is a *turning point*.

The value of y at A is called a *maximum value* and A is called a *maximum point*.

Moving through B from left to right the curve is falling, then turns at B and begins to rise, i.e. the gradient changes from negative to zero at B and then becomes positive.

At B there is a *turning point*.

The value of y at B is called a *minimum value* and B is called a *minimum point*.

> The tangent is always horizontal at a turning point.

Note that a maximum value of y is *not necessarily the greatest value of* y *overall*. The terms maximum and minimum apply only to the behaviour of the curve in the neighbourhood of a stationary point.

At C the curve does not turn. The gradient goes from positive, to zero at C and then becomes positive again, i.e. the gradient does not change sign at C.

C is not a turning point but, because there is a change in the sense in which the curve is turning (from clockwise to anti-clockwise), C is called a *point of inflexion*.

INVESTIGATING THE NATURE OF STATIONARY POINTS

We already know how to locate stationary points on a curve and now examine several ways of distinguishing between the different types of stationary point.

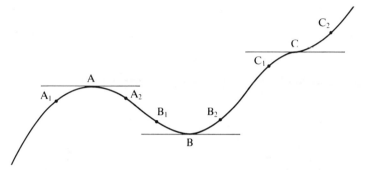

METHOD (i)

This method compares the value of y at the stationary point with values of y at points on either side of, and near to, the stationary point.

For a maximum value, e.g. at A y at $A_1 < y$ at A

                                                                y at $A_2 < y$ at A

For a minimum point, e.g. at B y at $B_1 > y$ at B

                                                                y at $B_2 > y$ at B

For a point of inflexion, e.g. at C y at $C_1 < y$ at C

                                                                y at $C_2 > y$ at C

Collecting these conclusions we have:

	Maximum	Minimum	Inflexion
y values on each side of the stationary point	both smaller	both larger	one larger and one smaller

Note that, between the points chosen on either side of the stationary point there must be no *other* stationary point, nor any discontinuity on the graph.

METHOD (ii)

This method examines the sign of the gradient, again at points close to, and on either side of, the stationary point where the gradient is zero.

For a maximum point, A $\dfrac{dy}{dx}$ at A_1 is +ve

$\dfrac{dy}{dx}$ at A_2 is −ve

For a minimum point, B $\dfrac{dy}{dx}$ at B_1 is −ve

$\dfrac{dy}{dx}$ at B_2 is +ve

For a point of inflexion, C $\dfrac{dy}{dx}$ at C_1 is +ve

$\dfrac{dy}{dx}$ at C_2 is +ve

Collecting these conclusions we have:

Sign of $\dfrac{dy}{dx}$	Passing through maximum + 0 −	Passing through minimum − 0 +	Passing through point of inflexion + 0 + or − 0 −
Gradient of tangent	/ ‾ \	\ _ /	/ ‾ / or \ ‾ \

Example 16c

Locate the stationary points on the curve $y = 4x^3 + 3x^2 - 6x - 1$ and determine the nature of each one.

$$y = 4x^3 + 3x^2 - 6x - 1 \quad \Rightarrow \quad \frac{dy}{dx} = 12x^2 + 6x - 6$$

At stationary points, $\dfrac{dy}{dx} = 0$

i.e. $12x^2 + 6x - 6 = 0 \quad \Rightarrow \quad 6(2x - 1)(x + 1) = 0$

∴ there are stationary points where $x = \frac{1}{2}$ and $x = -1$

When $x = \frac{1}{2}$, $y = -2\frac{3}{4}$ and when $x = -1$, $y = 4$

i.e. the stationary points are $(\frac{1}{2}, -2\frac{3}{4})$ and $(-1, 4)$

Considering the sign of $\dfrac{dy}{dx}$ when $x = 0$ and at $x = -1$

(i.e. on either side of $x = \frac{1}{2}$) gives

$$\left.\begin{array}{l} \dfrac{dy}{dx} = -6 \text{ when } x = 0 \\[3mm] \dfrac{dy}{dx} = 12 \text{ when } x = 1 \end{array}\right\} \;\Rightarrow\; (\tfrac{1}{2}, -2\tfrac{3}{4}) \text{ is a minimum point}$$

Considering the sign of $\dfrac{dy}{dx}$ when $x = -\frac{3}{2}$ and at $x = -\frac{1}{2}$

(i.e. on either side of $x = -1$) gives

$$\left.\begin{array}{l} \dfrac{dy}{dx} = 12 \text{ when } x = -\tfrac{3}{2} \\[3mm] \dfrac{dy}{dx} = -6 \text{ when } x = -\tfrac{1}{2} \end{array}\right\} \;\Rightarrow\; (-1, 4) \text{ is a maximum point}$$

EXERCISE 16c

Find the stationary points on the following curves and distinguish between them.

1. $y = 2x - x^2$

2. $y = 3x - x^3$

3. $y = 9/x + x$

4. $y = x^2(x - 5)$

5. $y = x^2$

6. $y = (2x - 5)^2$

7. $y = x^3$

8. $y = x^4$

9. $y = (2x + 1)(x - 3)$

10. $y = x^5 - 5x$

Find the stationary value(s) of each of the following functions and determine their character.

11. $x + 1/x$

12. $3 - x + x^2$

13. $27x - x^3$

14. $x^2(3x^2 - 2x - 3)$

PROBLEMS

Examples 16d _____

1. An open box is made from a square sheet of cardboard, with sides half a metre long, by cutting out a square from each corner, folding up the sides and joining the cut edges. Find the maximum capacity of the box.

The capacity of the box depends upon the unknown length of the side of the square cut from each corner so we denote this by x metres. The side of the cardboard sheet is $\frac{1}{2}$ m, so we know that $0 < x < \frac{1}{4}$

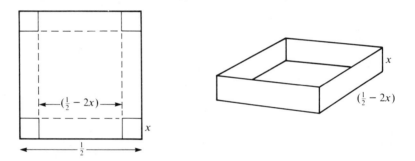

Using metres throughout,

the base of the box is a square of side $(\frac{1}{2} - 2x)$
and the height of the box is x

∴   the capacity, C, of the box is given by

$$C = x(\tfrac{1}{2} - 2x)^2 = \tfrac{1}{4}x - 2x^2 + 4x^3 \quad \text{for} \quad 0 < x < \tfrac{1}{4}$$

⇒ $$\frac{dC}{dx} = \tfrac{1}{4} - 4x + 12x^2$$

At a stationary value of C, $\dfrac{dC}{dx} = 0$

i.e. $12x^2 - 4x + \tfrac{1}{4} = 0$ ⇒ $48x^2 - 16x + 1 = 0$

$(4x - 1)(12x - 1) = 0$ ⇒ $x = \tfrac{1}{4}$ or $x = \tfrac{1}{12}$

there are stationary values of C when $x = \tfrac{1}{4}$ and when $x = \tfrac{1}{12}$

It is obvious that it is not possible to make a box if $x = \frac{1}{4}$ so it is tempting to *assume* that $x = \frac{1}{12}$ gives a maximum capacity. However we ought to check that this is the case by considering the signs of $\dfrac{dC}{dx}$ close to $x = \frac{1}{12}$ (i.e. $x \approx 0.08$).

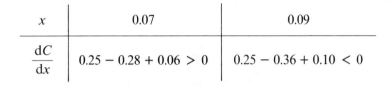

x	0.07	0.09
$\dfrac{dC}{dx}$	$0.25 - 0.28 + 0.06 > 0$	$0.25 - 0.36 + 0.10 < 0$

Therefore C has a maximum value of $\frac{1}{12}(\frac{1}{2} - \frac{1}{6})^2$,   i.e. $\frac{1}{108}$

i.e. the maximum capacity of the box is $\frac{1}{108}$ m^3

or, correct to 3 s.f., 9260 cm^3

Alternatively the nature of the stationary point where $x = \frac{1}{12}$ can be investigated by using the sketch given by a graphics calculator for the curve $C = \frac{1}{4}x - 2x^2 + 4x^3$ and looking at the section for which $0 < x < \frac{1}{4}$, i.e.

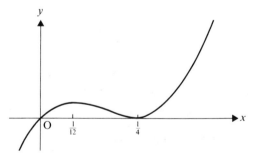

The sketch shows that there is a maximum point between $x = 0$ and $x = \frac{1}{4}$ so $x = \frac{1}{12}$ must give the maximum value of C.

The sketch also shows that there is a minimum point *on the curve* where $x = \frac{1}{4}$ but this is *not* a minimum value of the *capacity*, as a box cannot be made if $x = \frac{1}{4}$

2. The function $ax^2 + bx + c$ has a gradient function $4x + 2$ and a stationary value of 1. Find the values of a, b and c

$$f(x) = ax^2 + bx + c \quad \Rightarrow \quad f'(x) = 2ax + b$$

But we know that $f'(x) = 4x + 2$

$\therefore$ $\qquad\qquad\qquad$ $2ax + b$ is identical to $4x + 2$

i.e. $\qquad\qquad\qquad\qquad$ $a = 2$ and $b = 2$

The stationary value of $f(x)$ occurs when $f'(x) = 0$

i.e. when $\qquad\qquad$ $4x + 2 = 0 \quad \Rightarrow \quad x = -\frac{1}{2}$

the stationary value of $f(x)$ is $2(-\frac{1}{2})^2 + 2(-\frac{1}{2}) + c = -\frac{1}{2} + c$

But the stationary value of $f(x)$ is also 1

$\therefore$ $\qquad\qquad\qquad$ $-\frac{1}{2} + c = 1 \quad \Rightarrow \quad c = \frac{3}{2}$

---

3. A cylinder has a radius r metres and a height h metres. The sum of the radius and height is 2 m. Find an expression for the volume, V cubic metres, of the cylinder in terms of r only. Hence find the maximum volume.

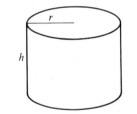

$$V = \pi r^2 h \quad \text{and} \quad r + h = 2$$

$\therefore$ $\qquad\qquad$ $V = \pi r^2 (2 - r) = \pi(2r^2 - r^3)$

Now for maximum volume, $\dfrac{dV}{dr} = 0,$

i.e. $\qquad\qquad$ $\pi(4r - 3r^2) = 0 \quad \Rightarrow \quad \pi r(4 - 3r) = 0$

Therefore there are stationary values of V when $r = 0$ and $r = \frac{4}{3}$

It is obvious that, when $r = 0$, $V = 0$ and no cylinder exists, so we check that $r = \frac{4}{3}$ does give the maximum volume

r	1	$\frac{5}{3}$
$\dfrac{dV}{dr}$	$\pi\,(> 0)$	$-\frac{5}{3}\,(< 0)$

Therefore the maximum value of V occurs when $r = \frac{4}{3}$
and is $\pi(\frac{4}{3})^2(2 - \frac{4}{3})$

i.e. the maximum volume is $\dfrac{32\pi}{27}$ m^3

Note that the solution of this problem depends fundamentally on
having an expression for V in terms of only one other variable. This
is true of *all* problems on stationary points so, if three or more
variables are involved initially, some of them must be replaced so that
we have a basic relationship containing only two variables.

EXERCISE 16d

1. A farmer has an 80 m length of fencing. He wants to use it to
   form three sides of a rectangular enclosure against an existing fence
   which provides the fourth side. Find the maximum area that he
   can enclose and give its dimensions.

2. A large number of open cardboard boxes are to be made and each
   box must have a square base and a capacity of 4000 cm^3. Find the
   dimensions of the box which contains the minimum area of
   cardboard.

3.

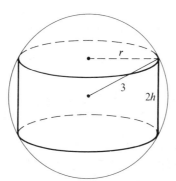

   The diagram shows a cylinder cut from a solid sphere of radius
   3 cm. Given that the cylinder has a height of $2h$, find its radius in
   terms of h. Hence show that the volume, V cubic metres, of the
   cylinder is given by

   $$V = 2\pi h(9 - h^2)$$

   Find the maximum volume of the cylinder as h varies.

4. A variable rectangle has a constant perimeter of 20 cm. Find the lengths of the sides when the area is maximum.

5. A variable rectangle has a constant area of 35 cm². Find the lengths of the sides when the perimeter is minimum.

6. The curve $y = ax^2 + bx + c$ crosses the y-axis at the point $(0, 3)$ and has a stationary point at $(1, 2)$. Find the values of a, b and c

7. The gradient of the tangent to the curve $y = px^2 - qx - r$ at the point $(1, -2)$ is 1 If the curve crosses the x-axis where $x = 2$, find the values of p, q and r. Find the other point of intersection with the x-axis and sketch the curve.

8. y is a quadratic function of x. The line $y = 2x$ is a tangent to the curve at the point $(3, 6)$. The turning point on the curve occurs where $x = -2$ Find the equation of the curve.

MIXED EXERCISE 16

1. Find the gradient of the curve with equation $y = 6x^2 - x$ at the point where $x = 1$ Find the equation of the tangent at this point. Where does this tangent meet the line $y = 2x$?

2. Find the equation of the normal to the curve $y = 1 - x^2$ at the point where the curve crosses the positive x-axis. Find also the coordinates of the point where the normal meets the curve again.

3. Find the coordinates of the points on the curve $y = x^3 + 3x$ where the gradient is 15

4. Find the equations of the tangents to the curve $y = x^3 - 6x^2 + 12x + 2$ which are parallel to the line $y = 3x$

5. Find the equation of the normal to the curve $y = x^2 - 6$ which is parallel to the line $x + 2y - 1 = 0$

6. Locate the turning points on the curve $y = x(x^2 - 12)$, determine their nature and draw a rough sketch of the curve.

7. Find the stationary values of the function $x + 1/x$ and sketch the function.

8. If the perimeter of a rectangle is fixed in length, show that the area of the rectangle is greatest when it is square.

9. A door is in the shape of a rectangle surmounted by a semicircle whose diameter is equal to the width of the rectangle. If the perimeter of the door is 7 m, and the radius of the semicircle is r metres, express the height of the rectangle in terms of r. Show that the area of the door has a maximum value when the width is $7/(4 + \pi)$.

10. An open tank is constructed, with a square base and vertical sides, to hold 32 cubic metres of water. Find the dimensions of the tank if the area of sheet metal used to make it is to have a minimum value.

11. Triangle ABC has a right angle at C. The shape of the triangle can vary but the sides BC and CA have a fixed total length of 10 cm. Find the maximum area of the triangle.

CHAPTER 17

TRIGONOMETRIC FUNCTIONS

THE TRIG RATIOS OF 30°, 45°, 60°

The sine, cosine and tangent of 30°, 45°, and 60°, can be expressed exactly in surd form and are worth remembering.

This triangle shows that

$$\sin 45° = \frac{1}{\sqrt{2}}$$

$$\cos 45° = \frac{1}{\sqrt{2}}$$

$$\tan 45° = 1$$

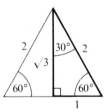

And this triangle gives

$$\sin 60° = \frac{\sqrt{3}}{2}, \quad \sin 30° = \frac{1}{2}$$

$$\cos 60° = \frac{1}{2}, \quad \cos 30° = \frac{\sqrt{3}}{2}$$

$$\tan 60° = \sqrt{3}, \quad \tan 30° = \frac{1}{\sqrt{3}}$$

THE GENERAL DEFINITION OF AN ANGLE

An angle is defined as a measure of the rotation of a line OP about a fixed point O. Taking Ox as the initial direction of OP, anticlockwise rotation describes a positive angle and clockwise rotation describes a negative angle. The rotation of OP is not limited to one revolution, so an angle can be as big as we choose to make it.

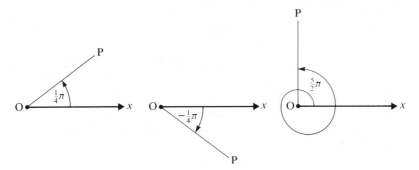

If θ is any angle, then θ can be measured
    either in degrees (one revolution $= 360°$)
    or   in radians (one revolution $= 2\pi$ radians)
and in either case we see that θ can take all real values.

THE TRIGONOMETRIC FUNCTIONS

The sine, cosine and tangent of *any* angle θ can now be defined as follows.

If OP is drawn on x and y axes as shown and if, for all values of θ, the length of OP is r and the coordinates of P are (x, y), then the sine, cosine and tangent functions are defined as follows.

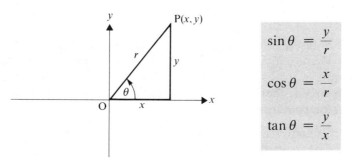

$$\sin \theta = \frac{y}{r}$$

$$\cos \theta = \frac{x}{r}$$

$$\tan \theta = \frac{y}{x}$$

We will now look at each of these functions in turn.

THE SINE FUNCTION

From the definition $f(\theta) = \sin \theta$, and measuring θ in radians, we can see that

for $0 \leqslant \theta \leqslant \frac{1}{2}\pi$, OP is in the first quadrant; y is positive and increases in value from 0 to r as θ increases from 0 to $\frac{1}{2}\pi$. Now r is always positive, so $\sin \theta$ increases from 0 to 1

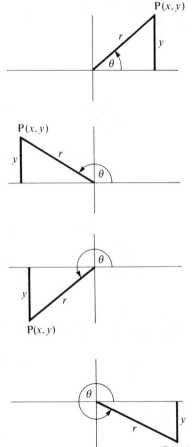

for $\frac{1}{2}\pi \leqslant \theta \leqslant \pi$, OP is in the second quadrant; again y is positive but decreases in value from r to 0, so $\sin \theta$ decreases from 1 to 0

for $\pi \leqslant \theta \leqslant \frac{3}{2}\pi$, OP is in the third quadrant; y is negative and decreases from 0 to $-r$, so $\sin \theta$ decreases from 0 to -1

for $\frac{3}{2}\pi \leqslant \theta \leqslant 2\pi$, OP is in the fourth quadrant; y is still negative but increases from $-r$ to 0, so $\sin \theta$ increases from -1 to 0

For $\theta \geqslant 2\pi$, the cycle repeats as OP travels round the quadrants again. For negative values of θ, OP rotates clockwise round the quadrants in the order 4th, 3rd, 2nd, 1st, etc. So $\sin \theta$ decreases from 0 to -1, then increases to 0 and on to 1 before decreasing to zero and repeating the pattern.

From this analysis we see that $\sin \theta$ is positive for $0 < \theta < \pi$ and negative when $\pi < \theta < 2\pi$

Further, $\sin \theta$ varies in value between -1 and 1 and the pattern repeats every revolution.

A plot of the graph of $f(\theta) = \sin\theta$ confirms these observations.

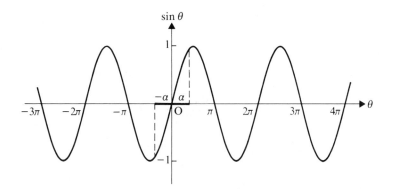

A graph of this shape is called, for obvious reasons, a *sine wave* and shows clearly the following characteristics of the sine function.

> The curve is continuous (i.e. it has no breaks).
> $$-1 \leqslant \sin\theta \leqslant 1$$

The shape of the curve from $\theta = 0$ to $\theta = 2\pi$ is repeated for each complete revolution. Any function with a repetitive pattern is called *periodic* or *cyclic*. The width of the repeating pattern, as measured on the horizontal scale, is called the *period*.

> The period of the sine function is 2π

Other properties of the sine function shown by the graph are as follows.

$$\sin\theta = 0 \quad \text{when} \quad \theta = n\pi \quad \text{where } n \text{ is an integer.}$$

The curve has rotational symmetry about the origin so the sine function is odd,

i.e. for any angle, α
$$\sin(-\alpha) = -\sin\alpha$$

$$\text{e.g.} \quad \sin(-30°) = -\sin(30°) = -\tfrac{1}{2}$$

$$\sin(-\tfrac{1}{2}\pi) = -\sin(\tfrac{1}{2}\pi) = -1$$

Also if OP′ is the reflection in the y-axis of OP, then OP′ represents a rotation from Ox of (180° − θ°).

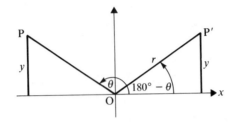

The values of y at P and P′ are equal so $\sin \theta = \sin(180° − θ°)$ and the sine function is symmetrical about $\theta = 90°$.

Taking an enlarged section of the graph for $0 \leqslant \theta \leqslant 2\pi$, we find further relationships.

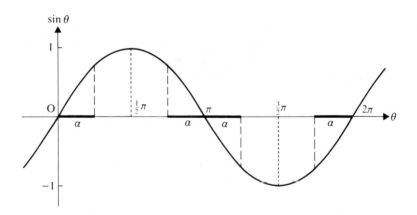

The curve is symmetrical about the line $\theta = \frac{1}{2}\pi$, so

$$\sin(\pi − \alpha) = \sin\alpha, \quad \text{e.g.} \quad \sin 130° = \sin(180° − 130°) = \sin 50°$$

The curve has rotational symmetry about $\theta = \pi$, so

$$\sin(\pi + \alpha) = −\sin\alpha \quad \text{and} \quad \sin(2\pi − \alpha) = −\sin\alpha$$

Examples 17a

1. Find the exact value of $\sin \frac{4}{3}\pi$.

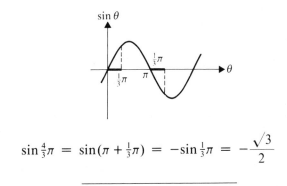

$$\sin \tfrac{4}{3}\pi = \sin(\pi + \tfrac{1}{3}\pi) = -\sin \tfrac{1}{3}\pi = -\frac{\sqrt{3}}{2}$$

2. Sketch the graph of $y = \sin(\theta - \tfrac{1}{4}\pi)$ for values of θ between 0 and 2π

Remember that the curve $y = f(x - a)$ is a translation of the curve $y = f(x)$ by a units in the positive direction of the x-axis.

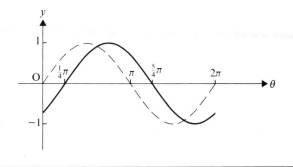

EXERCISE 17a

Find the exact value of

1. $\sin 120°$ 2. $\sin(-2\pi)$ 3. $\sin 300°$ 4. $\sin(-210°)$
5. $\sin(-60°)$ 6. $\sin \frac{13}{6}\pi$ 7. $\sin \frac{5}{4}\pi$ 8. $\sin(-\tfrac{2}{3}\pi)$

9. Write down all the values of θ between 0 and 6π for which $\sin \theta = 1$

10. Write down all the values of θ between 0 and -4π for which $\sin \theta = -1$

Express in terms of the sine of an acute angle.

11. $\sin 125°$ **12.** $\sin 290°$ **13.** $\sin(-120°)$ **14.** $\sin \frac{7}{6}\pi$

Sketch each of the following curves for values of θ in the range $0 \leqslant \theta \leqslant 3\pi$.

15. $y = \sin(\theta + \frac{1}{3}\pi)$ **16.** $y = -\sin\theta$ **17.** $y = \sin(-\theta)$

18. $y = 1 - \sin\theta$ **19.** $y = \sin(\pi - \theta)$ **20.** $y = \sin(\frac{1}{2}\pi - \theta)$

SG   Use a graphics calculator or computer for Questions 21 to 23 and set the range for θ as -2π to 4π.

21. On the same set of axes draw the graphs of $y = \sin\theta$, $y = 2\sin\theta$, and $y = 3\sin\theta$. What can you deduce about the relationship between the curves $y = \sin\theta$ and $y = a\sin\theta$?

22. On the same set of axes draw the curves $y = \sin\theta$ and $y = \sin 2\theta$.

23. On the same set of axes draw the curves $y = \sin\theta$ and $y = \sin 3\theta$.
What can you deduce about the relationship between the two curves?

24. *Sketch* the curves   (a) $y = \sin 4\theta$ (b) $y = 4\sin\theta$

One-way Stretches

Questions 21 to 24 in the last exercise show examples of one-way stretches. For example, the curve $y = 2\sin\theta$ is seen to be a one-way stretch of the curve $y = \sin\theta$ by a factor 2 parallel to the y-axis.

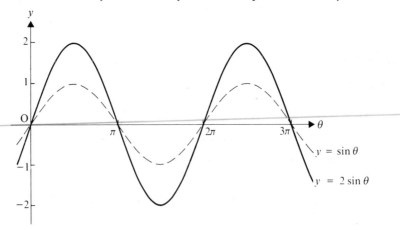

In general, if we compare points on the curves $y = f(x)$ and $y = af(x)$ with the same x-coordinate, then the y-coordinate of the point on $y = af(x)$ is a times the y-coordinate of the point on $y = f(x)$. Therefore

the curve $y = af(x)$ is a one-way stretch of the curve $y = f(x)$ by a factor a parallel to the y-axis.

Also, the curve $y = \sin 2\theta$ was seen to be a one-way stretch of the curve $y = \sin \theta$ by a factor $\frac{1}{2}$ parallel to the x-axis (or a one-way shrinkage by a factor 2).

Now consider points on the curves $y = f(x)$ and $y = f(ax)$ with the same y-coordinate. The x-coordinate on $y = f(ax)$ must be $\frac{1}{a}$ times the x-coordinate on $y = f(x)$. Therefore, in general,

the curve $y = f(ax)$ is a one-way stretch of the curve $y = f(x)$ by a factor $\frac{1}{a}$ parallel to the x-axis.

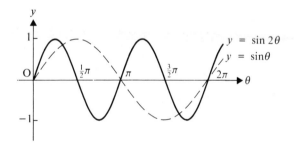

THE COSINE FUNCTION

For any position of P, $\cos \theta = \dfrac{x}{r}$

As OP rotates through the first quadrant, x decreases from r to 0 as θ increases, so $\cos \theta$ decreases from 1 to 0

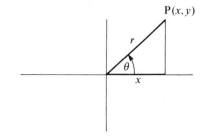

Similar observations of the behaviour of $\dfrac{x}{r}$ as OP continues to rotate show that

in the second quadrant x decreases from 0 to $-r$,
so $\cos \theta$ decreases from 0 to -1,
in the third quadrant $\cos \theta$ increases from -1 to 0,
and in the fourth quadrant $\cos \theta$ increases from 0 to 1

The cycle then repeats, and we get this graph of $f(\theta) = \cos \theta$

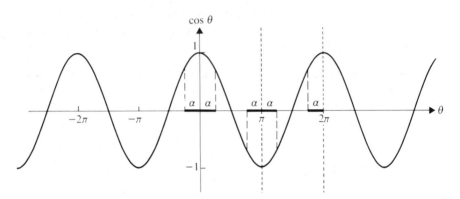

The characteristics of this graph are as follows.

> The curve is continuous
>
> $$-1 \leqslant \cos \theta \leqslant 1$$
>
> It is periodic with a period of 2π

It is the same shape as the sine wave but is translated a distance $\frac{1}{2}\pi$ to the left. Such a translation of a sine wave is called a *phase shift*.

$$\cos \theta = 0 \quad \text{when} \quad \theta = \ldots -\tfrac{1}{2}\pi, \tfrac{1}{2}\pi, \tfrac{3}{2}\pi, \tfrac{5}{2}\pi, \ldots$$

The curve is symmetric about $\theta = 0$, so the cosine function is an even function and $\qquad \cos(-\alpha) = \cos \alpha$

The curve has rotational symmetry about $\theta = \tfrac{1}{2}\pi$, so

$$\cos(\pi - \alpha) = -\cos \alpha$$

Further considerations of symmetry show that

$$\cos(\pi + \alpha) = -\cos \alpha \quad \text{and} \quad \cos(2\pi - \alpha) = \cos \alpha$$

Note that each of these properties can be proved from the definition of $\cos \theta$ with the help of a quadrant diagram but they are easier to see and remember from the graph.

EXERCISE 17b

1. Write in terms of the cosine of an acute angle
   (a) $\cos 123°$ (b) $\cos 250°$ (c) $\cos(-20°)$ (d) $\cos(-154°)$

2. Find the exact value of
   (a) $\cos 150°$ (b) $\cos \frac{3}{2}\pi$ (c) $\cos \frac{5}{4}\pi$ (d) $\cos 6\pi$
   (e) $\cos(-60°)$ (f) $\cos \frac{7}{3}\pi$ (g) $\cos(-\frac{5}{6}\pi)$ (h) $\cos 225°$

3. Sketch each of the following curves.
   (a) $y = \cos(\theta + \pi)$ (b) $y = \cos(\theta - \frac{1}{3}\pi)$ (c) $y = \cos(-\theta)$

4. Sketch the graph of $y = \cos(\theta - \frac{1}{4}\pi)$ for values of θ between $-\pi$
   and π. Use the graph to find the values of θ in this range for which
   (a) $\cos(\theta - \frac{1}{4}\pi) = 1$ (b) $\cos(\theta - \frac{1}{4}\pi) = -1$
   (c) $\cos(\theta - \frac{1}{4}\pi) = 0$

5. On the same set of axes, sketch the graphs $y = \cos \theta$ and
   $y = 3 \cos \theta$.

6. On the same set of axes, sketch the graphs $y = \cos \theta$ and
   $y = \cos 3\theta$.

7. Sketch the graph of $f(\theta) = \cos 4\theta$ for $0 \leqslant \theta \leqslant \pi$.
   Hence find the values of θ in this range for which $f(\theta) = 0$

8. Sketch the graph of $y = \cos(\frac{1}{2}\pi - \theta)$
   What relationship does this suggest between $\sin \theta$ and $\cos(\frac{1}{2}\pi - \theta)$?
   Is there a similar relationship between $\cos \theta$ and $\sin(\frac{1}{2}\pi - \theta)$?

9. Sketch the graph of $f(\theta) = \cos 2\theta$ for $0 \leqslant \theta \leqslant 2\pi$.
   Hence find the values of θ in this range for which
   (a) $\cos 2\theta = 0$ (b) $\cos 2\theta = -1$

10. On the same set of axes *sketch* the graphs of $y = 2 \cos \theta$ and
    $y = \cos 2\theta$ for $0 \leqslant \theta \leqslant \pi$. Use your sketch to *estimate* the value
    of θ for which $2 \cos \theta = \cos 2\theta$

THE TANGENT FUNCTION

For any position of P, $\tan \theta = \dfrac{y}{x}$

As OP rotates through the first
quadrant, x decreases from r to 0
while y increases from 0 to r. This
means that the fraction y/x increases
from 0 to very large values indeed.
In fact, as $\theta \to \frac{1}{2}\pi$, $\tan \theta \to \infty$

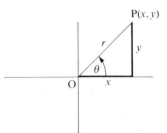

Looking at the behaviour of y/x in the other quadrants shows that
in the second quadrant, $\tan \theta$ is negative and increases from $-\infty$ to 0,
in the third quadrant, $\tan \theta$ is positive and increases from 0 to ∞,
and in the fourth quadrant, $\tan \theta$ is negative and increases from $-\infty$ to 0
The cycle then repeats and we can draw the graph of $f(\theta) = \tan \theta$

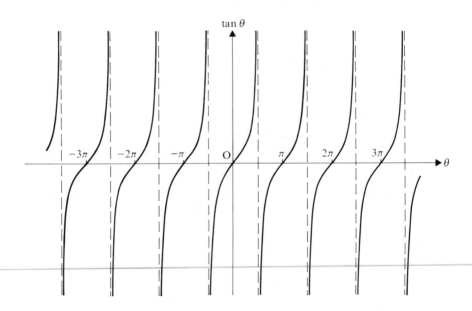

From the graph we can see that the characteristics of the tangent
function are different from those of the sine and cosine functions in
several respects.

It is not continuous, being *undefined* when $\theta = \ldots -\frac{1}{2}\pi, \frac{1}{2}\pi, \frac{2}{3}\pi, \ldots$

The range of values of $\tan\theta$ is unlimited.

It is periodic with a period of π (not 2π as in the other cases).

The graph has rotational symmetry about $\theta = 0$, so

$$\tan(-\alpha) = -\tan\alpha$$

The graph has rotational symmetry about $\theta = \frac{1}{2}\pi$, giving

$$\tan(\pi - \alpha) = -\tan\alpha$$

As the cycle repeats itself from $\theta = \pi$ to 2π, we have

$$\tan(\pi + \alpha) = \tan\alpha \quad \text{and} \quad \tan(2\pi - \alpha) = -\tan\alpha$$

Example 17c _____

Express $\tan\frac{11}{4}\pi$ as the tangent of an acute angle.

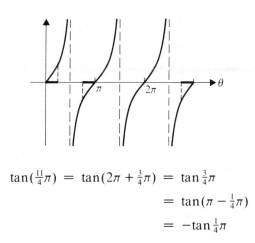

$$\tan(\tfrac{11}{4}\pi) = \tan(2\pi + \tfrac{3}{4}\pi) = \tan\tfrac{3}{4}\pi$$
$$= \tan(\pi - \tfrac{1}{4}\pi)$$
$$= -\tan\tfrac{1}{4}\pi$$

EXERCISE 17c

1. Find the exact value of
   (a) $\tan\frac{9}{4}\pi$ (b) $\tan 120°$ (c) $\tan -\frac{2}{3}\pi$ (d) $\tan\frac{7}{4}\pi$

2. Write in terms of the tangent of an acute angle
   (a) $\tan 220°$ (b) $\tan\frac{12}{7}\pi$ (c) $\tan 310°$ (d) $\tan -\frac{7}{5}\pi$

3. Sketch the graph of $y = \tan\theta$ for values of θ in the range 0 to 2π. From this sketch find the values of θ in this range for which

(a) $\tan\theta = 1$ (b) $\tan\theta = -1$ (c) $\tan\theta = 0$ (d) $\tan\theta = \infty$

4. Using the basic definitions of $\sin\theta$, $\cos\theta$ and $\tan\theta$, show that

$$\tan\theta = \frac{\sin\theta}{\cos\theta}$$

for all values of θ

RELATIONSHIPS BETWEEN SIN θ, COS θ AND TAN θ

Because each trig ratio is a ratio of two of the three quantities x, y and r, we would expect to find several relationships between $\sin\theta$, $\cos\theta$ and $\tan\theta$. Some of these relationships will be investigated in a later chapter, but here is a summary of the results from the various exercises so far.

The graph of $\cos\theta$ shifted by $\frac{1}{2}\pi$ to the right gives the graph of $\sin\theta$.

Two angles which add up to $\frac{1}{2}\pi$ (90°) are called *complementary* angles,

so   the sine of an angle is equal to
        the cosine of the complementary angle and vice-versa.

Now $\sin\theta = \dfrac{y}{r}$, $\cos\theta = \dfrac{x}{r}$ and $\tan\theta = \dfrac{y}{x}$

$\therefore$ $\dfrac{\sin\theta}{\cos\theta} = \dfrac{y/r}{x/r} = \dfrac{y}{x} = \tan\theta$

i.e. for all values of θ, $\tan\theta = \dfrac{\sin\theta}{\cos\theta}$

We have also seen that the sign of each trig ratio depends on the size of the angle, i.e. the quadrant in which P is. So we can summarise the sign of each ratio in a quadrant diagram:

sin +ve	All +ve
tan +ve	cos +ve

Examples 17d _____

1. Give all the values of x between 0 and 360° for which
   $\sin x = -0.3$

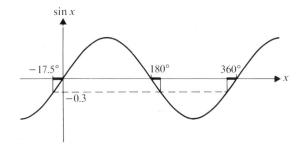

The value given for x by a calculator is $-17.5°$

From the graph, we see that, when $\sin x = -0.3$, the values of x in the specified range are $180° + 17.5°$ and $360° - 17.5°$

When $\sin x = -0.3$, $x = 197.5°$ and $342.5°$

(Note that when the range of values is given in degrees, the answer should also be given in degrees and the same applies for radians.)

_____

2. Find the smallest positive value of θ for which $\cos \theta = 0.7$ and $\tan \theta$ is negative.

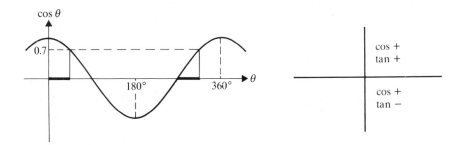

If $\cos \theta = 0.7$, the possible values of θ are 45.6°, 314.4°,...
Now $\tan \theta$ is positive if θ is in the first quadrant and negative if θ is in the fourth quadrant.

Therefore the required value of θ is 314.4°

EXERCISE 17d

1. Within the range $-2\pi \leqslant \theta \leqslant 2\pi$, give all the values of θ for which
   (a) $\sin\theta = 0.4$ (b) $\cos\theta = -0.5$ (c) $\tan\theta = 1.2$

2. Within the range $0 \leqslant \theta \leqslant 720°$, give all the values of θ for which
   (a) $\tan\theta = -0.8$ (b) $\sin\theta = -0.2$ (c) $\cos\theta = 0.1$

3. Find the smallest angle (positive or negative) for which
   (a) $\cos\theta = 0.8$ and $\sin\theta \geqslant 0$
   (b) $\sin\theta = -0.6$ and $\tan\theta \leqslant 0$
   (c) $\tan\theta = \sin\frac{1}{6}\pi$

4. Using $\tan\theta = \dfrac{\sin\theta}{\cos\theta}$, show that the equation $\tan\theta = \sin\theta$ can
   be written as $\sin\theta(\cos\theta - 1) = 0$, provided that $\cos\theta \neq 0$
   Hence find the values of θ between 0 and 2π for which
   $\tan\theta = \sin\theta$.

5. Sketch the graph of $y = \sin 2\theta$. Use your sketch to help find the
   values of θ in the range $0 \leqslant \theta \leqslant 360°$ for which $\sin 2\theta = 0.4$

6. Sketch the graph of $y = \cos 3\theta$. Hence find the values of θ in
   the range $0 \leqslant \theta \leqslant 2\pi$ for which $\cos 3\theta = -1$

THE RECIPROCAL TRIGONOMETRIC FUNCTIONS

The reciprocals of the three main trig functions have their own names
and are sometimes referred to as the *minor* trig ratios.

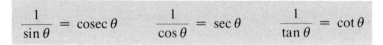

$$\frac{1}{\sin\theta} = \operatorname{cosec}\theta \qquad \frac{1}{\cos\theta} = \sec\theta \qquad \frac{1}{\tan\theta} = \cot\theta$$

The names given above are abbreviations of cosecant, secant and
cotangent respectively.

The graph of $f(\theta) = \operatorname{cosec} \theta$ is given below.

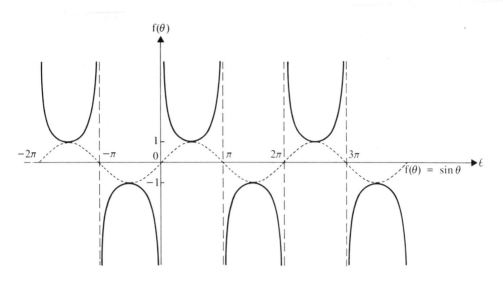

From this graph we can see that

the cosec function is not continuous, being undefined when θ is any integral multiple of π (we would expect this because these are values of θ where $\sin \theta = 0$ and the reciprocal of 0 is $\pm\infty$).

The pattern of the graph of $f(\theta) = \sec \theta$ is similar to that of the cosec graph, as would be expected.

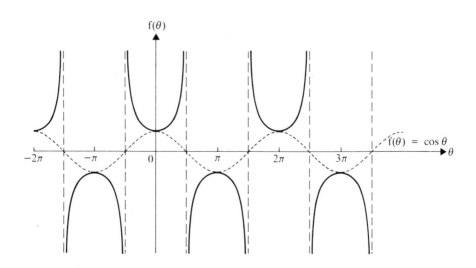

The graph of $f(\theta) = \cot\theta$ is given below.

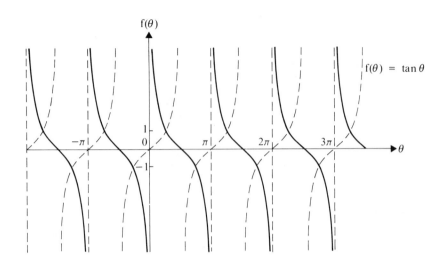

Example 17e _____

For $0 \leqslant \theta \leqslant 360°$, find the values of θ for which $\operatorname{cosec}\theta = -8$

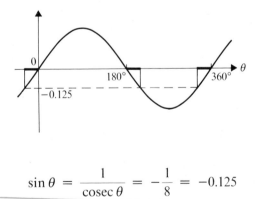

$$\sin\theta = \frac{1}{\operatorname{cosec}\theta} = -\frac{1}{8} = -0.125$$

∴ from a calculator $\theta = -7.2°$

From the sketch, the required values of θ are $187.2°$ and $352.8°$

EXERCISE 17e

1. Find, for values of θ in the range $0 \leqslant \theta \leqslant 360°$, the values of θ for which

   (a) $\sec \theta = 2$ (b) $\cot \theta = 0.6$ (c) $\operatorname{cosec} \theta = 1.5$

2. Within the range $-180° \leqslant \theta \leqslant 180°$ find the values of θ for which

   (a) $\cot \theta = 1.2$ (b) $\sec \theta = -1.5$ (c) $\operatorname{cosec} \theta = -2$

3. Given that $\tan \theta = \dfrac{\sin \theta}{\cos \theta}$, write $\cot \theta$ in terms of $\sin \theta$ and $\cos \theta$. Hence show that $\cot \theta - \cos \theta = 0$ can be written in the form $\cos \theta(1 - \sin \theta) = 0$, provided that $\sin \theta \neq 0$. Thus find the values in the range $-\pi \leqslant \theta \leqslant \pi$ for which $\cot \theta - \cos \theta = 0$

4. Find, in surd form, the values of

   (a) $\cot \frac{1}{4}\pi$ (b) $\sec \frac{5}{4}\pi$ (c) $\operatorname{cosec} \frac{11}{6}\pi$

5. Sketch the graph of $f(\theta) = \sec(\theta - \frac{1}{4}\pi)$ for $0 \leqslant \theta \leqslant 2\pi$ and give the values of θ for which $f(\theta) = 1$

6. Sketch the graph of $f(\theta) = \cot(\theta + \frac{1}{3}\pi)$ for $-\pi \leqslant \theta \leqslant \pi$. Hence give the values of θ in this range for which $f(\theta) = 1$

7. Use this diagram to show that the cotangent of an angle is equal to the tangent of its complement.

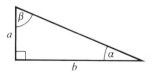

8. Sketch the graph of $f(\theta) = \tan(\frac{1}{2}\pi - \theta)$ for $0 < \theta < \pi$ and compare your sketch with the graph of $f(\theta) = \cot \theta$. What can you deduce about the relationship between $\tan(\frac{1}{2}\pi - \theta)$ and $\cot \theta$?

GRAPHICAL SOLUTIONS OF TRIG EQUATIONS

If you have worked through all the exercises in this chapter, you will already have solved several trig equations with the help of sketch graphs. In this section we are going to look at more complicated equations which require accurate plots of the graphs to solve them.

Consider the equation $\theta = 3 \sin \theta$

The values of θ for which $\theta = 3 \sin \theta$ can be found by plotting the graphs of $y = \theta$ and $y = 3 \sin \theta$ on the same axes and hence finding the values of θ at points of intersection.

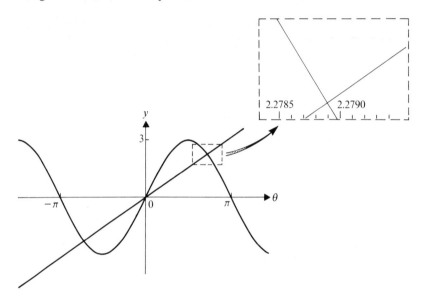

From the enlarged section of the graph $\theta = 2.2789$ rad

Therefore the three points of intersection occur where

$$\theta = -2.2789 \text{ rad, } 0, \ 2.2789 \text{ rad} \quad \text{correct to 4 d.p.}$$

SG If these graphs are produced on a graphics calculator or on a computer using suitable software, then it is possible to zoom in on the points of intersection and get very accurate values for θ. If the graphs are hand drawn, the accuracy of the results will depend on the patience and accuracy of the drawer! Plotting accurate graphs manually is tedious, so if you do not have either of the tools mentioned above, try just Question 1 in the following exercise and do not attempt to get answers correct to more than 2 s.f.

EXERCISE 17f

1. Plot the graphs of $y = \theta$ and $y = 2\cos\theta$ for values of θ in the range $-\pi \leqslant \theta \leqslant \pi$. Hence find the values of θ for which $\theta = 2\cos\theta$

2. Repeat Question 1 using *sketch* graphs and measuring θ in degrees. If this was plotted accurately, would it give the same solutions as when the angle is measured in radians?

3. Measuring the angle in radians throughout, find graphically the values of θ for which

   (a) $2\theta = 4\sin\theta$ (b) $\sin\theta = \theta^2$ (c) $\cos\theta = \theta - 1$

CHAPTER 18

TRIGONOMETRIC IDENTITIES AND EQUATIONS

IDENTITIES

At this stage it is important to know the difference between identities and equations.

This is an equation: $(x - 1)^2 = 4$
The equality is true only when $x = 3$ or when $x = -1$
In any equation, the equality is valid only for a restricted set of values.

This is an identity: $(x - 1)^2 = x^2 - 2x + 1$
The RHS is a different way of expressing the LHS, and the equality is true for all values of x.

In an identity, the equality is true for *any* value of the variable.
The symbol $\equiv$ means 'is identical to', so, strictly, we should write

$$(x - 1)^2 \equiv x^2 - 2x + 1$$

In practice however the symbol $=$ is often used in an algebraic identity.

In this chapter we concentrate on some trigonometric identities and some of their uses.
One such identity, introduced in Chapter 17, is

$$\tan \theta \equiv \frac{\sin \theta}{\cos \theta}$$

THE PYTHAGOREAN IDENTITIES

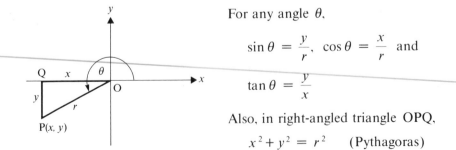

For any angle θ,

$$\sin \theta = \frac{y}{r}, \quad \cos \theta = \frac{x}{r} \quad \text{and}$$

$$\tan \theta = \frac{y}{x}$$

Also, in right-angled triangle OPQ,

$$x^2 + y^2 = r^2 \quad \text{(Pythagoras)}$$

284

Therefore, $(\cos \theta)^2 + (\sin \theta)^2 = \left(\dfrac{x}{r}\right)^2 + \left(\dfrac{y}{r}\right)^2 = \dfrac{x^2 + y^2}{r^2} = 1$

Using the notation $\cos^2 \theta$ to mean $(\cos \theta)^2$, etc., we have

$$\cos^2 \theta + \sin^2 \theta \equiv 1 \qquad\qquad\qquad [1]$$

Using the identity $\tan \theta \equiv \dfrac{\sin \theta}{\cos \theta}$ we can write [1] in two other forms.

$[1] \div \cos^2 \theta \Rightarrow$ $1 + \dfrac{\sin^2 \theta}{\cos^2 \theta} \equiv \dfrac{1}{\cos^2 \theta}$

$\Rightarrow$ $1 + \tan^2 \theta \equiv \sec^2 \theta$

$[1] \div \sin^2 \theta \Rightarrow$ $\dfrac{\cos^2 \theta}{\sin^2 \theta} + 1 \equiv \dfrac{1}{\sin^2 \theta}$

$\Rightarrow$ $\cot^2 \theta + 1 \equiv \operatorname{cosec}^2 \theta$

These identities can be used to
     simplify trig expressions,
     eliminate trig terms from pairs of equations,
     derive a variety of further trig relationships,
     calculate other trig ratios of any angle for which one trig ratio is
     known.

These identities are also very useful in the solution of certain types of trig equations and we will look at this application later in this chapter.

Examples 18a _____

1. Simplify $\dfrac{\sin \theta}{1 + \cot^2 \theta}$

$$\dfrac{\sin \theta}{1 + \cot^2 \theta} \equiv \dfrac{\sin \theta}{\operatorname{cosec}^2 \theta} \equiv \sin^3 \theta$$

Using $1 + \cot^2 \theta \equiv \operatorname{cosec}^2 \theta$ and $\operatorname{cosec} \theta \equiv \dfrac{1}{\sin \theta}$

2. Eliminate θ from the equations $x = 2\cos\theta$ and $y = 3\sin\theta$

$$\cos\theta = \tfrac{1}{2}x \quad \text{and} \quad \sin\theta = \tfrac{1}{3}y$$

Using $\cos^2\theta + \sin^2\theta \equiv 1$ gives

$$\left(\frac{x}{2}\right)^2 + \left(\frac{y}{3}\right)^2 = 1$$

$\Rightarrow \qquad\qquad\qquad 9x^2 + 4y^2 = 36$

In Example 2, both x and y initially depend on θ, a variable angle. Used in this way, θ is called a *parameter*, and is a type of variable that plays an important part in the analysis of curves and functions.

---

3. If $\sin A = -\tfrac{1}{3}$ and A is in the third quadrant, find $\cos A$ without using a calculator.

There are two ways of doing this problem. The first method involves drawing a quadrant diagram and working out the remaining side of the triangle, using Pythagoras theorem.

From the diagram, $x = -2\sqrt{2}$

$\therefore \qquad \cos A = \dfrac{x}{r} = -\dfrac{2\sqrt{2}}{3}$

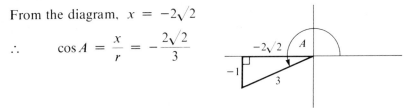

The second method uses the identity $\cos^2 A + \sin^2 A \equiv 1$ giving

$$\cos^2 A + \frac{1}{3} = 1 \quad \Rightarrow \quad \cos A = \pm\sqrt{\frac{8}{3}} = \pm\frac{2\sqrt{2}}{3}$$

As A is between π and $\tfrac{3}{2}\pi$, $\cos A$ is negative, i.e.

$$\cos A = -\frac{2\sqrt{2}}{3}$$

---

4. Prove that $(1 - \cos A)(1 + \sec A) \equiv \sin A \tan A$

Because the relationship has yet to be proved, we must not assume its truth by using the complete identity in our working. The left and right hand sides must be isolated throughout the proof, preferably by working on only one of these sides.

Consider the LHS:

$$(1 - \cos A)(1 + \sec A) \equiv 1 + \sec A - \cos A - \cos A \sec A$$

$$\equiv 1 + \sec A - \cos A - \cos A \left(\frac{1}{\cos A} \right)$$

$$\equiv \sec A - \cos A$$

$$\equiv \frac{1 - \cos^2 A}{\cos A}$$

$$\equiv \frac{\sin^2 A}{\cos A} \qquad (\cos^2 A + \sin^2 A \equiv 1)$$

$$\equiv \sin A \left[\frac{\sin A}{\cos A} \right]$$

$$\equiv \sin A \tan A \equiv \text{RHS}$$

EXERCISE 18a

1. Without using a calculator, complete the following table.

	$\sin \theta$	$\cos \theta$	$\tan \theta$	type of angle
(a)		$-\frac{5}{13}$		reflex
(b)	$\frac{3}{5}$			obtuse
(c)			$\frac{7}{24}$	acute
(d)				straight line

Simplify the following expressions.

2. $\dfrac{1 - \sec^2 A}{1 - \csc^2 A}$

3. $\dfrac{\sin\theta}{\sqrt{(1 - \cos^2\theta)}}$

4. $\dfrac{\sin\theta}{\cos\theta} + \dfrac{\cos\theta}{\sin\theta}$

5. $\dfrac{\sqrt{(1 + \tan^2\theta)}}{\sqrt{(1 - \sin^2\theta)}}$

6. $\dfrac{1}{\cos\theta\sqrt{(1 + \cot^2\theta)}}$

7. $\dfrac{\sin\theta}{1 + \cot^2\theta}$

Eliminate θ from the following pairs of equations.

8. $x = 4\sec\theta$
   $y = 4\tan\theta$

9. $x = a\csc\theta$
   $y = b\cot\theta$

10. $x = 2\tan\theta$
    $y = 3\cos\theta$

11. $x = 1 - \sin\theta$
    $y = 1 + \cos\theta$

12. $x = 2 + \tan\theta$
    $y = 2\cos\theta$

13. $x = a\sec\theta$
    $y = b\sin\theta$

Prove the following identities.

14. $\cot\theta + \tan\theta \equiv \sec\theta\csc\theta$

15. $\dfrac{\cos A}{1 - \tan A} + \dfrac{\sin A}{1 - \cot A} \equiv \sin A + \cos A$

16. $\tan^2\theta + \cot^2\theta \equiv \sec^2\theta + \csc^2\theta - 2$

17. $\dfrac{\sin A}{1 + \cos A} \equiv \dfrac{1 - \cos A}{\sin A}$ (Hint. Multiply top and bottom of LHS by $(1 - \cos A)$.)

18. $\dfrac{\sin A}{1 + \cos A} + \dfrac{1 + \cos A}{\sin A} \equiv \dfrac{2}{\sin A}$

19. $\sec^2 A \equiv \dfrac{\csc A}{\csc A - \sin A}$

20. $(1 + \sin A + \cos A)^2 = 2(1 + \sin A)(1 + \cos A)$

SOLVING EQUATIONS

We have already solved some simple trig equations in Chapter 17. We can now solve a greater variety of equations using the Pythagorean identities.

Examples 18b

1. Solve the equation $2\cos^2\theta - \sin\theta = 1$ for values of θ in the range 0 to 2π

The given equation is quadratic, but it involves the sine and the cosine of θ, so we use $\cos^2\theta + \sin^2\theta \equiv 1$ to express the equation in terms of $\sin\theta$ only.

$$2\cos^2\theta - \sin\theta = 1$$

$\Rightarrow \qquad 2(1 - \sin^2\theta) - \sin\theta = 1$

$\Rightarrow \qquad 2\sin^2\theta + \sin\theta - 1 = 0$

$\Rightarrow \qquad (2\sin\theta - 1)(\sin\theta + 1) = 0 \qquad \Rightarrow \qquad \sin\theta = \tfrac{1}{2} \text{ or } -1$

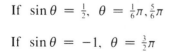

If $\sin\theta = \tfrac{1}{2}$, $\theta = \tfrac{1}{6}\pi, \tfrac{5}{6}\pi$

If $\sin\theta = -1$, $\theta = \tfrac{3}{2}\pi$

Therefore the solution of the equation is $\theta = \tfrac{1}{6}\pi, \tfrac{5}{6}\pi, \tfrac{3}{2}\pi$

2. Solve the equation $\cot x = \sin x$ for values of x from 0 to 360°

Using $\cot x \equiv \dfrac{\cos x}{\sin x}$ gives $\dfrac{\cos x}{\sin x} = \sin x$

We can now multiply the equation by $\sin x$ provided that $\sin x \neq 0$ Thus we must exclude any values of x for which $\sin x = 0$ from the solution set.

$\Rightarrow$ $\cos x = \sin^2 x$

$\Rightarrow$ $\cos^2 x + \cos x - 1 = 0$

This equation does not factorise, so we use the formula, giving

$$\cos x = \tfrac{1}{2}(-1 \pm \sqrt{5})$$

∴ $\cos x = -1.618$ and there is no value of x for which this is true.

or $\cos x = 0.618$

$\Rightarrow$ $x = 51.8°$ or $308.2°$

EXERCISE 18b

Solve the following equations for angles in the range $0 \leqslant \theta \leqslant 360°$

1. $\sec^2 \theta + \tan^2 \theta = 6$

2. $4 \cos^2 \theta + 5 \sin \theta = 3$

3. $\cot^2 \theta = \operatorname{cosec} \theta$

4. $\tan \theta + \cot \theta = 2 \sec \theta$

5. $\tan \theta + 3 \cot \theta = 5 \sec \theta$

6. $\sec \theta = 1 - 2 \tan^2 \theta$

Solve the following equations for angles in the range $-\pi \leqslant \theta \leqslant \pi$

7. $5 \cos \theta - 4 \sin^2 \theta = 2$

8. $4 \cot^2 \theta + 12 \operatorname{cosec} \theta + 1 = 0$

9. $4 \sec^2 \theta - 3 \tan \theta = 5$

10. $2 \cos \theta - 4 \sin^2 \theta + 2 = 0$

11. $2 \sin \theta \cos \theta + \sin \theta = 0$

12. $2 \cos \theta = \cos \theta \operatorname{cosec} \theta$

13. $\sqrt{3} \tan \theta = 2 \sin \theta$

14. $\cot \theta = \cos \theta$

EQUATIONS INVOLVING MULTIPLE ANGLES

Many trig equations involve ratios of a multiple of θ, for example

$$\cos 2\theta = \tfrac{1}{2} \qquad \tan 3\theta = -2$$

Simple equations of this type can be solved by finding first the values of the multiple angle and then, by division, the corresponding values of θ

However it must be remembered that if values of θ are required in the range $\alpha \leqslant \theta \leqslant \beta$ then values of 2θ will be needed in double that range, i.e. $2\alpha \leqslant 2\theta \leqslant 2\beta$ and similarly for other multiples of θ

Examples 18c _____

1. Find the angles in the interval $[-\tfrac{1}{2}\pi, \tfrac{1}{2}\pi]$ which satisfy the equation $\cos 2\theta = \tfrac{1}{2}$

As values of θ in the interval $[-\tfrac{1}{2}\pi, \tfrac{1}{2}\pi]$ are required, we need to find values of 2θ in the interval $[-\pi, \pi]$.

In the interval $[-\pi, \pi]$, the
solutions of $\cos 2\theta = \tfrac{1}{2}$ are
$2\theta = \pm\tfrac{1}{3}\pi$

Hence $\theta = \pm\tfrac{1}{6}\pi$

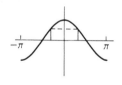

_____

EQUATIONS INVOLVING COMPOUND ANGLES

When a compound angle appears in a trig equation such as

$$\cos\left(\theta - \tfrac{1}{4}\pi\right) = \tfrac{1}{2}$$

the equation can be solved by first finding values of the compound angle. The values of θ can then be found from a simple linear equation. If values of θ are required in the range $0 \leqslant \theta \leqslant \pi$ say, then we must find the values of $\theta + \alpha$ in the interval $[0 + \alpha, \pi + \alpha]$

Examples 18c (continued) _____

    2. Solve the equation $\cos(\theta - 20°) = -\frac{1}{2}$ for values of θ in the range $-180° \leqslant \theta \leqslant 180°$

As values of θ are required in the interval $[-180°, 180°]$, we need values of $(\theta - 20°)$ in the interval $[-200°, 160°]$.

In the interval $[-200°, 160°]$, the
solutions of $\cos(\theta - 20°) = -\frac{1}{2}$
are $\theta - 20° = \pm 120°$

Hence

$\theta = 120° + 20°$ or $-120° + 20°$

    $= 140°$ or $-100°$

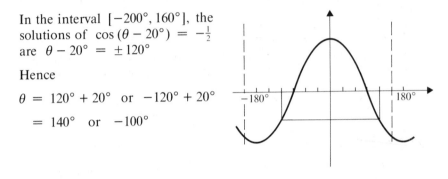

EXERCISE 18c

Find the values of θ in the range $0 \leqslant \theta \leqslant 360°$ which satisfy the
following equations.

    1. $\tan 2\theta = 1$ **2.** $\cos 3\theta = -0.5$

    3. $\sin\frac{1}{2}\theta = -\frac{\sqrt{2}}{2}$ **4.** $\sec 5\theta = 2$

    5. $\cot\frac{1}{3}\theta = -4$ **6.** $\cos 2\theta = 0.63$

    7. $\cos(\theta - 45°) = 0$ **8.** $\sin(\theta + 30°) = -1$

    9. $\tan(\theta - 60°) = 0$ **10.** $\cos(\theta + 60°) = \frac{1}{2}$

Solve the following equations for values of θ in the range $-\pi \leqslant \theta \leqslant \pi$

   11. $\cos(\theta + \frac{1}{4}\pi) = \frac{1}{2}$ **12.** $\tan(\theta - \frac{1}{3}\pi) = -1$

   13. $\sin(\theta + \frac{1}{6}\pi) = \frac{1}{2}$ **14.** $\cos(\theta - \frac{1}{3}\pi) = -\frac{1}{2}$

Solve these equations for values of θ in the range $-180° \leqslant \theta \leqslant 180°$

   15. $\tan 2\theta = 1.8$ **16.** $\sin 3\theta = 0.7$ **17.** $\cos\frac{1}{2}\theta = 0.85$

Solve these equations for values of θ in the range $0 \leqslant \theta \leqslant 2\pi$

   18. $\tan 4\theta = -\sqrt{3}$ **19.** $\sec 5\theta = 2$ **20.** $\cot\frac{1}{2}\theta = -1$

MIXED EXERCISE 18

1. Eliminate α from the equations $x = \cos\alpha,\ y = \operatorname{cosec}\alpha$

2. If $\cos\beta = 0.5$, find possible values for $\sin\beta$ and $\tan\beta$, giving your answers in exact form.

3. Simplify the expression $\dfrac{1}{1 + \cos\theta} + \dfrac{1}{1 - \cos\theta}$. Hence solve the equation $\dfrac{1}{1 + \cos\theta} + \dfrac{1}{1 - \cos\theta} = 4$ for values of θ in the range $0 \leqslant \theta \leqslant 2\pi$

4. Solve the equation $\sec\theta + \tan^2\theta = 5$
    Give the answer in degrees, in the range $0 \leqslant \theta \leqslant 180°$

5. Prove that $(\cot\theta + \operatorname{cosec}\theta)^2 \equiv \dfrac{1 + \cos\theta}{1 - \cos\theta}$

6. Find the values of θ for which $\tan(3\theta - \frac{1}{3}\pi) = 1$ in the interval $[-\pi, \pi]$

7. Eliminate θ from the equations
    (a) $x - 2 = \sin\theta,\ y + 1 = \cos\theta$
    (b) $x = \sec\theta - 3,\ y = 2 - \tan\theta$

8. Solve the equation $\tan 2\alpha = \cot 2\alpha$ for $0 \leqslant \alpha \leqslant \pi$

9. Prove that $(\cos A + \sin A)^2 + (\cos A - \sin A)^2 \equiv 2$

10. Simplify $(1 + \cos A)(1 - \cos A)$

11. Find in degrees the solutions of the equation $\tan\theta = 3\sin\theta$ for which $0 \leqslant \theta \leqslant 360°$

12. Simplify $\sec^4\theta - \sec^2\theta$

CONSOLIDATION D

SUMMARY

CURVES

A chord is a straight line joining two points on a curve.

A tangent to a curve is a line that touches the curve at one point, called the point of contact.

A normal to a curve is the line perpendicular to a tangent and through its point of contact.

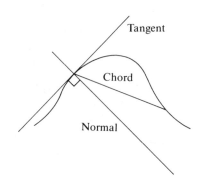

The gradient of a curve at a point on the curve is the gradient of the tangent at that point.

$y = af(x)$ is a one-way stretch of $y = f(x)$ by a factor, a, parallel to Oy

$y = f(ax)$ is a one-way stretch of $y = f(x)$ by a factor, $1/a$, parallel to Ox

DIFFERENTIATION

Differentiation is the process of finding a general expression for the gradient of a curve at any point on the curve.
This general expression is called the gradient function, or the derived function or the derivative.

The derivative is denoted by $\dfrac{dy}{dx}$ or by $f'(x)$ where

$$\frac{dy}{dx} = \lim_{\delta x \to 0} \left[\frac{f(x + \delta x) - f(x)}{\delta x} \right]$$

When $y = x^n$, $\qquad \dfrac{dy}{dx} = nx^{n-1}$

When $y = ax^n$, $\qquad \dfrac{dy}{dx} = anx^{n-1}$

When $y = c$, $\qquad \dfrac{dy}{dx} = 0$

294

Stationary Values

A stationary value of $f(x)$ is its value where $f'(x) = 0$

The point on the curve $y = f(x)$ where $f(x)$ has a stationary value is called a stationary point.

At all stationary points, the tangents to the curve $y = f(x)$ are parallel to the x-axis.

Turning Points

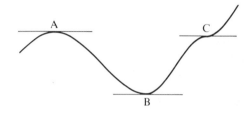

A, B and C are stationary points on the curve $y = f(x)$
The points A and B are called turning points.
The point C is called a point of inflexion.

At A, $f(x)$ has a maximum value and A is called a maximum point.
At B, $f(x)$ has a minimum value and B is called a minimum point.

There are two methods for distinguishing stationary points:

	Max	Min	Inflexion
1. Find value of y on each side of stationary value	Both smaller	Both larger	One smaller One larger
2. Find sign of $\dfrac{dy}{dx}$ on each side of stationary value	+ 0 −	− 0 +	+ 0 + or − 0 −
Gradient	╱‾╲	╲_╱	╱—╱ or ╲—╲

TRIGONOMETRIC FUNCTIONS

The sine function, $f(x) = \sin x$,

is defined for all values of x

is periodic with a period 2π

has a maximum value of 1 when $x = (2n + \frac{1}{2})\pi$

and a minimum value of -1 when $x = (2n + \frac{3}{2})\pi$

is zero when $x = n\pi$

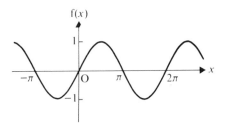

The cosine function, $f(x) = \cos x$,

is defined for all values of x

is periodic with a period 2π

has a maximum value of 1 when $x = 2n\pi$

and a minimum value of -1 when $x = (2n - 1)\pi$

is zero when $x = \frac{1}{2}(2n + 1)\pi$

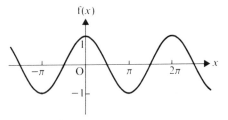

The tangent function, $y = \tan x$,

is undefined for some values of x

these values being all odd multiples of $\frac{1}{2}\pi$

is periodic with period π

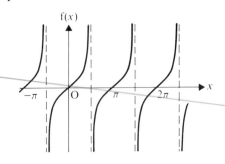

The reciprocal trig functions are

$$\sec x \; \left(= \frac{1}{\cos x} \right) \qquad \operatorname{cosec} x \; \left(= \frac{1}{\sin x} \right) \qquad \cot x \; \left(= \frac{1}{\tan x} \right)$$

TRIGONOMETRIC IDENTITIES

$$\sin \theta \equiv \cos \left(\tfrac{1}{2}\pi - \theta \right)$$

$$\cos \theta \equiv \cos \left(\tfrac{1}{2}\pi - \theta \right)$$

$$\tan \theta \equiv \frac{\sin \theta}{\cos \theta}$$

Pythagorean Identities

$$\cos^2\theta + \sin^2\theta \equiv 1$$

$$1 + \tan^2\theta \equiv \sec^2\theta$$

$$\cot^2\theta + 1 \equiv \operatorname{cosec}^2\theta$$

MULTIPLE CHOICE EXERCISE D

TYPE I

1. The minimum value of $(x + 1)(x - 5)$ is where x equals

   A 1   **B** 5   **C** 2   **D** 0   **E** −2

2. The curve $y = \sin(x - 30°)$ could be

A

C

E

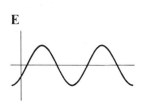

B

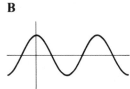

D

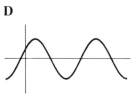

3. The function $x^3 - 12x + 5$ has a stationary value when

   A $x = \sqrt{6}$ C $x = 0$ E $x = 1$

   B $x = -2$ D $x = 4$

4. When $x = 1$ the function $x^3 - 3x^2 + 7$ is

   A stationary   C maximum   E minimum

   B increasing   D decreasing

5. The rate of increase w.r.t. x of the function $x^2 - \dfrac{1}{x^2}$ is

   A $2x + \dfrac{2}{x^3}$ C $2x - \dfrac{2}{x^3}$ E $2x + \dfrac{1}{2x}$

   B $2x - \dfrac{2}{2x}$ D $2x + \dfrac{3}{x^3}$

6. The gradient function of $y = (x - 3)(x^2 + 2)$ is

   A $2x$ C $3x^2 - 6x + 2$ E $x^3 - 3x^2 + 2x - 6$

   B $2x - 3$ D $-3(2x + 2)$

7. The graph of the function $f(\theta) \equiv \cos 2\theta$ has a period

   A 2π C $\frac{1}{2}\pi$ E none of these

   B π D -2π

TYPE II

8. When $\tan\theta = 1$

   A $\cos\theta = \sqrt{2}$

   B $\theta = \frac{1}{4}\pi$

   C θ lies in quadrants 1 and 2

9. $f(x) = x + \dfrac{1}{x}$

   A $f(x)$ is stationary when $x = -1$

   B $\dfrac{d}{dx} f(x) = 1 - \dfrac{1}{x^2}$

   C $y = f(x)$ has no turning points.

10. $y = x^3 - 4x + 5$

    A y is a quadratic function of x
    B The curve has two turning points.
    C y is increasing when $x = 2$

11. $y = x^4$

    A y is decreasing when $x = 1$
    B x^4 has only one stationary value
    C $\dfrac{dy}{dx} = 4x^3$

12. An angle θ is such that $\tan\theta = 1$ and $\cos\theta$ is negative.

    A $\sin\theta$ is positive   **B** $\cos\theta = -\tfrac{1}{2}\sqrt{2}$ **C** $\cot\theta = -1$

13. $f(\theta) = \cos\theta$

    A For $-\tfrac{1}{2}\pi < \theta < \tfrac{1}{2}\pi$, $f(\theta) > 0$
    B $f(\theta)$ is undefined when $\theta = \tfrac{1}{2}\pi$
    C $-1 \leqslant f(\theta) \leqslant 1$

14. The graph of $f(\theta) = \cos\theta$ compared with the graph of $f(\theta) = \sin\theta$ is

    A inverted   **B** $90°$ to the left   **C** $90°$ to the right

15. The solution of the equation $\cos 2\theta = \tfrac{1}{2}$ is the same as the solution of the equation

    A $\tan 2\theta = \sqrt{3}$ **B** $\cos(-2\theta) = \tfrac{1}{2}$ **C** $\cos\theta = \tfrac{1}{4}$

16. If $x = 1 - \tan\theta$ and $y = \sec\theta$ the Cartesian equation given by eliminating θ is

    A $x^2 - y^2 = 2x$
    B $x^2 - y^2 + 2 = 2x$
    C $(1 - x)^2 = (y - 1)(y + 1)$

TYPE III

17. The function $f(\theta) = \cos\theta$ is such that $-1 \leqslant \theta \leqslant 1$

18. $\sin\theta = 0$ when θ is a multiple of π

19. A solution of the equation $\cos\theta = -1$ is $\theta = \pi$

20. $y = \dfrac{1}{x^2} \Rightarrow \dfrac{dy}{dx} = \dfrac{-1}{2x}$

MISCELLANEOUS EXERCISE D

1. Given that $f(x) = 2x^3 - 5x^2 - 4x + 3$, find the stationary values of $f(x)$. Show that $f(x) = (x + 1)(2x - 1)(x - 3)$

   Hence sketch the curve with equation $y = f(x)$, marking on your sketch the coordinates of the points where the curve crosses the coordinate axes.   (AEB)p

2. Sketch the graph of the curve with equation $y = x(1 - x)$. Determine the greatest and least values of y when $-1 \leqslant x \leqslant 1$
                                             (C)

3. Find all the solutions in the interval $-\pi \leqslant \theta \leqslant \pi$ of the equation
   $$2 \sin^2\theta + 5 \cos \theta + 1 = 0$$
   giving each solution in terms of π (JMB)

4. Find the equation of the normal to the curve $y = -3x^3 + 5$ at the point $(-1, 8)$.   (C)

5. Differentiate with respect to x

   (a) $(4x + 1)^3$ (b) $\dfrac{1}{\sqrt{(2 - 3x)}}$ (c) $x^{2/3} + \dfrac{1}{x^{2/3}}$

6. Show that the gradient of the curve with equation $y = x^3 + x + 6$ at the point $A(1, 8)$ is 4

   Hence, or otherwise, find

   (a) an equation of the tangent to the curve at A,

   (b) the x-coordinate of the point where this tangent meets the curve again.   (U of L)

7. The production cost per kilogram, C (in thousands of pounds), when x kilograms of a chemical are made is given by
   $$C = 3x + \frac{100}{x}, \quad x > 0$$
   Find the value of x for which the cost is a minimum, and the minimum cost. Show, also, that this cost is a minimum rather than a maximum.   (O/C, SU & C)

8. The profit, P, in thousands of pounds made by a company t months after its launch, where $0 \leqslant t \leqslant 12$, is given by

$$P = 2t^3 - 15t^2 + 24t + 2$$

Find $\dfrac{dP}{dt}$, and hence determine the range of values of t for which the profit is decreasing.   (U of L 89)

9. A lorry is to be driven 200 km at a steady speed of x km/hour. Petrol is consumed at the rate of $\left(3 + \dfrac{x^2}{240}\right)$ litres per hour.

If petrol costs 50 p per litre and the driver earns £6 an hour, form an expression for the cost, £C, of the journey.
Find the most economical speed and the total cost when this speed is maintained.

10. A cylinder, with an open top, is to be designed to hold 6000 cc of liquid. Show that the surface area is a minimum when the radius of the circular base is $10\left(\dfrac{6}{\pi}\right)^{1/3}$ cm.   (U of L)

11. (a)

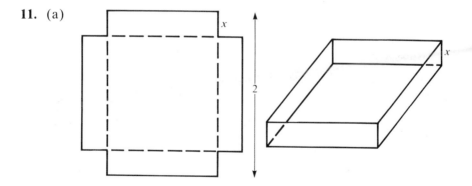

The diagrams show how an open metal box may be formed.
From a square thin sheet of metal measuring 2 units by 2 units, four equal squares of side x are removed; the projecting pieces are then folded upwards to make the sides of the box, which has depth x.
Show that the volume V of the box is given by $V = 4x(1 - x)^2$.
Find the value of x for which V is a maximum, and find also the maximum value of V.

(b) Sketch the graph of $y = 4x(1 - x)^2$, and indicate on your sketch that part of the curve where x can represent the depth of a box as described in (a) above.   (O/C, SU & C)

12. The rate of working, P watts, of an engine which is travelling at a speed $v\,\mathrm{m\,s^{-1}}$ is given by

$$P = 10v + \frac{4000}{v}, \quad v > 0$$

Find the speed at which the rate of working is least.   (U of L)

13. A manufacturer produces closed cylindrical tin cans of radius r cm and height h cm. Each can has a total surface area of $54\pi\,\mathrm{cm^2}$.

(a) Show that $h = \dfrac{27 - r^2}{r}$ and hence find an expression for the volume, $V\,\mathrm{cm^3}$, of each can in terms of r.

(b) Find the value of r for which the cans have their maximum possible volume.   (O/C, SU & C)

14. Find all the solutions in the interval $0 \leqslant \theta \leqslant 360°$ of the equation

$$\cos^2\theta + 5\sin\theta = 3$$

15. Solve the following equations, giving angles correct to 1 decimal place in the range $0 \leqslant x \leqslant 180°$.

(a) $\tan \frac{1}{2}x = \frac{1}{2}$ (b) $\sin(x - 30°) = 0.4$

16. Use the remainder theorem to show that $x - 2$ is a factor of the expression $6x^3 - 13x^2 + 4$ and find the other linear factors.

Hence find all solutions of the equation $6\sin^3\theta - 13\sin^2\theta + 4 = 0$ in the interval $0 \leqslant \theta \leqslant 360°$ (WJEC)

17. Find the values of θ, in the range $-\frac{1}{2}\pi \leqslant \theta \leqslant \frac{1}{2}\pi$, which satisfy the equations

(a) $2\sin^2\theta + \cos\theta = 2$ (b) $\sec^2\theta = 1 + \tan\theta$

CHAPTER 19

INTEGRATION

DIFFERENTIATION REVERSED

When x^2 is differentiated with respect to x the derivative is $2x$

Conversely, if the derivative of an unknown function is $2x$ then it is clear that the unknown function could be x^2

This process of finding a function from its derivative, which reverses the operation of differentiating, is called *integration*.

The Constant of Integration

As seen above, $2x$ is the derivative of x^2,
but it is also the derivative of $x^2 + 3$, $x^2 - 9$, and, in fact,
the derivative of $x^2 +$ any constant.

Therefore the result of integrating $2x$, which is called *the integral of $2x$*, is not a unique function but is of the form

$$x^2 + K \quad \text{where } K \text{ is any constant}$$

K is called *the constant of integration*.

This is written

$$\int 2x \, dx = x^2 + K$$

where $\int \ldots dx$ means the integral of ... w.r.t. x

Integrating *any* function reverses the process of differentiating so, for any function $f(x)$ we have

$$\int \frac{d}{dx} f(x)\, dx = f(x) + K$$

e.g. because differentiating x^3 w.r.t. x gives $3x^2$ we have

$$\int 3x^2\, dx = x^3 + K$$

and it follows that

$$\int x^2\, dx = \tfrac{1}{3}x^3 + K$$

Note that it is not necessary to write $\tfrac{1}{3}K$ in the second form, as K represents *any* constant in either expression.

In general, the derivative of x^{n+1} is $(n+1)x^n$ so

$$\int x^n\, dx = \frac{1}{(n+1)} x^{n+1} + K$$

i.e.   to integrate a power of x,

increase the power by 1 and *divide* by the new power.

This rule can be used to integrate any power of x *except* -1

EXERCISE 19a

Integrate with respect to x,

1. x^4 2. x^7 3. x^3 4. x^{11}

5. $\dfrac{1}{x^2}$ 6. $\dfrac{1}{x^5}$ 7. $x^{1/2}$ 8. $x^{3/4}$

9. $\sqrt[3]{x}$ 10. x^{-3} 11. $x^{-1/2}$ 12. x

Integrating a Sum or Difference of Functions

We saw in Chapter 15 that a function can be differentiated term by term. Therefore, as integration reverses differentiation, integration also can be done term by term.

Example 19b _____

Find the integral of $x^7 + \dfrac{1}{x^2} - \sqrt{x}$

$$\int \left(x^7 + \frac{1}{x^2} - \sqrt{x} \right) dx = \int (x^7 + x^{-2} - x^{1/2})\, dx$$

$$= \int x^7\, dx + \int x^{-2}\, dx - \int x^{1/2}\, dx$$

$$= \frac{1}{8}x^8 + \frac{1}{-1}x^{-1} - \frac{1}{\frac{3}{2}}x^{3/2} + K$$

$$= \frac{1}{8}x^8 - \frac{1}{x} - \frac{2}{3}x^{3/2} + K$$

EXERCISE 19b

Integrate with respect to x,

1. $x^5 + \sqrt{x}$ 2. $\dfrac{1}{x^5} - x^2$ 3. $\sqrt[4]{x} + x^4$ 4. $x^{-3} - x^3$

5. $\dfrac{1}{x^{5/2}} + x^{2/5}$ 6. $x^{-1/2} + x^{1/2}$ 7. $x - \dfrac{1}{x^2}$ 8. $\dfrac{1}{\sqrt[3]{x}} + \sqrt{(x^3)}$

Integrating $(ax + b)^n$

First consider the function $f(x) = (2x + 3)^4$

To differentiate $f(x)$ we make a substitution

i.e. $u = 2x + 3$ $\Rightarrow$ $f(x) = u^4$

giving $\dfrac{d}{dx}(2x + 3)^4 = (4)(2)(2x + 3)^3$

Hence $\int (4)(2)(2x + 3)^3\, dx = (2x + 3)^4 + K$

or $\int (2x + 3)^3\, dx = \dfrac{1}{(2)(4)}(2x + 3)^4 + K$

Considering $f(x) = (ax + b)^{n+1}$ in a similar way gives the general result

$$\int (ax + b)^n\, dx = \frac{1}{(a)(n + 1)}(ax + b)^{n+1}$$

EXERCISE 19c

Integrate with respect to x,

1. $(4x + 1)^3$ 2. $(2 + 3x)^7$ 3. $(5x - 4)^5$

4. $(6x - 2)^3$ 5. $(4 - x)^2$ 6. $(3 - 2x)^3$

7. $(3x + 1)^{-2}$ 8. $(2x - 5)^{-4}$ 9. $(4 - 3x)^{-2}$

10. $(2x + 3)^{1/2}$ 11. $(3x - 1)^{1/3}$ 12. $(2x - 1)^{-1/2}$

13. $(8 - 3x)^{-1/2}$ 14. $\sqrt{(1 - 4x)}$ 15. $(3 - 2x)^{3/2}$

16. $\dfrac{1}{(2 - 7x)^2}$ 17. $\dfrac{1}{\sqrt{(3 - x)}}$ 18. $\dfrac{1}{(1 - 5x)^{1/3}}$

USING INTEGRATION TO FIND AN AREA

The area shown in the diagram is bounded by the curve $y = f(x)$, the x-axis and the lines $x = a$ and $x = b$

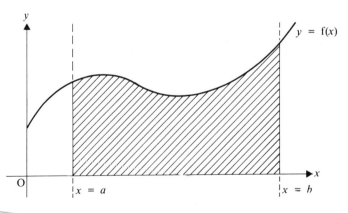

There are several elementary ways in which this area can be estimated, e.g. by counting squares on graph paper. A better method is to divide the area into thin vertical strips and treat each strip, or *element*, as being approximately rectangular.

118

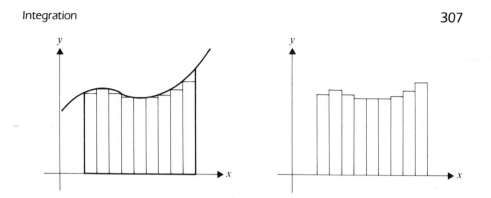

The sum of the areas of the rectangular strips then gives an approximate value for the required area. The thinner the strips are, the better is the approximation.

Note that every strip has one end on the x-axis, one end on the curve and two vertical sides, i.e., they all have the same type of boundaries.

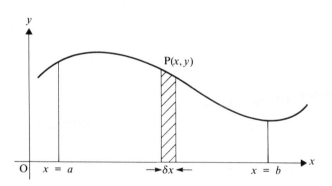

Now, considering a typical element bounded on the left by the ordinate through a general point $P(x, y)$, we see that

> the width of the element represents a small increase in the value of x and so can be called δx

Also, if A represents the part of the area up to the ordinate through P, then

> the area of the element represents a small increase in the value of A and so can be called δA

The shape of a typical strip is approximately a rectangle of height y and width δx

Therefore, for any element

$$\delta A \approx y\delta x \qquad\qquad\qquad\qquad\qquad [1]$$

The required area can now be found by adding the areas of all the strips from $x = a$ to $x = b$

The notation for a summation of this kind is $\displaystyle\sum_{x=a}^{x=b} \delta A$

so,
$$\text{total area} = \sum_{x=a}^{x=b} \delta A$$

$\Rightarrow$
$$\text{total area} \approx \sum_{x=a}^{x=b} y\,\delta x$$

As δx gets smaller the accuracy of the results increases until, in the limiting case,

$$\text{total area} = \lim_{\delta x \to 0} \sum_{x=a}^{x=b} y\,\delta x$$

Equation [1] above can also be written in the alternative form

$$\frac{\delta A}{\delta x} \approx y$$

This form too becomes more accurate as δx gets smaller giving, in the limiting case,

$$\lim_{\delta x \to 0} \frac{\delta A}{\delta x} = y$$

But
$$\lim_{\delta x \to 0} \frac{\delta A}{\delta x} \quad \text{is} \quad \frac{dA}{dx} \quad \text{so} \quad \frac{dA}{dx} = y$$

Hence
$$A = \int y\,dx$$

The boundary values of x defining the total area are $x = a$ and $x = b$ and we indicate this by writing

$$\text{total area} = \int_a^b y\,dx$$

The total area can therefore be found in two ways, either as the limit of a sum or by integration

i.e.
$$\lim_{\delta x \to 0} \sum_{x=a}^{x=b} y\,\delta x = \int_a^b y\,dx$$

and we conclude that   integration is a process of summation.

At this stage we will use integration only to find areas bounded by straight lines and a curve, but first we must investigate the meaning of $\int_a^b y\,dx$

DEFINITE INTEGRATION

Suppose that we wish to find the area bounded by the x-axis, the lines $x = a$ and $x = b$ and the curve $y = 3x^2$

Using the method above we find that $A = \int 3x^2\,dx$

i.e. $A = x^3 + K$

From this area function we can find the value of A corresponding to a particular value of x.

Hence using $x = a$ gives $A_a = a^3 + K$

and   using $x = b$ gives $A_b = b^3 + K$

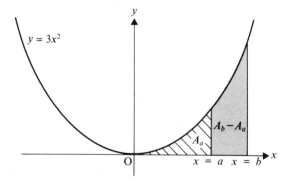

Then the area between $x = a$ and $x = b$ is given by $A_b - A_a$

where $A_b - A_a = (b^3 + K) - (a^3 + K) = b^3 - a^3$

Now $A_b - A_a$ is referred to as

the definite integral from a to b of $3x^2$

and is denoted by $\displaystyle\int_a^b 3x^2\,dx$

i.e. $\displaystyle\int_a^b 3x^2\,dx = (x^3)_{x=b} - (x^3)_{x=a}$

The RHS of this equation is usually written in the form $\left[x^3\right]_a^b$ where a and b are called the *boundary values* or *limits of integration*; b is the *upper limit* and a is the *lower limit*.

Whenever a definite integral is calculated, the constant of integration disappears.

Note. A definite integral can be found in this way only if the function to be integrated is defined for every value of x from a to b, e.g. $\displaystyle\int_{-1}^1 \frac{1}{x^2}\,dx$ cannot be found directly as $\dfrac{1}{x^2}$ is undefined when $x = 0$.

Examples 19d _____

1. Find the value of $\displaystyle\int_0^9 x^{3/2}\,dx$

$$\int_0^9 x^{3/2}\,dx = \left[\tfrac{2}{5}x^{5/2}\right]_0^9$$
$$= \tfrac{2}{5}\{(9)^{5/2} - 0\}$$
$$= 97\tfrac{1}{5}$$

2. Evaluate $\displaystyle\int_1^4 \frac{1}{(x+3)^2}\,dx$

$$\int_1^4 \frac{1}{(x+3)^2}\,dx \equiv \int_1^4 (x+3)^{-2}\,dx$$
$$= \left[-(x+3)^{-1}\right]_1^4$$
$$= \{-(4+3)^{-1}\} - \{-(1+3)^{-1}\}$$
$$= -\tfrac{1}{7} + \tfrac{1}{4} = \tfrac{3}{28}$$

EXERCISE 19d

Evaluate each of the following definite integrals.

1. $\int_0^1 x^4 \, dx$

2. $\int_1^2 x^{-2} \, dx$

3. $\int_1^4 x^{1/2} \, dx$

4. $\int_1^2 \frac{1}{x^5} \, dx$

5. $\int_4^9 \frac{1}{\sqrt{x}} \, dx$

6. $\int_0^1 (2x + 3)^2 \, dx$

7. $\int_0^2 x^3 \, dx$

8. $\int_1^2 \sqrt{x^5} \, dx$

9. $\int_2^4 (x^2 + 4) \, dx$

10. $\int_3^8 \sqrt{(1 + x)} \, dx$

11. $\int_0^3 (x^2 + 2x - 1) \, dx$

12. $\int_0^2 (x^3 - 3x) \, dx$

13. $\int_{-1}^0 \frac{1}{(1 - x)^2} \, dx$

14. $\int_{-1}^2 \frac{3}{\sqrt{(x + 2)}} \, dx$

15. $\int_{-1}^0 (2 + 3x)^6 \, dx$

16. $\int_{1/2}^7 (4x - 1)^{1/3} \, dx$

FINDING AREA BY DEFINITE INTEGRATION

As we have seen, the area bounded by a curve $y = f(x)$, the lines $x = a$, $x = b$, and the x-axis, can be found from the definite integral

$$\int_a^b f(x) \, dx$$

It is recommended, however, that this is not regarded as a *formula* but that the required area is first considered as the summation of the areas of elements, a typical element being shown in a diagram.

Example 19e _____

Find the area in the first quadrant bounded by the x and y axes and the curve $y = 1 - x^2$

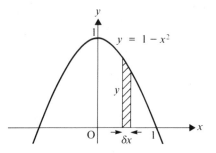

The required area starts at the y-axis, i.e. at $x = 0$ and ends where the curve crosses the x-axis, i.e. where $x = 1$. So it is given by

$$\lim_{\delta x \to 0} \sum_{x=0}^{x=1} y\,\delta x = \int_0^1 (1 - x^2)\,dx = \left[x - \frac{x^3}{3}\right]_0^1$$

$$= (1 - \tfrac{1}{3}) - (0 - 0) = \tfrac{2}{3}$$

The required area is $\tfrac{2}{3}$ of a square unit.

_____

EXERCISE 19e

In each question find the area with the given boundaries.

1. The x-axis, the curve $y = x^2 + 3$ and the lines $x = 1, x = 2$

2. The curve $y = \sqrt{x}$, the x-axis and the lines $x = 4, x = 9$

3. The x-axis, the lines $x = -1, x = 1$, and the curve $x^2 + 1$

4. The curve $y = x^2 + x$, the x-axis and the line $x = 3$

5. The positive x and y axes and the curve $y = 4 - x^2$

6. The lines $x = 2, x = 4$, the x-axis and the curve $y = x^3$

7. The curve $y = 4 - x^2$, the positive y-axis and the negative x-axis.

8. The x-axis, the lines $x = 1$ and $x = 2$, and the curve $y = \tfrac{1}{2}x^3 + 2x$

9. The x-axis and the lines $x = 1, x = 5$, and $y = 2x$
   Check the result by sketching the required area and finding it by mensuration.

The Meaning of a Negative Result

Consider the area bounded by $y = 4x^3$ and the x-axis if the other boundaries are the lines
(a) $x = -2$ and $x = -1$ (b) $x = 1$ and $x = 2$

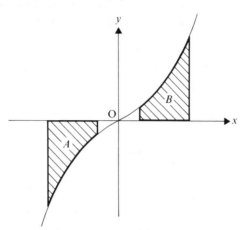

This curve is symmetrical about the origin so the two shaded areas are equal.

(a) Considering A

$$\lim_{\delta x \to 0} \sum_{x = -2}^{x = -1} y\, \delta x = \int_{-2}^{-1} y\, dx$$

$$= \int_{-2}^{-1} 4x^3\, dx$$

$$= \left[x^4 \right]_{-2}^{-1}$$

$$= 1 - 16 = -15$$

(b) Considering B

$$\lim_{\delta x \to 0} \sum_{x = 1}^{x = 2} y\, \delta x = \int_{1}^{2} y\, dx$$

$$= \int_{1}^{2} 4x^3\, dx$$

$$= \left[x^4 \right]_{1}^{2}$$

$$= 16 - 1 = 15$$

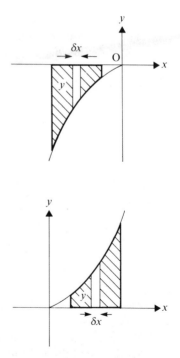

So we see that, while the magnitudes of the two areas are equal, the result for the area of A, which is below the x-axis, is negative. This is explained by the fact that the length of a strip in A was taken as y, which is negative for the part of the curve bounding A.

Note. Care must be taken with problems involving a curve that crosses the x-axis between the boundary values.

Example 19f

Find the area enclosed between the curve $y = x(x - 1)(x - 2)$ and the x-axis.

The area enclosed between the curve and the x-axis is the sum of the areas A and B.

For A we use

$$\int_0^1 y \, dx = \int_0^1 (x^3 - 3x^2 + 2x) \, dx$$

$$= \left[\frac{x^4}{4} - x^3 + x^2 \right]_0^1$$

$$= \tfrac{1}{4}$$

For B we use

$$\int_1^2 (x^3 - 3x^2 + 2x) \, dx = \left[\frac{x^4}{4} - x^3 + x^2 \right]_1^2$$

$$= (4 - 8 + 4) - (\tfrac{1}{4} - 1 + 1)$$

$$= -\tfrac{1}{4}$$

The minus sign refers only to the *position* of area B relative to the x-axis.
The actual area is $\tfrac{1}{4}$ of a square unit.
So the total shaded area is $\tfrac{1}{4} + \tfrac{1}{4} = \tfrac{1}{2}$ of a square unit.

EXERCISE 19f

1. Find the area enclosed by the curve $y = x^2$, the x-axis and the lines $x = 1$ and $x = 2$

2. Find the area bounded by the x and y axes, the line $x = 1$ and part of the curve $y = x^3 + 2$

3. Find the area of the region of the xy-plane which is enclosed by the x and y axes, the line $x = 3$ and part of the curve with equation $y = x^2 + 3x + 2$

4. Find the area bounded by the curve $y = 1 - x^3$, the x-axis and the lines $x = 2$, $x = 3$

5. The boundaries of a region in the xy plane are the lines $x = 1$, $x = 2$, the x-axis and part of the curve $y = x(3 - x)$. Find the area of this region.

6. Find the area below the x-axis and above the curve $y = x^2 - 1$

7. Sketch the curve with equation $y = (x - 1)^2$. Calculate the area of the region enclosed by this curve and the x and y axes.

8. Sketch the curve $y = x(x^2 - 1)$, showing where it crosses the x-axis. Find
   (a) the area enclosed above the x-axis and below the curve
   (b) the area enclosed below the x-axis and above the curve
   (c) the total area between the curve and the x-axis.

9. Repeat Question 8 for the curve $y = x(4 - x^2)$

10. Evaluate

    (a) $\int_0^2 (x - 2)\,dx$ (b) $\int_2^4 (x - 2)\,dx$ (c) $\int_0^4 (x - 2)\,dx$

    Interpret your results by means of a sketch.

USING HORIZONTAL ELEMENTS

Suppose that area between the curve $x = y(4 - y)$ and the y-axis is required.

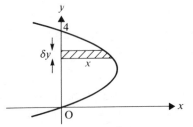

The curve crosses the y-axis where $y = 0$ and $y = 4$ as shown.

A vertical element is not suitable in this case because it has *both ends on the curve* and its length is therefore not easily found.

However it is easy to find the approximate area of a *horizontal* strip, by treating it as a rectangle with length x and width δy

i.e. area of element $\approx x\,\delta y$ and the required area is therefore given by

$$\lim_{\delta y \to 0} \sum_{y=0}^{y=4} x\,\delta y = \int_0^4 x\,dy = \int_0^4 y(y - 4)\,dy$$

EXERCISE 19g

1. Evaluate

   (a) $\displaystyle\int_3^6 (y^2 - y)\,dy$ (b) $\displaystyle\int_2^{10} \frac{1}{\sqrt{(2y + 5)}}\,dy$ (c) $\displaystyle\int_{-4}^3 (4 - y)^{-2/3}\,dy$

   In Questions 2 to 5 find the area specified by the given boundaries.

2. The curve $x = y^2$, the y-axis and the lines $y = 1$, $y = 2$

3. The y-axis, the curve $x = \sqrt{y}$ and the line $y = 4$

4. The y-axis and the curve $x = 9 - y^2$

5. The y-axis and the curve $x = (y - 2)(y + 1)$

6. Find the area in the first quadrant bounded by the x and y axes and the curve $y = 16 - x^2$
   (a) by using vertical elements
   (b) by using horizontal elements and the equation of the curve in the form $x = \sqrt{(16 - y)}$

7. If $y = x^2$, show by means of sketch graphs and *not* by evaluating the integrals, that $\displaystyle\int_0^1 y\,dx = 1 - \int_0^1 x\,dy$

MIXED EXERCISE 19

Integrate with respect to x

1. $x^2 - 1/x^2$ **2.** $\sqrt{(3x + 7)}$ **3.** $\sqrt{x} + 1/\sqrt{x}$

4. $\dfrac{1}{(4x - 3)}$ **5.** $x^3 - \dfrac{1}{(1 - x)^3}$ **6.** $\dfrac{x^2 - 1}{\sqrt{x}}$

Evaluate

7. $\displaystyle\int_5^6 (6 - x)^4 \, dx$ **8.** $\displaystyle\int_{-1}^{12} \dfrac{3}{\sqrt[3]{(2y + 3)}} \, dy$ **9.** $\displaystyle\int_1^{32} \left(\sqrt[5]{x} - \dfrac{1}{\sqrt[5]{x}} \right) dx$

Find the areas specified in Questions 10 to 15

10. Between the curve $y = 6 + x - x^2$ and the x-axis.

11. Bounded by part of the curve $y = x^2 + 2,$ the x-axis and the lines $x = 1, \ x = 4$

12. Bounded by the x and y axes and the curve $y = 1 - x^3$

13. Bounded by the y-axis, the lines $y = 1, \ y = 3$ and part of the curve $3y^2 = 2x - 3$

14. Bounded by the curve $x = y^2 - 4$ and the y-axis.

15. The *total* area between the curve $y = (x - 1)(x - 2)(x - 3)$ and the x-axis.

16. (a) Sketch the curve with equation $y = (x + 2)(x - 5)$ marking the x-coordinates of the points where the curve crosses the x-axis.

     (b) Evaluate $\displaystyle\int_{-2}^5 (x^2 - 3x - 10) \, dx$

     (c) Write down the value of the area between the curve and the x-axis.

CHAPTER 20

KINEMATICS

GENERAL RATES OF INCREASE

We have already seen that

$$\frac{dy}{dx} \text{ represents the rate at which } y \text{ increases compared with } x$$

Whenever the variation in one quantity, p say, depends upon the changing value of another quantity, q, then the rate of increase of p compared with q can be expressed as $\frac{dp}{dq}$

There are many every-day situations where such relationships exist, e.g.

1) liquid expands when it is heated so, if V is the volume of a quantity of liquid and T is the temperature, then the rate at which the volume increases with temperature can be written $\frac{dV}{dT}$

2) if the profit, P, made by a company selling radios depends upon the number, n, of radios sold, then $\frac{dP}{dn}$ represents the rate of increase of profit compared with the increase in sales.

EXERCISE 20a

1. Express, in the form $\frac{dy}{dx}$, the rate at which:

 (a) the quantity, q, of petrol consumed increases with the number, n, of miles that a car is driven,

 (b) the girth, g, of a tree increases with the age, a, of the tree in years,

 (c) the surface area, A, of a crystal increases with time, t.

318

MOTION IN A STRAIGHT LINE

Suppose that we begin to observe a particle P moving along a straight line. If P is initially at a fixed point O on the line then t seconds later P will be at a distance s metres from O.

When this distance is measured in a specified direction it is called displacement, i.e. *displacement is a vector quantity.*
(Any quantity that has both magnitude and direction is a vector.)

The speed of any moving object tells us the rate at which it covers the ground but it does not give any indication of the direction of motion, i.e. speed is not a vector quantity.

The rate at which the *displacement* of the particle P increases with time, however, is a vector called *velocity*, represented by v and measured in metres per second, $m\,s^{-1}$ (or m/s).

Because velocity is a vector it can be negative, indicating a direction opposite to the specified direction of motion. Speed, on the other hand, is the magnitude of the velocity and is therefore always positive.

Unless the velocity is constant it will change with time. The rate at which the velocity of P increases with time is the *acceleration*. This is another vector quantity, represented by a and measured in metres per second per second, $m\,s^{-2}$.
A negative acceleration is a retardation.

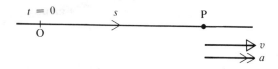

As velocity is the rate at which the displacement is increasing with time,

$$v = \frac{ds}{dt} \quad \text{and hence} \quad \int v\, dt = s$$

As acceleration is the rate at which the velocity is increasing with time,

$$a = \frac{dv}{dt} \quad \text{and hence} \quad \int a\, dt = v$$

Examples 20b

1. A particle P is moving along a straight line. When $t = 0$, P is at a fixed point O on that line and t seconds later its displacement from O is given by s metres where $s = 3t^2 - 5t$

(a) Find the velocity of P 5 seconds after passing through O.
(b) Show that the acceleration of P is constant.

(a)
$$s = 3t^2 - 5t$$

$$v = \frac{ds}{dt} = \frac{d}{dt}(3t^2 - 5t)$$

$$\Rightarrow \qquad v = 6t - 5$$

When $t = 5$, $v = 30 - 5 = 25$

So the velocity of P after 5 seconds is $25\ \mathrm{m\,s^{-1}}$

(b)
$$a = \frac{dv}{dt} = \frac{d}{dt}(6t - 5)$$

$$\Rightarrow \qquad a = 6$$

Therefore the acceleration has a constant value of $6\ \mathrm{m\,s^{-2}}$

2. A is a fixed point on a straight line AB. The displacement from A of a particle P moving on this line, is given by

$$s = t^3 - 6t^2 + 9t$$

at any time t.
Find an expression for the velocity of P at time t and hence, using metres and seconds, find

(a) the initial velocity,
(b) the velocity when $t = 2$,
(c) the speed when $t = 2$,
(d) the times when P is momentarily at rest.

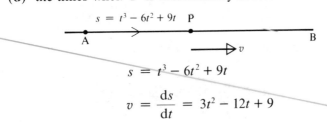

$$s = t^3 - 6t^2 + 9t$$

$$v = \frac{ds}{dt} = 3t^2 - 12t + 9$$

(a) The initial velocity is the velocity when $t = 0$

i.e.   initial velocity $= 9 \, \text{m s}^{-1}$

(b) When $t = 2$, $v = 12 - 24 + 9 = -3$

i.e.   the velocity when $t = 2$ is $-3 \, \text{m s}^{-1}$

(P is moving in the direction BA)

(c) When $t = 2$ the speed of P is $3 \, \text{m s}^{-1}$

(d) P is momentarily at rest when $v = 0$

$$v = 3t^2 - 12t + 9 = 3(t - 3)(t - 1)$$

So P is momentarily at rest when $t = 1$ and $t = 3$

---

3. A plane begins its take-off when $t = 0$ and moves down the runway so that its velocity after t seconds is given by $\frac{9}{4}t \, \text{m s}^{-1}$. Find an expression for s, where s metres is the displacement of the plane from its starting point t seconds after beginning to move. If the plane's lift-off velocity is $45 \, \text{m s}^{-1}$, find the minimum length of the runway.

$$v = \tfrac{9}{4}t$$

$$s = \int v \, dt = \int \tfrac{9}{4}t \, dt = \tfrac{9}{8}t^2 + K$$

$s = 0$ when $t = 0$ $\Rightarrow$ $K = 0$

therefore $s = \tfrac{9}{8}t^2$

The plane lifts off when $v = 45$

i.e.   when $\tfrac{9}{4}t = 45$ $\Rightarrow$ $t = 20$

When $t = 20$, $s = 450$

therefore the runway must be at least $450 \, \text{m}$ long.

---

4. A particle starts from rest and travels in a straight line so that its acceleration is $(t + 3)\,\mathrm{m\,s^{-2}}$ at any time t. Find the distance travelled in the interval of time from $t = 1$ to $t = 3$

$t = 0$ $\longrightarrow a = (t + 3)$

$v = 0$

$$a = t + 3$$

$$v = \int (t + 3)\,\mathrm{d}t = \tfrac{1}{2}t^2 + 3t + K$$

The particle starts from rest so $v = 0$ when $t = 0$ $\Rightarrow$ $K = 0$

$\therefore$
$$v = \tfrac{1}{2}t^2 + 3t$$

$$s = \int (\tfrac{1}{2}t^2 + 3t)\,\mathrm{d}t = \tfrac{1}{6}t^3 + \tfrac{3}{2}t^2 + K'$$

where s is the displacement at time t from a fixed point.

For positive values of t, s is always positive so the *distance* travelled in any time interval is given by the difference of the displacements at the beginning and end of that interval.

When $t = 1$, $s_1 = \tfrac{1}{6} + \tfrac{3}{2} + K' = \tfrac{5}{3} + K'$

When $t = 3$, $s_3 = \tfrac{1}{6}(27) + \tfrac{3}{2}(9) + K' = 18 + K'$

$\therefore$ $s_3 - s_1 = (18 + K') - (\tfrac{5}{3} + K') = 16\tfrac{1}{3}$

i.e. the distance travelled between $t = 1$ and $t = 3$ is $16\tfrac{1}{3}$ m.

EXERCISE 20b

1. A particle P moves in a straight line so that its displacement, s metres, from a fixed point A on the line, is given by $s = (3t^2 + 4)$ at time t seconds.

(a) Find the velocity and the acceleration of P when $t = 2$

(b) What was the initial distance of P from A?

2. A particle P is moving on a straight line and O is a fixed point on the line. The displacement of P from O at time t seconds is s metres where $s = 6 + 10t - t^2$. Find

(a) the distance travelled from $t = 1$ to $t = 4$

(b) the velocity when $t = 6$

(c) the speed when $t = 6$

3. The displacement, x metres, from O of an object P moving on a straight line Ox, is given by $x = 42t - 3t^2$ at time t seconds.

(a) Find the velocity when $t = 6$, when $t = 7$ and when $t = 8$. Explain what happens during this interval of time.

(b) Find the maximum distance of P from O during this time.

4. The velocity of a particle, which starts from rest at a point O and travels in a straight line, is $v \, \mathrm{m\,s^{-1}}$ at any time t seconds. If v is given by $v = 2\sqrt{t}$, find

(a) the acceleration when $t = 4$

(b) the displacement of the particle from O when $t = 9$

5. The velocity after t seconds of a particle moving in a straight line is $v \, \mathrm{m\,s^{-1}}$ where $v = 6t^2 + 1$. Find how far the particle travels in the interval from $t = 1$ to $t = 4$

6. A particle moving in a straight line starts from rest at a point O on the line, and, t seconds later, has an acceleration $(t - 6) \, \mathrm{m\,s^{-2}}$. Find expressions for the velocity and the displacement of the particle from O after t seconds. Find also the velocity and the displacement from O after 6 seconds.

7. The velocity of a particle moving in a straight line is $3t^2 \, \mathrm{m\,s^{-1}}$ at any time, t seconds. What is the acceleration of the particle at time t? If the particle passes through a fixed point A on the line, with velocity $12 \, \mathrm{m\,s^{-1}}$, find the displacement from A, 6 seconds later.

8. The acceleration of a particle travelling in a straight line is given by $15\sqrt{(t - 1)}$ at any time t seconds. When $t = 2$ the particle passes through a fixed point O on the line with velocity $2 \, \mathrm{m\,s^{-1}}$. Find expressions for the velocity of the particle and for its displacement from O at time t.

9. A cyclist is riding at $4 \, \mathrm{m\,s^{-1}}$ as he passes a kiosk at the top of a long straight downhill slope. He immediately starts to free-wheel and finds that he has a constant acceleration of $\frac{1}{3} \, \mathrm{m\,s^{-2}}$. Find expressions for his velocity, $v \, \mathrm{m\,s^{-1}}$, and his displacement, s m, from the kiosk after t seconds. If his speed at the foot of the slope is $14 \, \mathrm{m\,s^{-1}}$, find the length of the slope.

10. A particle, P, moves in a straight line with an acceleration of $1/t^3 \, \mathrm{m\,s^{-2}}$ after t seconds. When $t = 1$, P is at rest at a point O on the line of motion. Find expressions for its velocity and its displacement from O at time t and when $t = 2$

11. The velocity at time, t, of a particle P travelling in a straight line, is $v \, \mathrm{m\,s^{-1}}$ where $v = 2 + 1/t^2$. What is the acceleration of P when $t = 1$? Find the distance travelled by P during the time interval between $t = 2$ and $t = 6$

12. A particle moves in a straight line with velocity $v \, \mathrm{m\,s^{-1}}$ where, after t seconds, $v = 2 + 1/(t + 1)^2$. When $t = 0$ the particle passes through a fixed point O on the line.

   (a) Find an expression for the displacement of the particle from O after t seconds.

   (b) Show that the velocity has a limiting value (i.e. v approaches a fixed value as t approaches infinity), and state this value.

CHAPTER 21

NUMBER SERIES

SEQUENCES

Consider the following sets of numbers,

$$2, 4, 6, 8, 10, \ldots$$
$$1, 2, 4, 8, 16, \ldots$$
$$4, 9, 16, 25, 36, \ldots$$

Each set of numbers, in the order given, has a pattern and there is an obvious rule for obtaining the next number and as many subsequent numbers as we wish to find.

Such sets are called *sequences* and each member of the set is a term of the sequence.

SERIES

When the terms of a sequence are added, a series is formed,
e.g., $1 + 2 + 4 + 8 + 16 + \ldots$ is a series.

If the series stops after a finite number of terms it is called a finite series,
e.g., $1 + 2 + 4 + 8 + 16 + 32 + 64$ is a finite series of seven terms.

If the series continues indefinitely it is called an infinite series,
e.g., $1 + \frac{1}{2} + \frac{1}{4} + \frac{1}{8} + \frac{1}{16} + \frac{1}{32} + \ldots + \frac{1}{1024} + \ldots$ is an infinite series.

Consider again the series $1 + 2 + 4 + 8 + 16 + 32 + 64$

As each term is a power of 2 we can write this series in the form

$$2^0 + 2^1 + 2^2 + 2^3 + 2^4 + 2^5 + 2^6$$

All the terms of this series are of the form 2^r, so 2^r is a general term. We can then define the series as the sum of terms of the form 2^r where r takes all integral values in order from 0 to 6 inclusive.

Using Σ as a symbol for 'the sum of terms such as' we can redefine our series more concisely as $\Sigma\, 2^r$, r taking all integral values from 0 to 6 inclusive, or, even more briefly,

$$\sum_{r=0}^{6} 2^r$$

Placing the lowest and highest value that r takes below and above the sigma symbol respectively, indicates that r takes all integral values between these extreme values.

Thus $\displaystyle\sum_{r=2}^{10} r^3$ means 'the sum of all terms of the form r^3 where r takes all integral values from 2 to 10 inclusive',

i.e. $\displaystyle\sum_{r=2}^{10} r^3 = 2^3 + 3^3 + 4^3 + 5^3 + 6^3 + 7^3 + 8^3 + 9^3 + 10^3$

Note that a finite series, when written out, should always end with the last term even if intermediate terms are omitted, e.g. $3 + 6 + 9 + \ldots + 99$

The infinite series $\qquad\qquad 1 + \frac{1}{2} + \frac{1}{4} + \frac{1}{8} + \frac{1}{16} + \ldots$

may also be written in the sigma notation. The continuing dots after the last written term indicate that the series is infinite, i.e. there is *no* last term. Each term of this series is a power of $\frac{1}{2}$ so a general term can be written $(\frac{1}{2})^r$. The first term is 1 or $(\frac{1}{2})^0$, so the first value that r takes is zero. There is no last term of this series, so there is no upper limit for the value of r.

Therefore $\quad 1 + \frac{1}{2} + \frac{1}{4} + \frac{1}{8} + \frac{1}{16} + \ldots\quad$ may be written as $\displaystyle\sum_{r=0}^{\infty} (\tfrac{1}{2})^r$

Note that when a given series is rewritten in the sigma notation it is as well to check that the first few values of r give the correct first few terms of the series.

Writing a series in the sigma notation, apart from the obvious advantage of brevity, allows us to select a particular term of a series without having to write down all the earlier terms.

For example, in the series $\sum\limits_{r=3}^{10} (2r + 5)$,

the first term is the value of $2r + 5$ when $r = 3$, i.e. $2 \times 3 + 5 = 11$
the last term is the value of $2r + 5$ when $r = 10$, i.e. 25
the fourth term is the value of $2r + 5$ when r takes its fourth value in
order from $r = 3$, i.e. when $r = 6$

Thus the fourth term of $\sum\limits_{r=3}^{10} (2r + 5)$ is $2 \times 6 + 5 = 17$

Example 21a

Write the following series in the sigma notation,

(a) $1 - x + x^2 - x^3 + \ldots$ (b) $2 - 4 + 8 - 16 + \ldots + 128$

(a) A general term of this series is $\pm x^r$, having a positive sign when r
is even and a negative sign when r is odd.
Because $(-1)^r$ is positive when r is even and negative when r is odd,
the general term can be written $(-1)^r x^r$

The first term of this series is 1, or x^0

Hence $1 - x + x^2 - x^3 + \ldots = \sum\limits_{r=0}^{\infty} (-1)^r x^r$

(b) $2 - 4 + 8 - 16 + \ldots + 128 = 2 - (2)^2 + (2)^3 - (2)^4 + \ldots + (2)^7$

So a general term is of the form $\pm 2^r$, being positive when r is odd
and negative when r is even,
i.e. the general term is $(-1)^{r+1} 2^r$

Hence $2 - 4 + 8 - 16 + \ldots + 128 = \sum\limits_{r=1}^{7} (-1)^{r+1} 2^r$

EXERCISE 21a

1. Write the following series in the sigma notation:

(a) $1 + 8 + 27 + 64 + 125$ (d) $1 + \frac{1}{3} + \frac{1}{9} + \frac{1}{27} + \ldots$

(b) $2 + 4 + 6 + 8 + \ldots + 20$ (e) $-4 - 1 + 2 + 5 \ldots + 17$

(c) $\frac{1}{2} + \frac{1}{3} + \frac{1}{4} + \frac{1}{5} + \ldots + \frac{1}{50}$ (f) $8 + 4 + 2 + 1 + \frac{1}{2} + \ldots$

2. Write down the first three terms and, where there is one, the last term of each of the following series:

(a) $\displaystyle\sum_{r=1}^{\infty} \frac{1}{r}$

(b) $\displaystyle\sum_{r=0}^{5} r(r+1)$

(c) $\displaystyle\sum_{r=0}^{20} \frac{r+2}{(r+1)(2r+1)}$

(d) $\displaystyle\sum_{r=0}^{\infty} \frac{1}{(r^2+1)}$

(e) $\displaystyle\sum_{r=-1}^{8} r(r+1)(r+2)$

(f) $\displaystyle\sum_{r=0}^{\infty} a^r(-1)^{r+1}$

3. For the following series, write down the term indicated, and the number of terms in the series.

(a) $\displaystyle\sum_{r=1}^{9} 2^r$,   3rd term

(b) $\displaystyle\sum_{r=-1}^{8} (2r+3)$,   5th term

(c) $\displaystyle\sum_{r=-6}^{-1} \frac{1}{(2r+1)}$,   last term

(d) $\displaystyle\sum_{r=0}^{\infty} \frac{1}{(r+1)(r+2)}$,   20th term

(e) $\displaystyle\sum_{r=1}^{\infty} \left(\frac{1}{2}\right)^r$, nth term

(f) $8 + 4 + 0 - 4 - 8 - 12 \ldots - 80$ 15th term

(g) $\frac{1}{16} + \frac{1}{8} + \frac{1}{4} + \frac{1}{2} + \ldots + 32$,   7th term

ARITHMETIC PROGRESSION

Consider the sequence 5, 8, 11, 14, 17, ..., 29
Each term of this sequence exceeds the previous term by 3, so the sequence can be written in the form

5, $(5 + 3)$, $(5 + 2 \times 3)$, $(5 + 3 \times 3)$, $(5 + 4 \times 3)$, ..., $(5 + 8 \times 3)$

This sequence is an example of an arithmetic progression (AP) which is a sequence where any term differs from the preceding term by a constant, called the *common difference*.
The common difference may be positive or negative. For example, the first six terms of an AP whose first term is 8 and whose common difference is -3, are 8, 5, 2, -1, -4, -7

In general, if an AP has a first term a, and a common difference d, the first four terms are a, $(a + d)$, $(a + 2d)$, $(a + 3d)$, and the nth term, u_n, is $a + (n - 1)d$

Thus
> an AP with n terms can be written as
>
> a, $(a + d)$, $(a + 2d)$, ..., $[a + (n - 1)d]$

Examples 21b

1. The 8th term of an AP is 11 and the 15th term is 21. Find the common difference, the first term of the series, and the nth term.

If the first term of the series is a and the common difference is d, then the 8th term is $a + 7d$,

$\therefore$ $$a + 7d = 11 \tag{1}$$

and the 15th term is $a + 14d$,

$\therefore$ $$a + 14d = 21 \tag{2}$$

$[2] - [1]$ gives $\qquad 7d = 10 \quad \Rightarrow \quad d = \frac{10}{7}$

and $\qquad\qquad\qquad\qquad\qquad a = 1$

so the first term is 1 and the common difference is $\frac{10}{7}$

Hence the nth term is $a + (n - 1)d = 1 + (n - 1)\frac{10}{7} = \frac{1}{7}(10n - 3)$

---

2. The nth term of an AP is $12 - 4n$. Find the first term and the common difference.

If the nth term is $12 - 4n$, the first term $(n = 1)$ is 8
The second term $(n = 2)$ is 4

Therefore the common difference is -4

The Sum of an Arithmetic Progression

Consider the sum of the first ten even numbers, which is an AP.

Writing it first in normal, then in reverse, order we have

$$S = 2 + 4 + 6 + 8 + \ldots + 18 + 20$$
$$S = 20 + 18 + 16 + 14 + \ldots + 4 + 2$$

Adding gives $\quad 2S = 22 + 22 + 22 + 22 + \ldots + 22 + 22$

As there are ten terms in this series, we have

$$2S = 10 \times 22 \quad \Rightarrow \quad S = 110$$

This process is known as finding the sum from first principles.
Applying it to a general AP gives formulae for the sum, which may be quoted and used.

If S_n is the sum of the first n terms of an AP with last term l,

then $\qquad S_n = \quad a \quad + (a + d) + (a + 2d) + \ldots + (l - d) + \quad l$

reversing $\quad S_n = \quad l \quad + (l - d) + (l - 2d) + \ldots + (a + d) + \quad a$

adding $\qquad 2S_n = (a + l) + (a + l) + (a + l) \quad + \ldots + (a + l) + (a + l)$

as there are n terms we have $\qquad 2S_n = n(a + l)$

$$\Rightarrow \quad S_n = \tfrac{1}{2}n(a + l) \qquad \text{i.e.} \quad S_n = (\text{number of terms}) \times (\text{average term})$$

Also, because the nth term, l, is equal to $a + (n - 1)d$, we have

$$S_n = \tfrac{1}{2}n[a + a + (n - 1)d]$$

i.e.
$$S_n = \tfrac{1}{2}n[2a + (n - 1)d]$$

Either of these formulae can now be used to find the sum of the first n terms of an AP.

Examples 21b (continued) _____

3. Find the sum of the following series,

   (a) an AP of eleven terms whose first term is 1 and whose last term is 6

   (b) $\displaystyle\sum_{r=1}^{8}\left(2 - \frac{2r}{3}\right)$

(a) We know the first and last terms, and the number of terms so we use
$S_n = \frac{1}{2}n(a + l)$

$\Rightarrow$ $\qquad\qquad\qquad\qquad S_{11} = \frac{11}{2}(1 + 6) = \frac{77}{2}$

(b) $\displaystyle\sum_{r=1}^{8}\left(2 - \frac{2r}{3}\right) = \frac{4}{3} + \frac{2}{3} + 0 - \frac{2}{3} - \ldots - \frac{10}{3}$

This is an AP with 8 terms where $a = \frac{4}{3}$, $d = -\frac{2}{3}$

Using $S_n = \frac{1}{2}n[2a + (n - 1)d]$ gives

$\qquad\qquad\qquad S_8 = 4\left[\frac{8}{3} + 7\left(-\frac{2}{3}\right)\right] = -8$

_____

4. In an AP the sum of the first ten terms is 50 and the 5th term is three times the 2nd term. Find the first term and the sum of the first 20 terms.

If a is the first term and d is the common difference, and there are n terms, using $S_n = \frac{1}{2}n[2a + (n - 1)d]$ gives

$\qquad\qquad\qquad S_{10} = 50 = 5(2a + 9d)$ $\qquad\qquad$ [1]

Now using $u_n = a + (n - 1)d$ gives

$\qquad\qquad\qquad u_5 = a + 4d$ and $u_2 = a + d$

Therefore $\qquad\qquad\qquad a + 4d = 3(a + d)$ $\qquad\qquad$ [2]

From [1] and [2] we get $d = 1$ and $a = \frac{1}{2}$
so the first term is $\frac{1}{2}$ and the sum of the first 20 terms is S_{20} where

$\qquad\qquad\qquad S_{20} = 10(1 + 19 \times 1) = 200$

_____

5. The sum of the first n terms of a series is given by $S_n = n(n+3)$ Find the fourth term of the series and show that the terms are in arithmetic progression.

If the terms of the series are $a_1, a_2, a_3, \ldots a_n$

then $S_n = a_1 + a_2 + \ldots + a_n = n(n+3)$

So $S_4 = a_1 + a_2 + a_3 + a_4 = 28$

and $S_3 = a_1 + a_2 + a_3 \quad\quad = 18$

Hence the fourth term of the series, a_4, is 10

Now $S_n = a_1 + a_2 + \ldots + a_{n-1} + a_n = n(n+3)$

and $S_{n-1} = a_1 + a_2 + \ldots + a_{n-1} \quad\quad = (n-1)(n+2)$

Hence the nth term of the series, a_n, is given by

$$a_n = n(n+3) - (n-1)(n+2) = 2n+2$$

Replacing n by $n-1$ gives the $(n-1)$th term

i.e. $a_{n-1} = 2(n-1) + 2 = 2n$

Then $a_n - a_{n-1} = (2n+2) - 2n = 2$

i.e. there is a common difference of 2 between successive terms, showing that the series is an AP.

EXERCISE 21b

1. Write down the fifth term and the nth term of the following APs.

(a) $\displaystyle\sum_{r=1}^{n} (2r-1)$ (b) $\displaystyle\sum_{r=1}^{n} 4(r-1)$ (c) $\displaystyle\sum_{r=0}^{n} (3r+3)$

(d) first term 5, common difference 3

(e) first term 6, common difference -2

(f) first term p, common difference q

(g) first term 10, last term 30, 11 terms

(h) 1, 5, ...   (i) 2, $1\frac{1}{2}$, ...   (j) $-4, -1, \ldots$

2. Find the sum of the first ten terms of each of the series given in Question (1).

3. The 9th term of an AP is 8 and the 4th term is 20. Find the first term and the common difference.

4. The 6th term of an AP is twice the 3rd term and the first term is 3. Find the common difference and the 10th term.

5. The nth term of an AP is $\frac{1}{2}(3 - n)$. Write down the first three terms and the 20th term.

6. Find the sum, to the number of terms indicated, of each of the following APs.

   (a) $1 + 2\frac{1}{2} + \ldots$, 6 terms   (b) $3 + 5 + \ldots$, 8 terms

   (c) the first twenty odd   (d) $a_1 + a_2 + a_3 + \ldots + a_8$
        integers   where $a_n = 2n + 1$

   (e) $4 + 6 + 8 + \ldots + 20$ (f) $\displaystyle\sum_{r=1}^{3n} (3 - 4r)$

   (g) $S_n = n^2 - 3n$, 8 terms   (h) $S_n = 2n(n + 3)$, m terms

7. The sum of the first n terms of an AP is S_n where $S_n = n^2 - 3n$. Write down the fourth term and the nth term.

8. The sum of the first n terms of a series is given by S_n where $S_n = n(3n - 4)$. Show that the terms of the series are in arithmetic progression.

9. In an arithmetic progression, the 8th term is twice the 4th term and the 20th term is 40. Find the common difference and the sum of the terms from the 8th to the 20th inclusive.

10. How many terms of the AP, $1 + 3 + 5 + \ldots$ are required to make a sum of 1521?

11. Find the least number of terms of the AP, $1 + 3 + 5 + \ldots$ that are required to make a sum exceeding 4000.

12. If the sum of the first n terms of a series is S_n where $S_n = 2n^2 - n$,

    (a) prove that the series is an AP, stating the first term and the common difference,

    (b) find the sum of the terms from the 3rd to the 12th inclusive.

13. In an AP the 6th term is half the 4th term and the 3rd term is 15.

    (a) Find the first term and the common difference.

    (b) How many terms are needed to give a sum that is less than 65?

GEOMETRIC PROGRESSIONS

Consider the sequence

$$12, 6, 3, 1.5, 0.75, 0.375, \ldots$$

Each term of this sequence is half the preceeding term so the sequence may be written

$$12, 12(\tfrac{1}{2}), 12(\tfrac{1}{2})^2, 12(\tfrac{1}{2})^3, 12(\tfrac{1}{2})^4, 12(\tfrac{1}{2})^5, \ldots$$

Such a sequence is called a geometric progression (GP) which is a sequence where each term is a constant multiple of the preceding term. This constant multiplying factor is called the common ratio, and it may have any real value.

Hence, if a GP has a first term of 3 and a common ratio of -2 the first four terms are

$$3, 3(-2), 3(-2)^2, 3(-2)^3$$

or

$$3, -6, 12, -24$$

In general if a GP has a first term a, and a common ratio r, the first four terms are

$$a, ar, ar^2, ar^3$$

and the nth term, u_n, is ar^{n-1}, thus

> a GP with n terms can be written $a, ar, ar^2, \ldots, ar^{n-1}$

The Sum of a Geometric Progression

Consider the sum of the first eight terms, S_8, of the GP with first term 1 and common ratio 3

i.e.
$$S_8 = 1 + 1(3) + 1(3)^2 + 1(3)^3 + \ldots + 1(3)^7$$

$$\Rightarrow \quad 3S_8 = \qquad 3 + 3^2 + 3^3 + \ldots + 3^7 + 3^8$$

Hence $\quad S_8 - 3S_8 = 1 + 0 + 0 + 0 + \ldots + 0 - 3^8$

So $\quad S_8(1 - 3) = 1 - 3^8$

$$\Rightarrow \qquad S_8 = \frac{1 - 3^8}{1 - 3} = \frac{3^8 - 1}{2}$$

This process can be applied to a general GP.

Consider the sum, S_n, of the first n terms of a GP with first term a and common ratio r,

i.e. $S_n = a + ar + \ldots + ar^{n-2} + ar^{n-1}$

Multiplying by r gives

$$rS_n = \quad ar + ar^2 + \ldots \quad + ar^{n-1} + ar^n$$

Hence $S_n - rS_n = a - ar^n$

$\Rightarrow$ $S_n(1 - r) = a(1 - r^n)$

$\Rightarrow$ $$S_n = \frac{a(1 - r^n)}{1 - r}$$

If $r > 1$ the formula may be written $\dfrac{a(r^n - 1)}{r - 1}$

Examples 21c

1. The 5th term of a GP is 8, the third term is 4, and the sum of the first ten terms is positive. Find the first term, the common ratio, and the sum of the first ten terms.

For a first term a and common ratio r, the nth term is ar^{n-1}

Thus we have $ar^4 = 8$ $(n = 5)$

and $ar^2 = 4$ $(n = 3)$

dividing gives $r^2 = 2$

$\Rightarrow$ $r = \pm\sqrt{2}$ and $a = 2$

Using the formula $S_n = \dfrac{a(r^n - 1)}{r - 1}$ gives,

when $r = \sqrt{2}$, $S_{10} = \dfrac{2[(\sqrt{2})^{10} - 1]}{\sqrt{2} - 1} = \dfrac{62}{\sqrt{2} - 1}$

when $r = -\sqrt{2}$, $S_{10} = \dfrac{2[(-\sqrt{2})^{10} - 1]}{-\sqrt{2} - 1} = \dfrac{-62}{\sqrt{2} + 1}$

But we are told that $S_{10} > 0$, so we deduce that

$r = \sqrt{2}$ and $S_{10} = \dfrac{62}{\sqrt{2} - 1} = 62(\sqrt{2} + 1)$

2. A "Save As You Earn" scheme involves paying in £50 on the first day of each month. Interest of 1% of the total in the scheme is added at the end of each month. If the first payment is made on January 1st, find the total amount in the scheme on December 31st of the same year.

If P_1 is the total amount at the end of January, P_2 is the total amount at the end of February, and so on,

then $P_1 = £\{50 + 50(0.01)\} = £(50)(1.01)$

$$P_2 = £(50 + P_1)(1.01) = £\{(50)(1.01) + (50)(1.01)^2\}$$

$$P_3 = £(50 + P_2)(1.01) = £\{(50)(1.01) + (50)(1.01)^2 + (50)(1.01)^3\}$$

and so on, giving

$$P_{12} = £\{(50)(1.01) + (50)(1.01)^2 + \ldots + (50)(1.01)^{12}\}$$

$$= £(50)(1.01)[1 + (1.01) + (1.01)^2 + \ldots + (1.01)^{11}]$$

The expression in square brackets is a GP of 12 terms with $a = 1$ and $r = 1.01$, hence

$$P_{12} = £(50)(1.01)\left[\frac{(1.01)^{12} - 1}{1.01 - 1}\right]$$

$$= £640.47 \quad \text{to the nearest penny.}$$

3. The sum of the first n terms of a series is 3^{n-1}. Show that the terms of this series are in geometric progression and find the first term, the common ratio and the sum of the second n terms of this series.

If the series is $\qquad a_1 + a_2 + \ldots + a_n$

then $\qquad S_n = a_1 + a_2 + \ldots + a_{n-1} + a_n = 3^n - 1$

and $\qquad S_{n-1} = a_1 + a_2 + \ldots + a_{n-1} = 3^{n-1} - 1$

therefore $\qquad a_n = 3^n - 1 - (3^{n-1} - 1)$

i.e. the nth term is $3^n - 3^{n-1} = 3^{n-1}(3 - 1) = (2)3^{n-1}$

Similarly $\qquad a_{n-1} = (2)3^{n-2}$ so $a_n \div a_{n-1} = 3$

showing that successive terms in the series have a constant ratio of 3 Hence this series is a GP with first term 2 and common ratio 3

The sum of the second n terms is

(the sum of the first $2n$ terms) $-$ (the sum of the first n terms)

$$= S_{2n} - S_n$$
$$= (3^{2n} - 1) - (3^n - 1)$$
$$= 3^n(3^n - 1)$$

EXERCISE 21c

1. Write down the fifth term and the nth term of the following GPs:
   (a) 2, 4, 8, ...   (b) 2, 1, $\frac{1}{2}$, ...   (c) 3, -6, 12, ...
   (d) first term 8, common ratio $-\frac{1}{2}$
   (e) first term 3, last term $\frac{1}{81}$, 6 terms

2. Find the sum, to the number of terms given, of the following GPs.
   (a) $3 + 6 + ...$, 6 terms   (b) $3 - 6 + ...$, 8 terms
   (c) $1 + \frac{1}{2} + \frac{1}{4} + ...$, 20 terms
   (d) first term 5, common ratio $\frac{1}{5}$, 5 terms
   (e) first term $\frac{1}{2}$, common ratio $-\frac{1}{2}$, 10 terms
   (f) first term 1, common ratio -1, 2001 terms.

3. The 6th term of a GP is 16 and the 3rd term is 2. Find the first term and the common ratio.

4. Find the common ratio, given that it is negative, of a GP whose first term is 8 and whose 5th term is $\frac{1}{2}$

5. The nth term of a GP is $(-\frac{1}{2})^n$. Write down the first term and the 10th term.

6. Evaluate $\displaystyle\sum_{r=1}^{10} (1.05)^r$

7. Find the sum to n terms of the following series.

   (a) $x + x^2 + x^3 + ...$ (b) $x + 1 + \dfrac{1}{x} + ...$ (c) $1 - y + y^2 - ...$

   (d) $x + \dfrac{x^2}{2} + \dfrac{x^3}{4} + \dfrac{x^4}{8} + ...$ (e) $1 - 2x + 4x^2 - 8x^3 + ...$

8. Find the sum of the first n terms of the GP $2 + \frac{1}{2} + \frac{1}{8} + ...$ and find the least value of n for which this sum exceeds 2.65

9. The sum of the first 3 terms of a GP is 14. If the first term is 2, find the possible values of the sum of the first 5 terms.

10. Evaluate $\displaystyle\sum_{r=1}^{10} 3(3/4)^r$

11. A mortgage is taken out for £10 000 and is repaid by annual instalments of £2000. Interest is charged on the outstanding debt at 10%, calculated annually. If the first repayment is made one year after the mortgage is taken out find the number of years it takes for the mortgage to be repaid.

12. A bank loan of £500 is arranged to be repaid in two years by equal monthly instalments. Interest, *calculated monthly*, is charged at 11% p.a. on the remaining debt. Calculate the monthly repayment if the first repayment is to be made one month after the loan is granted.

CONVERGENCE OF SERIES

If a piece of string, of length l, is cut up by first cutting it in half and keeping one piece, then cutting the remainder in half and keeping one piece, then cutting the remainder in half and keeping one piece, and so on, the sum of the lengths retained is

$$\frac{l}{2} + \frac{l}{4} + \frac{l}{8} + \frac{l}{16} + \ldots$$

As this process can (in theory) be carried on indefinitely, the series formed above is infinite.

After several cuts have been made the remaining part of the string will be very small indeed, so the sum of the cut lengths will be very nearly equal to the total length, l, of the original piece of string. The more cuts that are made the closer to l this sum becomes, i.e. if after n cuts, the sum of the cut lengths is

$$\frac{l}{2} + \frac{l}{2^2} + \frac{l}{2^3} + \ldots + \frac{l}{2^n}$$

then, as $n \to \infty$, $\qquad \dfrac{l}{2} + \dfrac{l}{2^2} + \ldots + \dfrac{l}{2^n} \to l$

or $\qquad \displaystyle\lim_{n \to \infty} \left[\frac{l}{2} + \frac{l}{2^2} + \ldots + \frac{l}{2^n} \right] = l$

l is called the sum to infinity of this series.

In general, if S_n is the sum of the first n terms of any series and if $\lim_{n \to \infty} [S_n]$ exists and is finite, the series is said to be *convergent*.

In this case the sum to infinity, S_∞, is given by

$$S_\infty = \lim_{n \to \infty} [S_n]$$

The series $1/2 + 1/2^2 + 1/2^3 + \ldots$, for example, is convergent as its sum to infinity is 1.

However, for the series $1 + 2 + 3 + \ldots + n$, we have $S_n = \frac{1}{2}n(n + 1)$ As $n \to \infty$, $S_n \to \infty$ so this series does not converge and is said to be divergent.

For any AP, $S_n = \frac{1}{2}n[2a + (n - 1)d]$, which always approaches infinity as $n \to \infty$. Therefore any AP is divergent.

THE SUM TO INFINITY OF A GP

Consider the general GP $a + ar + ar^2 + \ldots$

Now
$$S_n = \frac{a(1 - r^n)}{1 - r}$$

and if $|r| < 1$, then $\lim_{n \to \infty} r^n = 0$

So
$$\lim_{n \to \infty} S_n = \lim_{n \to \infty} \left[\frac{a(1 - r^n)}{1 - r} \right] = \frac{a}{1 - r}$$

If $|r| > 1$, $\lim_{n \to \infty} r^n = \infty$ and the series does not converge.

Therefore, provided that $|r| < 1$, a GP converges to a sum of $\dfrac{a}{1 - r}$

i.e.

for a GP $\quad S_\infty = \dfrac{a}{1 - r}$

provided that $|r| < 1$

Arithmetic Mean

If three numbers, p_1, p_2, p_3, are in arithmetic progression then p_2 is called the *arithmetic mean* of p_1 and p_3

If $p_1 = a$, we may write p_2, p_3 as $a + d$, $a + 2d$ respectively,

hence $p_1 + p_3 = 2a + 2d = 2(a + d) = 2p_2$

∴ $p_2 = \frac{1}{2}(p_1 + p_3)$

i.e.   the arithmetic mean of two numbers m and n is $\frac{1}{2}(m + n)$

Geometric Mean

If p_1, p_2, p_3 are in geometric progression, p_2 is called the *geometric mean* of p_1 and p_3

If $p_1 = a$, then we may write $p_2 = ar$, $p_3 = ar^2$

thus $p_1 p_3 = a^2 r^2 = p_2^2$ $\Rightarrow$ $p_2 = \sqrt{(p_1 p_3)}$

i.e.   the geometric mean of two numbers m and n is $\sqrt{(mn)}$

Examples 21d

1. Determine whether each series converges. If it does, give its sum to infinity.
   (a) $3 + 5 + 7 + \ldots$ (b) $1 - \frac{1}{4} + \frac{1}{16} - \frac{1}{64} + \ldots$ (c) $3 + \frac{9}{2} + \frac{27}{4} + \ldots$

(a) $3 + 5 + 7 + \ldots$ is an AP $(d = 2)$ and so does not converge.

(b) $1 - \frac{1}{4} + \frac{1}{16} - \frac{1}{64} + \ldots = 1 + (-\frac{1}{4}) + (-\frac{1}{4})^2 + (-\frac{1}{4})^3 + \ldots$
which is a GP where $r = -\frac{1}{4}$, i.e. $|r| < 1$

So this series converges and $S_\infty = \dfrac{a}{1 - r} = \dfrac{1}{1 - (-\frac{1}{4})} = \dfrac{4}{5}$

(c) $3 + \frac{9}{2} + \frac{27}{4} + \ldots = 3 + 3(\frac{3}{2}) + 3(\frac{9}{4}) + \ldots = 3 + 3(\frac{3}{2}) + 3(\frac{3}{2})^2 + \ldots$
This series is a GP where $r = \frac{3}{2}$ and, as $|r| > 1$, the series does not converge.

2. Express the recurring decimal $0.1\dot{5}7\dot{6}$ as a fraction in its lowest terms.

$$0.1\dot{5}7\dot{6} = 0.1\widetilde{57}\,\widetilde{657}\,\widetilde{657}\,\widetilde{657}\,6\ldots$$

$$= 0.1 + 0.0576 + 0.000\,0576 + 0.000\,000\,0576 + \ldots$$

$$= \frac{1}{10} + \frac{576}{10^4} + \frac{576}{10^7} + \frac{576}{10^{10}} + \ldots$$

$$= \frac{1}{10} + \frac{576}{10^4}\left[1 + \frac{1}{10^3} + \frac{1}{10^6} + \ldots\right]$$

$$= \frac{1}{10} + \frac{576}{10^4}\left[1 + \frac{1}{10^3} + \left(\frac{1}{10^3}\right)^2 + \ldots\right]$$

Now the series in the square bracket is a GP whose first term is 1, and whose common ratio is $\dfrac{1}{10^3}$.

Hence it has a sum to infinity of $\dfrac{1}{1 - 10^{-3}} = \dfrac{10^3}{999}$

$$\Rightarrow \quad 0.1\dot{5}7\dot{6} = \frac{1}{10} + \frac{576}{10^4} \times \frac{10^3}{999} = \frac{1}{10} + \frac{576}{9990} = \frac{1575}{9990} = \frac{35}{222}$$

---

3. The 3rd term of a convergent GP is the arithmetic mean of the 1st and 2nd terms.
Find the common ratio and, if the first term is 1, find, the sum to infinity.

If the series is $a + ar + ar^2 + ar^3 + \ldots$

then $\qquad\qquad\qquad\qquad\qquad ar^2 = \tfrac{1}{2}(a + ar)$

$a \neq 0$, so $\qquad\qquad\qquad\qquad 2r^2 - r - 1 = 0$

$\Rightarrow \qquad\qquad\qquad\qquad\qquad (2r + 1)(r - 1) = 0$

i.e. $\qquad\qquad\qquad\qquad\qquad\qquad r = -\tfrac{1}{2}$ or 1

As the series is convergent, the common ratio is $-\tfrac{1}{2}$

When $r = -\tfrac{1}{2}$ and $a = 1$,

$$S_\infty = \frac{1}{1 + \tfrac{1}{2}} = \tfrac{2}{3}$$

EXERCISE 21d

1. Determine whether each of the series given below converge.

   (a) $4 + \dfrac{4}{3} + \dfrac{4}{3^2} + \ldots$ (b) $9 + 7 + 5 + 3 + \ldots$

   (c) $20 - 10 + 5 - 2.5 + \ldots$ (d) $\dfrac{5}{10} + \dfrac{5}{100} + \dfrac{5}{1000} + \ldots$

   (e) $p + 2p + 3p + \ldots$ (f) $3 - 1 + \dfrac{1}{3} - \dfrac{1}{9} + \ldots$

2. Find the sum to infinity of those series in Question 1 that are convergent.

3. Express the following recurring decimals as fractions
   (a) $0.16\dot{2}$ (b) $0.\dot{3}\dot{4}$ (c) $0.0\dot{2}\dot{1}$

4. The sum to infinity of a GP is twice the first term. Find the common ratio.

5. The sum to infinity of a GP is 16 and the sum of the first 4 terms is 15. Find the first four terms.

6. If a, b and c are the first three terms of a GP, prove that $\sqrt{a}$, $\sqrt{b}$ and $\sqrt{c}$ form another GP.

MIXED EXERCISE 21

In Questions 1 to 9 find the sum of each series.

1. $1 - \frac{1}{2} + \frac{1}{4} - \frac{1}{8} + \ldots$

2. $2 - (2)(3) + (2)(3)^2 - (2)(3)^3 + \ldots + (2)(3)^{10}$

3. $\displaystyle\sum_{r=2}^{n} ab^{2r}$ 4. $\displaystyle\sum_{r=5}^{n} 4r$

5. $e + e^2 + e^3 + \ldots + e^n$ 6. $\displaystyle\sum_{r=1}^{n} (2 + 3r)$

7. $\displaystyle\sum_{r=n}^{2n} (1 - 2r)$ 8. $\displaystyle\sum_{r=1}^{\infty} \dfrac{1}{2^r}$

9. The sum of the first n even numbers.

In Questions 10 and 11, express each decimal as a fraction in its lowest terms.

10. $0.0\dot{5}\dot{1}$ **11.** $0.\dot{1}\dot{0}$

12. The sum of the first n terms of a series is n^3. Write down the first four terms and the nth term of the series.

13. The fourth term of an AP is 8 and the sum of the first ten terms is 40. Find the first term and the tenth term.

14. The second, fourth and eighth terms of an AP are the first three terms of a GP. Find the common ratio of the GP.

15. Find the value of x for which the numbers $x + 1$, $x + 3$, $x + 7$, are in geometric progression.

16. The second term of a GP is $\frac{1}{2}$ and the sum to infinity of the series is 4. Find the first term and the common ratio of the series.

17. £2000 is invested in a pension fund on January 1st each year. Interest at 9% p.a. is added to the fund on December 31st each year. Calculate the amount in the fund on December 31st 2020 if the first payment was made on January 1st 1980.

18. Jane Smith takes out an endowment policy which involves making a fixed payment each year for 10 years. At the end of the 10 years Jane receives a sum of money equal to her total payments together with interest which is added at the rate of 8% p.a. of the total sum in the fund. Jane will get a payout of £100 000. What is her annual payment?

CHAPTER 22

THE BINOMIAL EXPANSION

POWER SERIES

A series such as $x + x^2 + x^3 + \ldots$ is called a power series because the terms involve powers of a variable quantity. Series, such as those considered in Chapter 21, each of whose terms has a fixed numerical value, are called number series.

THE BINOMIAL THEOREM

We saw in Chapter 1 that when an expression such as $(1 + x)^4$ is expanded, the coefficients of the terms in the expansion can be obtained from Pascal's Triangle. Now $(1 + x)^{20}$ could be expanded in the same way but, as the construction of the triangular array would be tedious we need a more general method to expand powers of $(1 + x)$.

This general method uses the *binomial theorem* which states that

when n is a positive integer

$$(1 + x)^n = 1 + nx + \frac{n(n-1)}{2}x^2 + \frac{n(n-1)(n-2)}{(2)(3)}x^3 + \frac{n(n-1)(n-2)(n-3)}{(2)(3)(4)}x^4 + \ldots + x^n$$

The coefficients of the powers of x are called *binomial coefficients*. For powers of x greater than 3, the coefficients are lengthy to write out in full so we use the notation $\binom{n}{r}$ to represent the coefficient of x^r.

The pattern shown in the coefficients given above shows that

$$\binom{n}{r} = \frac{n(n-1)(n-2)\ldots(n-r+1)}{(2)(3)(4)\ldots(r)}$$

Notice that the expansion of $(1 + x)^n$ is a finite series which has $(n + 1)$ terms and that the term in x^2 is the third term, the term in x^3 is the fourth term and the *term in x^r is the $(r + 1)$th term*.

344

Using the Σ notation, the binomial theorem may be written as

$$(1+x)^n = \sum_{r=0}^{n} \binom{n}{r} x^r \quad \text{for} \quad n \in \mathbf{Z}^+$$

For those readers who are interested, a justification of the binomial theorem follows.

From our knowledge of multiplying brackets, we know that $(1+x)^n$ can be expanded as a series of ascending powers of x, starting with 1 and ending with x^n,

i.e. $\qquad (1+x)^n \equiv 1 + a_1 x + a_2 x^2 + a_3 x^3 + \ldots + a_r x^r + \ldots + x^n \qquad$ [1]

where $a_1, a_2, \ldots$ are constants.

Differentiating both sides of this identity w.r.t. x gives

$$n(1+x)^{n-1} \equiv a_1 + 2a_2 x + 3a_3 x^2 + \ldots + ra_r x^{r-1} + \ldots + nx^{n-1} \qquad [2]$$

As [2] is also an identity it is true for all values of x, so when $x = 0$ we have

$$n(1)^{n-1} = a_1 \quad \Rightarrow \quad a_1 = n$$

Differentiating both sides of [2] w.r.t. x gives

$$n(n-1)(1+x)^{n-2} \equiv 2a_2 + (2)(3)a_3 x + \ldots + r(r-1)a_r x^{r-2}$$

$$+ \ldots + n(n-1)x^{n-2} \qquad [3]$$

When $x = 0$, [3] gives

$$n(n-1)(1)^{n-2} = 2a_2 \quad \Rightarrow \quad a_2 = \frac{n(n-1)}{2}$$

Differentiating both sides of [3] w.r.t. x and then using $x = 0$ gives

$$n(n-1)(n-2)(1)^{n-3} = (2)(3)a_3 \quad \Rightarrow \quad a_3 = \frac{n(n-1)(n-2)}{(2)(3)}$$

This process can be repeated to give all the coefficients but as the pattern is now clear we can deduce that

$$a_r = \frac{n(n-1)(n-2)\ldots(n-r+1)}{(2)(3)(4)\ldots(r)}$$

where a_r is the coefficient of x^r.

Examples 22a

1. Write down the first three terms in the expansion in ascending powers of x of

   (a) $\left(1 - \dfrac{x}{2}\right)^{10}$ (b) $(3 - 6x)^8$

   (a) Using the binomial theorem and replacing x by $-\dfrac{x}{2}$ and n by 10 we have

   $$\left(1 - \frac{x}{2}\right)^{10} = 1 + (10)\left(-\frac{x}{2}\right) + \frac{10 \times 9}{2}\left(-\frac{x}{2}\right)^2 + \ldots$$

   $$= 1 - 5x + \tfrac{45}{4}x^2 + \ldots$$

   (b) Writing $(3 - 6x)^8$ as $3^8(1 - 2x)^8$ we can use the binomial theorem replacing x by $-2x$ and n by 8

   $$(3 - 6x)^8 = 3^8(1 - 2x)^8 = 3^8\left[1 + (8)(-2x) + \frac{(8)(7)}{2}(-2x)^2 + \ldots\right]$$

   Therefore the first three terms of the expansion are

   $$3^8 + (3^8)(8)(-2x) + \frac{(3^8)(8)(7)}{2}(4x^2)$$

   i.e. $$3^8 - (3^8)(16)x + (3^8)(112)x^2$$

2. Expand $(2x - 1)^5$

   Writing $(2x - 1)^5$ as $(-1)^5(1 - 2x)^5$ we can use the binomial theorem replacing x by $-2x$ and n by 5

   $$(2x - 1)^5 = (-1)^5(1 - 2x)^5$$

   $$= (-1)\left[1 + (5)(-2x) + \frac{(5)(4)}{2}(-2x)^2 + \frac{(5)(4)(3)}{(2)(3)}(-2x)^3\right.$$

   $$\left. + \frac{(5)(4)(3)(2)}{(2)(3)(4)}(-2x)^4 + (-2x)^5\right]$$

   $$= -1 + 10x - 40x^2 + 160x^3 - 80x^4 + 32x^5$$

3. Write down the first three terms in the binomial expansion of

$$(1 - 2x)(1 + \tfrac{1}{2}x)^{10}$$

The third term in the binomial expansion is the term containing x^2, so start by expanding $(1 + \tfrac{1}{2}x)^{10}$ as far as the term in x^2

$$(1 + \tfrac{1}{2}x)^{10} = 1 + (10)(\tfrac{1}{2}x) + \frac{(10)(9)}{2}(\tfrac{1}{2}x)^2 + \ldots$$

$$= 1 + 5x + \tfrac{45}{4}x^2 + \ldots$$

$$\therefore \quad (1 - 2x)(1 + \tfrac{1}{2}x)^{10} = (1 - 2x)(1 + 5x + \tfrac{45}{4}x^2 + \ldots)$$

$$= 1 + 5x + \tfrac{45}{4}x^2 + \ldots - 2x - 10x^2 + \ldots$$

$$= 1 + 3x + \tfrac{5}{4}x^2 + \ldots$$

Notice that we do not write down the product of $-2x$ and $\tfrac{45}{4}x^2$, as terms in x^3 are not required.

---

4. Find the sixth term in the expansion of $(a + b)^{20}$ as a series of ascending powers of b.

$$(a + b)^{20} = a^{20}\left(1 + \frac{b}{a}\right)^{20}$$

The sixth term in the binomial expansion of $(1 + x)^n$ is the term in x^5, i.e. $\dbinom{n}{6}x^5$.
Replacing x by $\frac{b}{a}$ and n by 20 gives

the sixth term in the expansion of $(a + b)^{20}$ is

$$(a)^{20}\frac{(20)(19)(18)(17)(16)}{(2)(3)(4)(5)}\left(\frac{b}{a}\right)^5 = 31\,008a^{15}b^5$$

---

5. If the first two terms in the expansion of $(2 - ax)^6$ are $b + 12x$, find the values of a and b.

$$(2 - ax)^6 = 2^6\left(1 - \frac{a}{2}x\right)^6$$

$$= 2^6\left[1 + (6)\left(\frac{a}{2}x\right) + \ldots\right]$$

$\therefore$
$$2^6 + (2^6)(6)\left(\frac{a}{2}x\right) \equiv b + 12x$$

$\Rightarrow$
$$b = 2^6 \quad \text{and} \quad (2^6)(6)\left(\frac{a}{2}\right) = 12$$

i.e.
$$b = 64 \quad \text{and} \quad a = \tfrac{1}{16}$$

EXERCISE 22a

1. Write down the first four terms in the binomial expansion of
   (a) $(1 + 3x)^{12}$ (b) $(1 - 2x)^9$ (c) $(1 + 5x)^7$
   (d) $\left(1 - \frac{x}{3}\right)^{20}$ (e) $(1 - \frac{2}{3}x)^6$ (f) $(1 + \frac{3}{5}x)^{20}$

2. Write down the first three terms in the binomial expansion of
   (a) $(2 + x)^{10}$ (b) $\left(2 - \frac{3}{2}x\right)^7$
   (c) $\left(\frac{3}{2} + 2x\right)^9$

3. Write down the term indicated in the binomial expansion of each of the following functions.
   (a) $(1 - 4x)^7$, 3rd term (b) $\left(1 - \frac{x}{2}\right)^{20}$, 2nd term
   (c) $(1 + 2x)^{12}$, 4th term (d) $(1 - \frac{1}{2}x)^9$, 3rd term
   (e) $(2 - x)^{15}$, 4th term (f) $(1 - 2x)^{12}$, the term in x^4
   (g) $\left(2 + \frac{x}{2}\right)^9$, the term in x^5 (h) $(p - 2q)^{10}$, 5th term
   (i) $(a + b)^8$, the term in a^3 (j) $(a + 2b)^8$, 2nd term

4. Write down the binomial expansion of each function as a series of ascending powers of x as far as, and including, the term in x^2.

   (a) $(1 + x)(1 - x)^9$

   (b) $(1 - x)(1 + 2x)^{10}$

   (c) $(2 + x)\left(1 - \dfrac{x}{2}\right)^{20}$

   (d) $(1 + x)^2(1 - 5x)^{14}$

5. Write down the full expansion of each of the following expressions.

   (a) $(2x - 1)^6$ (b) $(3x - 2)^5$ (c) $(x - 4)^9$

6. Expand $(1 - 3x)^2(1 + 2x)^5$ as a series of ascending powers of x as far as and including the term in x^3

7. Factorise $5 - 4x - x^2$. Hence expand $(5 - 4x - x^2)^7$ as a series of ascending powers of x as far as and including the term in x^2.

8. Use the binomial theorem to find the first three terms in the expansion of $(1 + 2x - x^2)^{20}$ as a series of ascending powers of x by first writing $(1 + 2x - x^2)$ in the form $(1 + X)$ where $X = f(x)$ and initially expanding $(1 + X)^{20}$.

USING SERIES TO FIND APPROXIMATIONS

17

Consider $(1 + x)^{20}$ and its binomial expansion,

$$(1 + x)^{20} = 1 + 20x + \frac{(20)(19)}{2}x^2 + \frac{(20)(19)(18)}{(2)(3)}x^3 + \ldots + x^{20}$$

This is valid for all values of x so if, for example, $x = 0.01$ we have

$$(1.01)^{20} = 1 + 20(0.01) + \frac{(20)(19)}{2}(0.01)^2 + \frac{(20)(19)(18)}{(2)(3)}(0.01)^3 + \ldots + (0.01)^{20}$$

i.e. $(1.01)^{20} = 1 + 0.2 + 0.019 + 0.001\,14 + 0.000\,048\,45 + \ldots + 10^{-40}$

Because the value of x (i.e. 0.01) is small, we see that adding successive terms of the series makes progressively smaller contributions to the accuracy of $(1.01)^{20}$.

In fact, taking only the first four terms gives $(1.01)^{20} \approx 1.22014$

This approximation is correct to three decimal places as the fifth and succeeding terms do not add anything to the first four decimal places.

In general, if x is small so that successive powers of x quickly become negligible in value, then the sum of the first few terms in the expansion of $(1 + x)^n$ gives an approximate value for $(1 + x)^n$

The number of terms required to obtain a good approximation depends on two considerations

1) the value of x (the smaller x is, the fewer are the terms needed to obtain a good approximation).

2) the accuracy required (an answer correct to 3 s.f. needs fewer terms than an answer correct to 6 s.f.)

When finding an approximation, the binomial expansion of $(1 + x)^n$ should be used, e.g. to find the approximate value of $(3.006)^5$ we use $3^5(1 + 0.002)^5$

Examples 22b

1. By substituting 0.001 for x in the expansion of $(1 - x)^7$ find the value of $(1.998)^7$ correct to five significant figures.

Now $(1.998)^7 = (2 - 0.002)^7 = 2^7(1 - 0.001)^7$

$$= 2^7(1 - x)^7 \quad \text{when} \quad x = 0.001$$

Hence

$$(1.998)^7 = 2^7\left[1 - 7(0.001) + \frac{(7)(6)}{2}(0.001)^2 - \frac{(7)(6)(5)}{(2)(3)}(0.001)^3 + \ldots\right]$$

To give an answer correct to 5 s.f. we will work to 7 s.f. so only the first three terms need be considered

$$\therefore \quad (1.998)^7 = 128(1 - 0.007 + 0.000\ 021\ 0) \quad \text{to 7 s.f.}$$

$$= 127.11 \quad \text{correct to 5 s.f.}$$

In the example above, a calculator will give the value of $(1.998)^7$ to about 8 s.f. (depending on the particular calculator). If, however, the value is required to, say, 15 s.f., the method used in the worked example will give the extra accuracy.

The next worked example illustrates how a series expansion enables us to find a simple function which can be used as an approximation to a given function when x has values that are close to zero.

---

2. If x is so small that x^2 and higher powers can be neglected show that

$$(1 - x)^5 \left(2 + \frac{x}{2}\right)^{10} \approx 2^9 (2 - 5x)$$

Using the binomial expansion of $(1 - x)^5$ and neglecting terms containing x^2 and higher powers of x we have

$$(1 - x)^5 \approx 1 - 5x$$

Similarly
$$\left(2 + \frac{x}{2}\right)^{10} \equiv 2^{10} \left(1 + \frac{x}{4}\right)^{10}$$

$$\approx 2^{10} \left[1 + 10\left(\frac{x}{4}\right)\right]$$

Therefore
$$(1 - x)^5 \left(2 + \frac{x}{2}\right)^{10} \approx 2^{10} (1 - 5x)\left(1 + \frac{5x}{2}\right)$$

$$= 2^9 (1 - 5x)(2 + 5x)$$

$$\approx 2^9 (2 - 5x)$$

again neglecting the term in x^2

The graphical significance of the approximation in the last example is interesting.

If
$$y = (1 - x)^5 \left(2 + \frac{x}{2}\right)^{10}$$

then, for values of x close to zero, $y \approx 2^9 (2 - 5x)$ which is the equation of a straight line,

i.e. $y = 2^9 (2 - 5x)$ is the tangent to $y = (1 - x)^5 \left(2 + \frac{x}{2}\right)^{10}$ at the point where $x = 0$

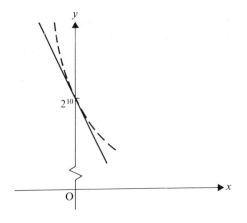

Note that the function $2^9(2 - 5x)$ is called a *linear approximation* for the function $(1 - x)^5\left(2 + \dfrac{x}{2}\right)^{10}$ in the region where $x \approx 0$

Examples 22b (continued)

3. The shape of the curve $y = (2x - 1)^9$ near the origin can be found by using $y \approx ax^2 + bx + c$. Find the values of a, b and c and hence find, correct to 2 s.f., the error involved in taking y as $ax^2 + bx + c$ when (a) $x = 0.01$ (b) $x = 0.1$

Using the binomial theorem gives

$$(2x - 1)^9 = (-1)^9(1 - 2x)^9 = (-1)(1 - 2x)^9$$

$$= -\left[1 + 9(-2x) + \frac{9 \times 8}{2}(-2x)^2 + \ldots\right]$$

$$= -1 + 18x - 144x^2 + \ldots$$

Near the origin, x is small so ignoring powers of x greater than 2 gives

$$(2x - 1)^9 \approx -144x^2 + 18x - 1$$

and

$$(2x - 1)^9 \approx ax^2 + bx + c$$

i.e. $a = -144$, $b = 18$ and $c = -1$

(a) When $x = 0.01$, $\quad y = (2x - 1)^9 = -0.833\,74\ldots$

$$y \approx -144x^2 + 18x - 1 = -0.8344$$

The error involved is $\quad -6.5 \times 10^{-4}$

(b) When $x = 0.1$, $\quad y = (2x - 1)^9 = -0.134\,21\ldots$

$$y \approx -144x^2 + 18x - 1 = -0.64$$

The error involved is $\quad -0.51$

This error is huge compared to the value of y, so we deduce that the approximation is reasonable only very close to the origin.

EXERCISE 22b

1. By substituting 0.01 for x in the binomial expansion of $(1 - 2x)^{10}$, find the value of $(0.98)^{10}$ correct to four decimal places.

2. By substituting 0.05 for x in the binomial expansion of $\left(1 + \dfrac{x}{5}\right)^6$, find the value of $(1.01)^6$ correct to four significant figures.

3. By using the binomial expansion of $(2 + x)^7$, show that, correct to 3 d.p., $(2.08)^7 = 168.439$

4. Show that, if x is small enough for x^2 and higher powers of x to be neglected, the function $(x - 2)(1 + 3x)^8$ has a linear approximation of $-2 - 47x$

5. If x is so small that x^3 and higher powers of x are negligible, show that $(2x + 3)(1 - 2x)^{10} \approx 3 - 58x + 500x^2$

6. By neglecting x^2 and higher powers of x find linear approximations for the following functions in the immediate neighbourhood of $x = 0$
   (a) $(1 - 5x)^{10}$ (b) $(2 - x)^8$ (c) $(1 + x)(1 - x)^{20}$

7. For values of x near the origin the curve $y = ax^2 + bx + c$ can be used as an approximation for the curve $(1 - 2x)^2(1 + x)^{20}$. Find the values of a, b and c.

8. Find the equation of the tangent to the curve $y = (1 - \frac{1}{2}x)^8$ at the point where the curve crosses the y-axis by

(a) differentiation

(b) using the binomial theorem to find a linear function that approximates to $(1 - \frac{1}{2}x)^8$ when x is small.

9. For small values of x, the value of y on the curve $y = (2x + 3)(1 - x)^{10}$ can be found approximately by using $y \approx ax^2 + bx + c$.

(a) Find the values of a, b and c.

(b) Find the error involved in using $y \approx ax^2 + bx + c$ when $x = 0.05$

(c) By trying different values of x, find correct to 2 s.f. a range of values of x for which the error in using the quadratic approximation is less than 5%.

CONSOLIDATION E

SUMMARY

INTEGRATION

Integration reverses differentiation.

$$\int x^n \, dx = \left(\frac{1}{n+1}\right) x^{n+1} + K$$

$$\int ax^n \, dx = \left(\frac{a}{n+1}\right) x^{n+1} + K$$

$$\int (a + bx)^n \, dx = \left(\frac{1}{b(n+1)}\right)(a + bx)^{n+1} + K$$

Integration can be performed across a sum or difference of functions, i.e.

$$\int [f(x) + g(x)] \, dx = \int f(x) \, dx + \int g(x) \, dx$$

Other combinations of functions can be integrated if they are expressed as a sum or difference of functions.

Integration as a Process of Summation

$$\lim_{\delta x \to 0} \sum_{x=a}^{x=b} f(x) \, \delta x = \int_a^b f(x) \, dx$$

where $\int_a^b f(x) \, dx = g(b) - g(a)$, such that $g(x) = \int f(x) \, dx$

PRACTICAL APPLICATIONS OF CALCULUS

Area

The area bounded by the x-axis, the two lines $x = a$ and $x = b$, and part of the curve $y = f(x)$ can be found by summing the areas of vertical strips of width δx and using

(i)

$$\text{Area} = \lim_{\delta x \to 0} \sum_{x = a}^{x = b} y \, \delta x = \int_a^b y \, dx$$

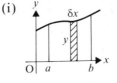

Similarly, for horizontal strips.

(ii)

$$\text{Area} = \lim_{\delta y \to 0} \sum_{y = a}^{y = b} x \, \delta y = \int_a^b x \, dy$$

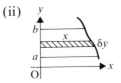

KINEMATICS

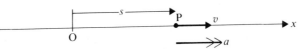

If a particle P is moving in a straight line and O is a fixed point on that line, then the *displacement*, s, of P from O is the directed distance of P from O,

the *velocity*, v, of P is the rate of increase of s w.r.t. time,

the acceleration, a, of P is the rate of increase of v w.r.t. time.

i.e.
$$v = \frac{ds}{dt} \quad \text{and} \quad a = \frac{dv}{dt}$$

$\Rightarrow$
$$s = \int v \, dt \quad \text{and} \quad v = \int a \, dt$$

NUMBER SERIES

Each term in a number series has a fixed numerical value.

A finite series has a finite number of terms,

e.g. $a_1 + a_2 + a_3 + \ldots + a_{10}$ is a finite series with ten terms.

The sum of the first n terms of a series is denoted by S_n, i.e.
$$S_n = a_1 + a_2 + a_3 + \ldots + a_n$$

An infinite series has no last term.

If, as $n \to \infty$, S_n tends to a finite value, S, then the series converges and S is called its sum to infinity.

If $S_n \to \infty$ as $n \to \infty$ then the series is divergent.

ARITHMETIC PROGRESSIONS

In an arithmetic progression, each term differs from the preceeding term by a constant (called the common difference).

An AP with first term a, common difference d and n terms, is

$$a, \; a + d, \; a + 2d, \; \ldots \; \{a + (n - 1)d\}$$

The sum of the first n terms of an A.P. is given by

$$S_n = \tfrac{1}{2}n(a + l) \qquad \text{where } l \text{ is the last term,}$$
$$= \tfrac{1}{2}n\{2a + (n - 1)d\}$$

GEOMETRIC PROGRESSIONS

In a geometric progression each term is a constant multiple of the preceeding term. This multiple is called the common ratio.

A GP with first term a, common ratio r and n terms is

$$a, \; ar, \; ar^2, \; \ldots \; ar^{n-1}$$

The sum of the first n terms is given by $S_n = \dfrac{a(1 - r^n)}{1 - r}$

The sum to infinity is given by $S = \dfrac{a}{1 - r}$ provided that $|r| < 1$

THE BINOMIAL THEOREM

If n is a positive integer then $(1 + x)^n$ can be expanded as a finite series,

where $\qquad (1 + x)^n = 1 + nx + \dbinom{n}{2}x^2 + \dbinom{n}{3}x^3 + \ldots + x^n$

and where $\qquad \dbinom{n}{r} = \dfrac{n(n - 1)(n - 2)\ldots(n - r + 1)}{(2)(3)(4)\ldots(r)}$

MULTIPLE CHOICE EXERCISE E

TYPE I

1. $\displaystyle\int 2x(x+1)\,dx$ is

   A $x^2(x+1)+c$ **C** $4x+2+c$ **E** 2

   B $x^2(\frac{1}{2}x^2+x)+c$ **D** $\frac{2}{3}x^3+x^2+c$

2. $\displaystyle\int_1^2 x\,dx$ is

   A 1 **B** $1\frac{1}{2}$ **C** 3 **D** $-1\frac{1}{2}$ **E** -1

3. The sum of the series $1+5+9+13+17+21+25+29$ is

   A 30 **B** 240 **C** 120 **D** 112 **E** 28

4.

The shaded area in the diagram is given by

   A $\displaystyle\int_{-3}^4 f(x)\,dx$ **C** $\displaystyle\int_{-3}^1 f(x)\,dx+\int_1^4 f(x)\,dx$

   B $\displaystyle\int_{-3}^0 f(x)\,dx+\int_0^4 f(x)\,dx$ **D** none of these

5. The first three terms of the series $\displaystyle\sum_{r=0}^{\infty}(-1)^{r+1}2^r x^{-r}$ are

   A $-1+\dfrac{2}{x}-\dfrac{4}{x^2}$ **C** $1+2x-4x^2$ **E** none of these

   B $1+\dfrac{2}{x}-\dfrac{4}{x^2}$ **D** $\dfrac{2}{x}-\dfrac{4}{x^2}+\dfrac{8}{x^3}$

6. 3 is the geometric mean of a and b. Possible values of a and b are

   A $5,4$ **B** $0,9$ **C** $3,1$ **D** $4,2$ **E** none of these

7. The series $1 - x + 2x^2 - 3x^3 + 4x^4 + \ldots$ may be written more briefly as

A $\displaystyle\sum_{r=0}^{\infty} (-1)^r rx^r$

D $1 - \displaystyle\sum_{r=0}^{\infty} (-1)^r rx^r$

B $1 + \displaystyle\sum_{r=0}^{\infty} (-1)^r rx^r$

E $\displaystyle\sum_{r=1}^{\infty} (-1)^{r+1} rx^r$

C $\displaystyle\sum_{r=1}^{\infty} rx^r$

8. The coefficient of x^3 in the binomial expansion of $(2 - x)^8$ is

A 1792   C -1792 E -448
B 56   D -2000

9. A particle moves along a straight line Ox such that at time t its displacement from O is given by $s = 3t^2 - 5$. When $t = 0$

A $s = 0$ B $a = 0$ C $s = -2$ D $v = -5$ E $v = 0$

10. The gradient at any point on the curve $y = f(x)$ is given by $\dfrac{dy}{dx} = 2x$. The point $(1, 0)$ is on the curve. The equation of the curve is

A $y = x^2 + 1$ C $y = x^2 - 1$ E $y = 2x^2 - \frac{1}{2}$
B $y = 2$ D $y = x^2$

TYPE 11

11. If $f(x) = (2x - 1)^2$ then

A $\displaystyle\int f(x)dx = \frac{1}{3}(x^2 - x)^3 + c$

B $\displaystyle\int_0^{1/2} f(x)dx$ gives the area enclosed by the curve $y = f(x)$ and the x and y axes

C $f(x)$ is an even function.

12. If $f(x) = 3x^2$ then $\int_0^1 f(x)dx$

    A represents the area under $y = f(x)$ from $x = 0$ to $x = 1$

    B is equal to x^3

    C has a value of 3

13. The sum of the first n terms, S_n, of a given series is given by

$$S_n = \frac{2n^2}{n^2 + 1}$$

    A The first two terms of the series are $1, \frac{8}{5}$

    B The sum of the third and fourth terms is $\frac{24}{85}$

    C The series converges.

14. $\displaystyle\sum_{r=2}^{12} \frac{2^r}{r}$

    A The series has eleven terms

    B The series is a G.P.

    C The third term of the series is $\dfrac{2^3}{3}$

15. $(1 + x)^3(1 - x)^{20}$ is expanded as a series of ascending powers of x

    A The series is finite

    B The first two terms of the series are $1 - 17x$

    C The last term of the series is x^{20}.

TYPE III

16. The third term in the binomial expansion of $(2 - 3x)^{10}$ is $(45)(2^7)(3^3)(x^3)$

17. If $2^n - 1$ is the sum of the first n terms of a series, $4^n - 1$ is the sum of the first $2n$ terms.

18. The fourth term of the series $\displaystyle\sum_{r=0}^{n} (-1)^{(r+1)}2^r$ is 16

19. The geometric mean of $3x$ and $12x$ is $6x$

20. The area between the curve $y = x^2$, the y-axis and the line $y = 1$ is given by $\displaystyle\int_0^1 x \, dy$

MISCELLANEOUS EXERCISE E

1. Given that $y = x(x + 2)$ find   (a) $\dfrac{dy}{dx}$ (b) $\displaystyle\int y \, dx$

2. A particle P moves in a straight line so that its displacement s metres, at time t seconds, from a fixed point O on the line is
$$s = 4t^3 - \frac{1}{t}.$$
Find the speed of P, correct to 3 significant figures, when $t = 1.25$.

3. Given that
$$(2x + 1)^5 \equiv ax^5 + bx^4 + 80x^3 + cx^2 + 10x + 1,$$
find the values of a, b and c.   (U of L)

4.

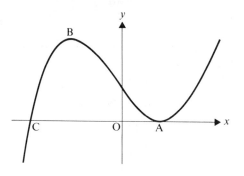

The figure shows a sketch of part of the curve with equation
$$y = x^3 - 12x + 16$$
Calculate
(a) the coordinates of the turning points A and B
(b) the coordinates of C, the point where the curve crosses the x-axis
(c) the area of the finite region enclosed by the curve and the x-axis.
   (U of L)

5. (a) Given that a, b and c are the first three terms respectively in an arithmetic series, and that $b = 3c$, find b and c in terms of a.

   (b) Given that e, f and g are the first three terms respectively in a geometric series, and that $f = g^3$, find f and g in terms of e.

   (U of L)

6. A particle moves along the x-axis such that at time t seconds, its displacement s metres from O is given by
$$s = \sqrt{(4 - t)}, \quad t \leqslant 4$$
   Find   (a) the initial speed of the particle

   (b) the acceleration of the particle when $t = 2$

7. The first three terms in the expansion of $(a - \frac{1}{3}x)^6$, in ascending powers of x are $64 - 64x + bx^2$, where a and b are constants. Find the value of a and the value of b.   (O/C, SU & C)

8. Write down the binomial expansions of $(2 + x)^4$ and $(1 - 3x)^4$ simplifying the coefficients as far as possible.
   Find the coefficient of x^2 in the expansion of $(2 - 5x - 3x^2)^4$

   (WJEC)

9. (a) Evaluate $\displaystyle\int_1^4 x(x - 1)\,dx$

   (b) Indicate on a sketch the region whose area is equal to the definite integral in part (a)

10. A particle moves in a straight line such that its velocity v m s^{-1} at time t seconds is given by
$$v = t^3 - 2t^2 + t - 1$$
   Find   (a) the initial velocity of the particle,

   (b) the value of t when the acceleration of the particle is zero.

11. A curve whose equation is $y = f(x)$ passes through the point $(2, 1)$. The gradient function is given by $f'(x) = \frac{1}{2}(1 - 2x)^2$. Find the equation of the curve.

12. A particle moves in a straight line such that its velocity, v m s^{-1} at time t seconds is given by
$$v = t(t - 2)$$
   (a) Find the values of t at which the particle is instantaneously at rest.

   (b) Find the distance travelled by the particle in the third second.

13. The coefficient of x in the binomial expansion of $(4 + x)^n$ equals the coefficient of x^3 in the binomial expansion of $(4 + x)^{n+1}$ where n is a positive integer. Find the value of n. (WJEC)

14. Evaluate (a) $\int_{-1}^{0} (x - 1)^3 dx$ (b) $\int_{0}^{\pi} \pi x^2 dx$

15. An arithmetic progression has first term a and common difference -1. The sum of the first n terms is equal to the sum of the first $3n$ terms. Express a in terms of n. (C)

16. Find the positive constants a and b such that 0.25, a, 9 are in geometric progression and 0.25, a, $9 - b$ are in arithmetic progression. (U of L)

17. Sketch the curve $y = (x - 1)^2(x - 2)$, showing the coordinates of the points where it meets the axes and of the turning points. Calculate the area enclosed by the curve and the x-axis.

18. A sum of money £P, is invested in a ten-year capital growth bond. Interest at the rate of r% p.a. is added to the value of the bond at the end of each year. Show that, at the end of the 10 year term, the value, £A, of the bond is given by

$$A = P\left(1 + \frac{r}{100}\right)^{10}$$

   (a) Find the first three terms in the expansion of $P\left(1 + \frac{r}{100}\right)^{10}$ as a series of ascending powers of r
   (b) Find the percentage error involved in using the answer to part (a) as an approximation for A when $r = 5$

19. Find the first three terms in the expansion of $(1 + x^2)^8$ as a series of ascending powers of x.
   Use this expansion to find an approximate value for

$$\int_{0}^{0.1} (1 + x^2)^8 dx$$

20. A geometric progression has first term a and common ratio $1/\sqrt{2}$. Show that the sum to infinity of the progression is $a(2 + \sqrt{2})$. (C)

21. Show that, provided $-1 < x < 1$, the sum to infinity of the series

$$1 - x + x^2 - x^3 + \ldots \text{ is } \frac{1}{1+x}.$$

Using the first three terms of the series, find an approximate value for

$$\int_0^{0.1} \frac{1}{1+x} \, dx.$$

22. Given that $f(x) = x(x-3)^2$, find the two values of x for which $f'(x) = 0$. Give the corresponding values of $f(x)$ and show that one of these is a maximum and the other is a minimum.

Sketch the curve $y = f(x)$.

Find the area of the region enclosed by the curve and the x-axis between $x = 0$ and $x = 3$

ANSWERS

Answers to questions taken from past examination papers are the sole responsibility of the authors and have not been approved by the Examining Boards.

CHAPTER 1

Exercise 1a – p. 2
1. $15x$
2. $2x^2$
3. $4x^2$
4. $10pq$
5. $8x^2$
6. $10p^2qr$
7. $9a^2$
8. $63ab$
9. $24st^2$
10. $8a^3$
11. $\frac{5}{3}x$
12. $2m$
13. $4ab^3$
14. $5xy$
15. $196p^4q^2$
16. $2a$
17. $6ax$
18. $2x$
19. $9b/5a$
20. $\frac{2}{3}x$
21. x^3/y^2

Exercise 1b – p. 3
1. $3x^2 - 4x$
2. $a - 12$
3. $2y - xy + y^2$
4. $5pq - 9p^2$
5. $3xy + y^2$
6. $x^3 - x^2 + x + 7$
7. $5 + t - t^2$
8. $a^2 - ab - 2b$
9. $7 - x$
10. $4x - 9$
11. $3x^2 + 18x - 20$
12. $ab - 2ac + cb$
13. $11cT - 2cT^2 - 55T^2$
14. $-x^3 + 7x^2 - 7x$
15. $-4y^2 + 24y - 10$
16. $5RS + 5RF - R^2$

Exercise 1c – p. 4
1. -7
2. 2
3. (a) 1 (b) -5 (c) -1
4. (a) 1 (b) 0 (c) -3

Exercise 1d – p. 4
1. $x^2 + 6x + 8$
2. $x^2 + 8x + 15$
3. $a^2 + 13a + 42$
4. $t^2 + 15t + 56$
5. $s^2 + 17s + 66$
6. $2x^2 + 11x + 5$
7. $5y^2 + 28y + 15$
8. $6a^2 + 17a + 12$
9. $35t^2 + 86t + 48$
10. $99s^2 + 49s + 6$
11. $x^2 - 5x + 6$
12. $y^2 - 5y + 4$
13. $a^2 - 11a + 24$
14. $b^2 - 17b + 72$
15. $p^2 - 15p + 36$
16. $2y^2 - 13y + 15$
17. $3x^2 - 13x + 4$
18. $6r^2 - 25r + 14$
19. $20x^2 - 19x + 3$
20. $6a^2 - 7ab + 2b^2$
21. $x^2 - x - 6$
22. $a^2 + a - 56$
23. $y^2 + 2y - 63$
24. $s^2 + s - 30$
25. $q^2 + 8q - 65$
26. $2t^2 + 3t - 20$
27. $4x^2 + 11x - 3$

28. $6q^2 - q - 15$
29. $x^2 - xy - 2y^2$
30. $2s^2 + st - 6t^2$

Exercise 1e – p. 5

1. $x^2 - 4$
2. $25 - x^2$
3. $x^2 - 9$
4. $4x^2 - 1$
5. $x^2 - 64$
6. $x^2 - a^2$
7. $x^2 - 1$
8. $9b^2 - 16$
9. $4y^2 - 9$
10. $a^2b^2 - 36$
11. $25x^2 - 1$
12. $x^2y^2 - 16$

Exercise 1f – p. 6

1. $x^2 + 8x + 16$
2. $x^2 + 4x + 4$
3. $4x^2 + 4x + 1$
4. $9x^2 + 30x + 25$
5. $4x^2 + 28x + 49$
6. $x^2 - 2x + 1$
7. $x^2 - 6x + 9$
8. $4x^2 - 4x + 1$
9. $16x^2 - 24x + 9$
10. $25x^2 - 20x + 4$
11. $9t^2 - 42t + 49$
12. $x^2 + 2xy + y^2$
13. $4p^2 + 36p + 81$
14. $9q^2 - 66q + 121$
15. $4x^2 - 20xy + 25y^2$

Exercise 1g – p. 7

1. $11x - 2x^2 - 12$
2. $x^2 - 49$
3. $6 - 25x + 4x^2$
4. $14p^2 - 3p - 2$
5. $9p^2 - 6p + 1$
6. $15t^2 + t - 2$
7. $16 - 8p + p^2$
8. $14t - 3 - 8t^2$
9. $x^2 + 4xy + 4y^2$
10. $16x^2 - 9$
11. $9x^2 + 42x + 49$
12. $15 - R - 2R^2$
13. $a^2 - 6ab + 9b^2$
14. $4x^2 - 20x + 25$
15. $49a^2 - 4b^2$
16. $9a^2 + 30ab + 25b^2$
17. (a) $6, -22$ (b) $15, 31$
    (c) $14, -31$ (d) $81, 18$

Exercise 1h – p. 9

1. $(x + 5)(x + 3)$
2. $(x + 7)(x + 4)$
3. $(x + 6)(x + 1)$
4. $(x + 4)(x + 3)$
1. $(x - 1)(x - 9)$
6. $(x - 3)^2$
7. $(x + 6)(x + 2)$
8. $(x - 8)(x - 1)$
9. $(x + 7)(x - 2)$
10. $(x + 4)(x - 3)$
11. $(x - 5)(x + 1)$
12. $(x - 12)(x + 2)$
13. $(x + 7)(x + 2)$
14. $(x - 1)^2$
15. $(x - 3)(x + 3)$
16. $(x + 8)(x - 3)$
17. $(x + 2)^2$
18. $(x - 1)(x + 1)$
19. $(x - 6)(x + 3)$
20. $(x + 5)^2$
21. $(x - 4)(x + 4)$
22. $(4 + x)(1 + x)$
23. $(2x - 1)(x - 1)$
24. $(3x + 1)(x + 1)$
25. $(3x - 1)^2$
26. $(3x + 1)(2x - 1)$
27. $(3 + x)^2$
28. $(2x - 3)(2x + 3)$
29. $(x + a)^2$
30. $(xy - 1)^2$

Exercise 1i – p. 10

1. $(3x - 4)(2x + 3)$
2. $(4x - 3)(x - 2)$
3. $(4x - 1)(x + 1)$
4. $(3x - 2)(x - 5)$
5. $(2x - 3)^2$
6. $(1 - 2x)(3 + x)$
7. $(5x - 4)(5x + 4)$
8. $(3 + x)(1 - x)$
9. $(5x - 1)(x - 12)$
10. $(3x + 5)^2$
11. $(3 - x)(1 + x)$
12. $(3 + 4x)(4 - 3x)$
13. $(1 + x)(1 - x)$
14. $(3x + 2)^2$
15. $(x + y)^2$
16. $(1 - 2x)(1 + 2x)$
17. $(2x - y)^2$
18. $(3 - 2x)(3 + 2x)$
19. $(6 + x)^2$
20. $(5x - 4)(8x + 3)$

21. $(7x + 30)(x - 5)$
22. $(6 - 5x)(6 + 5x)$
23. $(x - y)(x + y)$
24. $(9x - 2y)^2$
25. $(7 - 6x)^2$
26. $(5x - 2y)(5x + 2y)$
27. $(6x + 5y)^2$
28. $(2x - 3y)(2x + y)$
29. $(3x + 4y)(2x + y)$
30. $(7pq - 2)^2$

Exercise 1j – p. 11

1. not possible
2. $2(x + 1)^2$
3. $(x + 2)(x + 1)$
4. $3(x + 5)(x - 1)$
5. not possible
6. not possible
7. not possible
8. $2(x - 2)^2$
9. $3(x - 2)(x + 1)$
10. $2(x^2 - 3x + 4)$
11. $3(x - 4)(x + 2)$
12. $(x - 6)(x + 2)$
13. not possible
14. $4(x - 5)(x + 5)$
15. $5(x^2 - 5)$
16. not possible
17. not possible
18. not possible

Exercise 1k – p. 12

1. $x^3 - x^2 - x - 2$
2. $3x^3 - 5x^2 - x + 2$
3. $4x^3 - 8x^2 + 13x - 5$
4. $x^3 - 2x^2 + 1$
5. $2x^3 - 9x^2 - 24x - 9$
6. $x^3 + 6x^2 + 11x + 6$
7. $x^3 + 4x^2 - x - 4$
8. $x^3 - 4x^2 + x + 6$
9. $2x^3 + 7x^2 + 7x + 2$
10. $x^3 + 4x^2 + 5x + 2$
11. $4x^3 + 4x^2 - 7x + 2$
12. $27x^3 - 27x^2 + 9x - 1$
13. $4x^3 - 9x^2 - 25x - 12$
14. $4x^3 - 4x^2 - x + 1$
15. $6x^3 + 13x^2 + x - 2$
16. $x^3 + 3x^2 + 3x + 1$
17. $x^3 + x^2 - 4x - 4$
18. $2x^3 + 7x^2 - 9$
19. $24x^3 + 38x^2 - 51x + 10$
20. $4x^3 - 42x^2 + 68x + 210$
21. $3x^3 - 16x^2 + 28x - 16; -16, 28$

22. $6, -17$
23. $x^3 + 3x^2y + 3xy^2 + y^3$
24. $x^4 + 4x^3y + 6x^2y^2 + 4xy^3 + y^4$

Exercise 1l – p. 14

1. $x^3 + 9x^2 + 27x + 27$
2. $x^4 - 8x^3 + 24x^2 - 32x + 16$
3. $x^4 + 4x^3 + 6x^2 + 4x + 1$
4. $8x^3 + 12x^2 + 6x + 1$
5. $x^5 - 15x^4 + 90x^3 - 270x^2 + 405x - 243$
6. $p^4 - 4p^3q + 6p^2q^2 - 4pq^3 + q^4$
7. $8x^3 + 36x^2 + 54x + 27$
8. $x^5 - 20x^4 + 160x^3 - 640x^2 + 1280x$
$$- 1024$$
9. $81x^4 - 108x^3 + 54x^2 - 12x + 1$
10. $1 + 20a + 150a^2 + 500a^3 + 625a^4$
11. $64a^6 - 192a^5b + 240a^4b^2 - 160a^3b^3$
$$+ 60a^2b^4 - 12ab^5 + b^6$$
12. $8x^3 - 60x^2 + 150x - 125$

Mixed Exercise 1 – p. 15

1. -23
2. $115x - 105x^2 - 30$
3. 108
4. $3(x - 2)(x - 1)$
5. 250
6. $4(x - 3)(x + 3)$
7. $4x^3 - 4x^2 + x$
8. $(x - 5)^2$
9. 7
10. $12x^3 - 3x$
11. $(2x - 3y)^2$
12. 7
13. -12
14. $(3x + 2y)(2x - 5y)$
15. 24

CHAPTER 2

Exercise 2a – p. 17

1. $\frac{1}{4}$
2. $\frac{2(x + 2)}{3(x - 2)}$
3. $\frac{2}{3}$
4. $\frac{3}{5}$
5. $\frac{x}{y}$
6. not possible
7. not possible
8. $\frac{5x(x + y)}{5y + 2x}$

9. $\dfrac{2a - 6b}{6a + b}$

10. $\dfrac{b - 4}{3x(b + 4)}$

11. $\dfrac{x - 3}{x + 4}$

12. $\dfrac{4y^2 + 3}{(y + 3)(y - 3)}$

13. $\dfrac{1}{3(x + 3)}$

14. $\dfrac{x + 2}{2x + 1}$

15. $\dfrac{x - 2}{x - 1}$

16. $\dfrac{1}{2(a - 5)}$

17. $\dfrac{3}{p + 3q}$

18. $\dfrac{a^2 + 2a + 4}{(a + 5)(a + 2)}$; not possible

19. $\dfrac{x + 1}{3(x + 3)}$

20. $\dfrac{4(x - 3)}{(x + 1)^2}$

Exercise 2b – p. 18

1. $\dfrac{2x^2}{3y^2}$

2. $\dfrac{6t^2}{s}$

3. $\dfrac{8v^2}{3}$

4. $\dfrac{2r}{3}$

5. $3x$

6. $\dfrac{9x}{4y^2}$

7. $\dfrac{x^2}{24}$

8. $\dfrac{2}{a(a + b)}$

9. $\dfrac{1}{x + 1}$

10. $\dfrac{1}{2a}$

11. $\dfrac{1}{x - 1}$

12. $\dfrac{3}{2(x + 3)}$

13. $\dfrac{a^4}{27}$

14. $\frac{2}{3}$

15. $\dfrac{2r}{3s^2}$

16. $\dfrac{3x^2}{2(y - 2)}$

17. $\dfrac{b^2}{c^2}$

18. $2(x + 3)$

19. $\dfrac{x - 3}{3}$

20. $\dfrac{x(2x - 3)}{x - 1}$

Exercise 2c – p. 19

1. $\dfrac{b - a}{ab}$

2. $\dfrac{8}{15x}$

3. $\dfrac{q - p}{pq}$

4. $\dfrac{11}{10x}$

5. $\dfrac{x^2 + 1}{x}$

6. $\dfrac{x^2 - y^2}{xy}$

7. $\dfrac{2p^2 - 1}{p}$

8. $\dfrac{7x + 3}{12}$

9. $\dfrac{5x - 1}{6}$

10. $\dfrac{11 - 7x}{15}$

11. $\dfrac{\sin B + \sin A}{\sin A \, \sin B}$

12. $\dfrac{\sin A + \cos A}{\cos A \, \sin A}$

13. $\dfrac{12x^2 + 1}{4x}$

14. $\dfrac{2x^2 + x - 2}{2x + 1}$

15. $\dfrac{x^2 + 2x + 2}{x + 1}$

16. $\dfrac{2x + 3}{2x}$

17. $\dfrac{1 + x - x^2}{x}$

18. $\dfrac{n - 1}{n^2}$

19. $\dfrac{x(b^2 + a^2)}{a^2 b^2}$

20. $\dfrac{a^2 + 3a + 1}{a + 1}$

Exercise 2d – p. 21

1. $\dfrac{2x}{(x + 1)(x - 1)}$

2. $\dfrac{2x - 1}{(x + 1)(x - 2)}$

3. $\dfrac{7x + 18}{(x + 2)(x + 3)}$

4. $\dfrac{x}{(x - 1)(x + 1)}$

5. $\dfrac{-1 - 3a}{(a - 1)(a + 1)} = \dfrac{1 + 3a}{(1 - a)(1 + a)}$

6. $\dfrac{x + 2}{(x + 1)^2}$

7. $\dfrac{1 - 4x}{(2x + 1)^2}$

8. $\dfrac{-3x - 10}{(x + 1)(x + 4)} = -\dfrac{3x + 10}{(x + 1)(x + 4)}$

9. $\dfrac{2x + 6}{(x + 1)^2}$

10. $\dfrac{8 - x - x^2}{(x + 2)^2(x + 4)}$

11. $\dfrac{7x + 8}{6(x - 1)(x + 4)}$

12. $\dfrac{8 - 3x}{5(x + 2)(x + 4)}$

13. $\dfrac{15x - 58}{6(x + 1)(3x - 5)}$

14. $\dfrac{5x^2 - 9x - 32}{(x + 1)(x - 2)(x + 3)}$

15. $\dfrac{2x^2 + 6x + 6}{(x + 1)(x + 2)(x + 3)}$
$= \dfrac{2(x^2 + 3x + 3)}{(x + 1)(x + 2)(x + 3)}$

16. $-\dfrac{1}{x(x + 1)^2}$

17. $\dfrac{7t + 3}{(t + 1)^2}$

18. $\dfrac{-t^4 + 2t^3 - 2t^2 - 2t - 1}{(t^2 + 1)(t^2 - 1)}$

19. $\dfrac{1 + 3y - 3x}{(y - x)(y + x)}$

20. $\dfrac{n^3 + 6n^2 + 8n + 2}{n(n + 1)(n + 2)}$

Mixed Exercise 2 – p. 21

1. (a) $\dfrac{x^2 - 9}{2(x - 3)} = \dfrac{x + 3}{2}$ (b) $\dfrac{1}{x - 3}$

2. (a) $\dfrac{4a}{rp}$ (b) $\dfrac{2p^2 - 3r}{pr}$

3. (a) $\dfrac{2}{3(n + 2)}$ (b) $\dfrac{5x^2 + x - 1}{x(x + 1)(2x - 1)}$

4. (a) $\dfrac{4x^2 - x + 3}{4x - 1}$ (b) $\dfrac{x + 3}{x - 2}$

5. (a) $\dfrac{2(x - 1)}{(x + 1)(x - 2)}$

(b) $\dfrac{x(x^2 + x - 4)}{(x - 1)^2(x + 1)}$

6. (a) $\dfrac{(2x - 1)(x + 1)}{2x(x + 1)}$ (b) $(x - 1)^2$

7. (a) $\dfrac{2x - 5}{2x + 5}$ (b) $\dfrac{2t}{t^2 - 1}$

8. (a) $\dfrac{(x - 1)^3}{x + 1}$ (b) $\dfrac{ab + bc + ac}{abc}$

CHAPTER 3

Exercise 3a – p. 24

1. $2\sqrt{3}$
2. $4\sqrt{2}$
3. $3\sqrt{3}$
4. $5\sqrt{2}$
5. $10\sqrt{2}$
6. $6\sqrt{2}$
7. $9\sqrt{2}$
8. $12\sqrt{2}$
9. $5\sqrt{3}$
10. $4\sqrt{3}$
11. $10\sqrt{5}$
12. $2\sqrt{5}$

Exercise 3b – p. 26

1. $2\sqrt{3} - 3$
2. $5\sqrt{2} + 8$
3. $2\sqrt{5} + 5\sqrt{15}$
4. 4
5. $\sqrt{6} + \sqrt{2} - \sqrt{3} - 1$
6. $13 + 7\sqrt{3}$
7. 4
8. $5 - 3\sqrt{2}$

9. $22 - 10\sqrt{5}$

10. 9

11. $10 - 4\sqrt{6}$

12. $31 + 12\sqrt{3}$

13. $(4 + \sqrt{5})$

14. $(\sqrt{11} - 3)$

15. $(2\sqrt{3} + 4)$

16. $(\sqrt{6} + \sqrt{5})$

17. $(3 + 2\sqrt{3})$

18. $(2\sqrt{5} + \sqrt{2})$

Exercise 3c – p. 27

1. $\frac{3}{2}\sqrt{2}$

2. $\frac{1}{7}\sqrt{7}$

3. $\frac{2}{11}\sqrt{11}$

4. $\frac{3}{5}\sqrt{10}$

5. $\frac{1}{9}\sqrt{3}$

6. $\frac{1}{2}\sqrt{2}$

7. $\sqrt{2} + 1$

8. $\frac{1}{23}(15\sqrt{2} - 6)$

9. $\frac{1}{3}(4\sqrt{3} + 6)$

10. $-5(2 + \sqrt{5})$

11. $\frac{1}{4}(\sqrt{7} + \sqrt{3})$

12. $4(2 + \sqrt{3})$

13. $\sqrt{5} - 2$

14. $\frac{1}{13}(7\sqrt{3} + 2)$

15. $3 + \sqrt{5}$

16. $3(\sqrt{3} + \sqrt{2})$

17. $\frac{3}{19}(10 - \sqrt{5})$

18. $3 + 2\sqrt{2}$

19. $\frac{2}{3}(7 - 2\sqrt{7})$

20. $\frac{1}{2}(1 + \sqrt{5})$

21. $\frac{1}{4}(\sqrt{11} + \sqrt{7})$

22. $\frac{1}{6}(9 + \sqrt{3})$

23. $\frac{1}{14}(9\sqrt{2} - 20)$

24. $\frac{1}{6}(3\sqrt{2} + 2\sqrt{3})$

25. $\frac{1}{2}(2 + \sqrt{2})$

26. $\frac{1}{42}(3\sqrt{7} - \sqrt{21})$

27. $\frac{1}{9}(\sqrt{30} + 2\sqrt{3})$

Exercise 3d – p. 31

1. $\dfrac{1}{2^4}$

2. $\dfrac{1}{2^2}$

3. 3^2

4. x^2

5. 1

6. t^4

7. 1

8. 2

9. $y^{3/2}$

10. x^5

11. $\dfrac{1}{y^{3/4}}$

12. p

13. 3

14. $\frac{1}{32}$

15. $\frac{1}{2}$

16. 2

17. 27

18. $\frac{9}{4}$

19. 1

20. 16

21. $\frac{5}{4}$

22. -5

23. 1331

24. $\frac{3}{5}$

25. 6

26. $\frac{16}{27}$

27. 8

28. 5

29. 1

30. 1

Mixed Exercise 3 – p. 32

1. (a) $2\sqrt{21}$ (b) $10\sqrt{3}$ (c) $3\sqrt{5}$

2. (a) $7\sqrt{3} - 6$ (b) 8

3. (a) $8 - 2\sqrt{2}$ (b) $7 - 2\sqrt{10}$

4. (a) $(7 + \sqrt{3})(7 - \sqrt{3}) = 46$
      (b) $(2\sqrt{2} - 1)(2\sqrt{2} + 1) = 7$
      (c) $(\sqrt{7} + \sqrt{5})(\sqrt{7} - \sqrt{5}) = 2$

5. (a) $\frac{5}{7}\sqrt{7}$ (b) $\frac{1}{3}(\sqrt{13} + 2)$
      (c) $4(\sqrt{3} + \sqrt{2})$ (d) $2 - \sqrt{3}$

6. (a) 32   (b) 27   (c) 3

7. (a) 1   (b) 1

8. (a) $\frac{1}{4}$ (b) $\frac{4}{7}$ (c) $\frac{16}{243}\sqrt{6}$

9. (a) 1   (b) $\frac{1}{5}\sqrt{15}$

10. (a) 3   (b) $-\frac{3}{4}$ (c) 4

CHAPTER 4

Exercise 4a – p. 34

1. $x = -2$ or $x = -3$

2. $x = 2$ or $x = -3$

3. $x = 3$ or $x = -2$

4. $x = -2$ or $x = -4$

5. $x = 1$ or $x = 3$

6. $x = 1$ or $x = -3$

7. $x = -1$ or $x = -\frac{1}{2}$
8. $x = 2$ or $x = \frac{1}{4}$
9. $x = 1$ or $x = -5$
10. $x = 8$ or $x = -9$
11. $-1, 3$
12. $-1, -4$
13. $1, 5$
14. $2, -5$
15. $-2, 7$
16. $2, 7$

Exercise 4b – p. 36

1. $x = 2$ or $x = 5$
2. $x = 3$ or $x = -5$
3. $x = 4$ or $x = -1$
4. $x = 3$ or $x = 4$
5. $x = \frac{1}{3}$ or $x = -1$
6. $x = -1$ or $x = -6$
7. $x = 0$ or $x = 2$
8. $x = -1$ or $x = -\frac{1}{4}$
9. $x = \frac{2}{3}$ or $x = -1$
10. $x = 0$ or $x = -\frac{1}{2}$
11. $x = 0$ or $x = -6$
12. $x = 0$ or $x = 10$
13. $x = 0$ or $x = \frac{1}{2}$
14. $x = 5$ or $x = -4$
15. $x = 2$ or $x = -\frac{4}{3}$
16. $x = 2$ or $x = -1$
17. $x = 0$ or $x = 1$
18. $x = 0$ or $x = 2$
19. $x = 3$ or $x = -1$
20. $x = -1$ or $x = \frac{1}{2}$

Exercise 4c – p. 38

1. 4
2. 1
3. 9
4. 25
5. 2
6. $\frac{25}{4}$
7. 192
8. 81
9. 200
10. $\frac{1}{4}$
11. $\frac{1}{3}$
12. $\frac{9}{8}$
13. $x = -4 \pm \sqrt{17}$
14. $x = 1 \pm \sqrt{3}$
15. $x = -\frac{1}{2}(1 \pm \sqrt{5})$
16. $x = -\frac{1}{2}(1 \pm \sqrt{3})$

17. $x = -\frac{1}{2}(3 \pm \sqrt{5})$
18. $x = \frac{1}{4}(1 \pm \sqrt{17})$
19. $x = -2 \pm \sqrt{6}$
20. $x = -\frac{1}{6}(1 \pm \sqrt{13})$
21. $x = \frac{1}{2}(-2 \pm 3\sqrt{2})$
22. $x = \frac{1}{2}(1 \pm \sqrt{13})$
23. $x = -\frac{1}{8}(1 \pm \sqrt{17})$
24. $x = \frac{1}{4}(3 \pm \sqrt{41})$

Exercise 4d – p. 40

1. $x = -2 \pm \sqrt{2}$
2. $x = \frac{1}{4}(-1 \pm \sqrt{17})$
3. $x = \frac{1}{2}(-5 \pm \sqrt{21})$
4. $x = \frac{1}{4}(1 \pm \sqrt{33})$
5. $x = 2 \pm \sqrt{3}$
6. $x = \frac{1}{4}(1 \pm \sqrt{41})$
7. $x = \frac{1}{6}(1 \pm \sqrt{13})$
8. $x = -\frac{1}{6}(1 \pm \sqrt{13})$
9. $x = 1 \pm \sqrt{6}$
10. $x = \frac{1}{5}(1 \pm \sqrt{21})$
11. $x = -0.260$ or -1.540
12. $x = 2.781$ or 0.719
13. $x = 1.883$ or -0.133
14. $x = 0.804$ or -1.554
15. $x = 0.804$ or -1.554
16. $x = 0.724$ or 0.276
17. $x = 3.303$ or -0.303
18. $x = 7.873$ or 0.127
19. $x = 1.281, -0.781$
20. $x = 7.873, 0.127$

Exercise 4e – p. 42

1.

x	-2	1
y	-1	2

2. $x = -1, y = 3$
3. $x = 2, y = 3$

4.

x	-1	$\frac{1}{2}$
y	4	1

5.

x	2	$-\frac{1}{2}$
y	-3	2

6.

x	$\frac{7}{2}$	-2
y	$-\frac{1}{2}$	5

7.

x	1	2
y	2	1

8.

x	-1	3
y	-4	4

9. $x = 1, y = 5$

10.

x	6	-6
y	2	-4

11. $x = \frac{1}{2}, y = -1$

12.

x	1	0
y	$\frac{1}{3}$	$\frac{2}{3}$

13. $x = -1, y = -\frac{1}{2}$

14. $x = 1, y = -\frac{1}{3}$

15.

x	$-\frac{1}{3}$	$\frac{2}{3}$
y	$-\frac{1}{2}$	$\frac{1}{4}$

16. $x = 1, y = \frac{1}{2}$

17.

x	-3	6
y	-3	$\frac{3}{2}$

18.

x	1	$-\frac{1}{4}$
y	-2	3

19.

x	-1	2
y	$\frac{1}{3}$	$-\frac{1}{6}$

20.

x	$\frac{1}{2}$	0
y	$\frac{1}{2}$	1

21.

x	1	$3\frac{1}{2}$
y	1	-4

22.

x	-1	$7\frac{1}{2}$
y	2	$-\frac{7}{5}$

Exercise 4f – p. 46

1. 4

2. $-\frac{5}{3}$

3. 1

4. $\frac{4}{3}$

5. -3

6. $\frac{2}{5}$

7. real and different

8. not real

9. real and different

10. real and equal

11. real and different

12. real and equal

13. real and different

14. not real

15. real and different

16. real and equal

17. $k = \pm 12$

18. $a = 2\frac{1}{4}$

19. $p = 2$

22. $q^2 = 4p$

Mixed Exercise 4 – p. 48

1. (a) 5   (b) $-1, 6$ (c) 5

2. (a) 6   (b) $3 \pm \sqrt{14}$ (c) 6

3. (a) $-\frac{3}{2}$ (b) $\frac{1}{4}(-3 \pm \sqrt{17})$ (c) $-\frac{3}{2}$

4. (a) $-\frac{4}{3}$ (b) $\frac{1}{3}(-2 \pm \sqrt{19})$ (c) $-\frac{4}{3}$

5. (a) 2   (b) 1, 1   (c) 2

6. (a) $\frac{11}{4}$ (b) $-\frac{1}{4}, 3$ (c) $\frac{11}{4}$

7. (a) -1 (b) $\frac{1}{2}(-1 \pm \sqrt{13})$ (c) -1

8. (a) -4 (b) $-6, 2$ (c) -4

9. (a) -2 (b) $-1 \pm \sqrt{3}$ (c) -2

10. (a) -4 (b) $-2, -2$ (c) -4

11. $x = 0$ or 2

12. $x = -4$ or 1

13. $x = 2$ or $x \to \infty$

14. $x = 0$ or -1

15. $x = \frac{1}{2}(7 \pm \sqrt{89})$

16. $x = 1$ or $x \to \infty$

17. (a) not real   (b) real and different

(c) real and equal

(d) real and different

18. 4, -1

20. 2

21.

x	4	-22
y	5	31

22.

x	2	4
y	5	9

23. $x = 5$ or $\frac{2}{3}$;   (a) rational

(b) it factorises

CONSOLIDATION A

Multiple Choice Exercise A – p. 50

1. A

2. E

3. E

4. B

5. E

6. A

7. E

8. C
9. B, C
10. B
11. A
12. B, C
13. A, B
14. T
15. F
16. T
17. F

Miscellaneous Exercise A – p. 52

1. (a) $\frac{3}{2}$ (b) $\frac{7}{2}$
3. $3[(x + 1)^2 - \frac{7}{3}]$
4. -96
5. 1
6. $\frac{81}{4}$
7. $p = 8, q = 2$
9. $p = -4, q = \frac{1}{2}$
10. $a = 3, b = -36$
11. -4
12. $x = -5$ or 4
13. $x = -3, y = 0; x = \frac{9}{5}, y = \frac{12}{5}$
14. $k < 0, k > 3$

CHAPTER 5

Exercise 5a – p. 56

1. 2 cm
2. 90 cm
3. 20 cm
4. (a) 84 cm   (b) $7x$ cm
5. $5:7$
6. $3:1$
7. $y:(x - y)$
8. $\dfrac{ma}{n - m}$
9. $(a + b):b$

Exercise 5b – p. 58

1. 3.75 cm
2. TN $= 2.5$ cm, LN $= 4.5$ cm
3. BC $= 2.25$ cm
4. $\dfrac{yz}{x}$
5. (a) 1.25 cm   (b) $2:7$ externally
6. (a) Y   (b) Y   (c) Y
   (d) Y   (e) N   (f) Y

Exercise 5c – p. 63

2. $\frac{5}{26}$ cm
3. XZ $= \frac{7}{4}$ cm, QR $= 16$ cm
5. RT $= 12$ cm, AT $= 8$ cm
6. $9:4$
7. $10°$
8. BD $= \frac{24}{7}$, DC $= \frac{25}{7}$, AD $= \frac{120}{7}\sqrt{2}$
10. EC $= 3$ cm, AC $= \frac{25}{3}$ cm

CHAPTER 6

Exercise 6a – p. 67

1.

2. $(9, 5)$ and $(9, 1)$ or $(-3, 5)$ and $(-3, -1)$
3. $(-2, 2), (3, -3)$

Exercise 6b – p. 72

1. (a) 5   (b) $\sqrt{2}$ (c) $\sqrt{13}$
2. (a) $(\frac{5}{2}, 4)$ (b) $(\frac{5}{2}, \frac{1}{2})$ (c) $(3, \frac{7}{2})$
3. (a) $\sqrt{109}, (\frac{1}{2}, 1)$ (b) $\sqrt{5}, (-\frac{1}{2}, -1)$
   (c) $2\sqrt{2}, (-2, -3)$
4. $\sqrt{65}$
5. $\sqrt{13}$
6. $(2, -4)$
8. (b) $(-3\frac{1}{2}, -\frac{1}{2})$ (c) $17\frac{1}{2}$ sq units
9. (a) $\sqrt{5}(2 + \sqrt{2})$ (b) $(0, 4\frac{1}{2})$
   (c) $2\frac{1}{2}$
11. $(-5, -3)$

Exercise 6c – p. 77

1. (a) 3   (b) $\frac{3}{2}$ (c) $\frac{1}{3}$ (d) $\frac{3}{4}$
   (e) -4 (f) 6   (g) $-\frac{7}{3}$
   (h) $-\frac{3}{2}$ (i) $\dfrac{k}{h}$
3. (a) yes   (b) no   (c) yes
   (d) yes
3. (a) parallel   (b) perpendicular
   (c) perpendicular   (d) neither
   (e) parallel

Exercise 6d – p. 79

1. $a = 0, b = 4$
2. (b) $22\frac{1}{2}$ square units
5. $\sqrt{(a^2 + 4b^2)}$
8. $\left(\dfrac{p + q}{2}, \dfrac{p + q}{2}\right)$
9. $(a - 2)^2 + (b - 1)^2 = 9$
10. 8
11. $b(d - b) = ac$
12. $b^2 = 8a - 16$

CHAPTER 7

Exercise 7a – p. 81

1. $\sin A = \frac{12}{13}, \cos A = \frac{5}{13}$
2. $\tan X = \frac{3}{4}, \sin X = \frac{3}{5}$
3. $\cos P = \frac{9}{41}, \tan P = \frac{40}{9}$
4. $\sin A = \dfrac{1}{\sqrt{2}} = \cos A$
5. $\sin Y = \frac{1}{3}\sqrt{5}, \tan Y = \frac{1}{2}\sqrt{5}$
6. $\cos A = \frac{1}{2}\sqrt{3};\ 30°$
7. $\cos X = \frac{24}{25}$
8. $\cos X = \frac{4}{5}$
   $\cos^2 X + \sin^2 X = 1$
9. $\cos X = \frac{1}{2}\sqrt{3}$
   $\cos^2 X + \sin^2 X = 1$
10. $\cos X = \dfrac{1}{q}\sqrt{(q^2 - p^2)};\ 1$

Exercise 7b – p. 86

1. $\frac{3}{5}, -\frac{4}{5}, -\frac{3}{4}$
2. $\frac{5}{13}, -\frac{12}{13}, -\frac{5}{12}$
3. $\frac{3}{5}, -\frac{4}{5}, -\frac{3}{4}$
4. $\dfrac{2}{\sqrt{13}}, \dfrac{-3}{\sqrt{13}}, -\dfrac{2}{3}$
5. 80° or 100°
6. 60°
7. 105°
8. 52° or 128°
9. 135°
10. 150°
11. 81° or 99°
12. 57°
13. 90°
14. 80°
15. 89°
16. 180°
17. $\pm\frac{4}{5}$

18. $\frac{5}{13}$
19. 53° or 127°
20. 150°
21. 135°
22. (a) yes, 90°
    (b) yes, 0
    (c) no
    (d) yes, 0 or 180°
23. $A + B = 180°$
24.

25. (a) Yes   (b) No

Exercise 7c – p. 91

Answers are correct to 3 s.f.

1. 11.1
2. 13.7
3. 10.2
4. 8.83
5. 156
6. 113
7. 7.01
8. 89.1
9. 581
10. 141
11. 16.3
12. 28.1
13. 51.3
14. no; an angle and the side opposite to it are not known

Exercise 7d – p. 95

Answers are correct to the nearest degree.

1. 18°

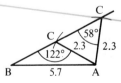

2. 58° or 122°

3. 17°

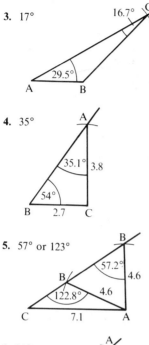

4. 35°

5. 57° or 123°

6. 30°

Exercise 7e – p. 98
1. 5.29
2. 12.9
3. 53.9
4. 4.04
5. 101
6. 12.0
7. 64.0
8. 31.8

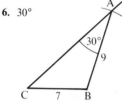

Exercise 7f – p. 100
1. 38°
2. 55°
3. 45°
4. 94°
5. (a) 18°   (b) 126°
6. 29°
7. 11.4 cm, 68°

Exercise 7g – p. 102
1. 87.4 cm
2. 23.8 cm
3. 17.5 cm
4. $\angle B = 81°$; $a = 112$ cm
5. $a = 164$ cm; $c = 272$ cm
6. $\angle B = 34°$; $a = 37.0$ cm
7. $\angle C = 43°$; $b = 19.4$ cm; $c = 13.5$ cm
8. $\angle A = 52°$; $a = 33.2$ cm; $c = 41.5$ cm
9. $\angle B = 43°$; $\angle C = 60°$; $a = 27.1$ cm
10. $\angle A = 22°$; $\angle C = 33°$; $b = 30.3$ cm
11. 14.1 m
12. 40°, 53°, 87°

Mixed Exercise 7 – p. 102
1. (a) 116°   (b) 86°
2. $-\frac{24}{25}$
3. (a) $\dfrac{5}{\sqrt{39}}$ (b) $\dfrac{-5}{\sqrt{39}}$
4. (a) $\dfrac{1}{\sqrt{2}}, \dfrac{-1}{\sqrt{2}}$ (b) $\dfrac{3}{\sqrt{13}}, \dfrac{-2}{\sqrt{13}}$
5. $-\frac{5}{13}$
6. 9.05 cm
7. 4.82
8. 83°
9. 54° or 126°
10. 108°, 50°, 22°

CHAPTER 8

Exercise 8a – p. 105
1. 12 300 cm²
2. 2190 cm²
3. 1680 cm²
4. 453 square units
5. 42.9 square units
6. 51°, 21.0 cm²
7. 10.6 cm, 59.8 cm²
8. 52°, 151 cm (or 150 cm)
9. 5.25 cm

Exercise 8b – p. 108
1. 11.0 cm, 67.7 cm² (or 67.8 cm²)
2. 58.5 km
3. 477 m
4. $\angle BAO = 74°$, $\angle CAO = 52°$; 22 cm²
5. (a) 60.8 cm   (b) 35°
   (c) 42.8 cm   (d) 2140 cm²
   (e) 427 000 cm³

Exercise 8c – p. 115

1. (a) $4\sqrt{2}$ cm   (b) $2\sqrt{29}$ cm
   (c) $2\sqrt{29}$ cm   (d) $2\sqrt{33}$ cm
2. (a) 6 cm, $6\sqrt{2}$ cm, $2\sqrt{34}$ cm
   (b) 53°   (c) 43°   (d) 53°
3. (a) $5\sqrt{2}$ cm   (b) $5\sqrt{5}$ cm
   (c) $\sqrt{109}$ cm   (d) 21°   (e) 34°
4. $\frac{1}{3}$
5. P = 55.4°, Q = 31.2°, R = 93.4°
6. 71°
7. (a) 6 m   (b) 9 m   (c) 48°
   (d) 240 m^2
8. (a) 15.3 m   (b) 2.29 cm
   (c) 15°   (d) 30°
9. 420 m; 31°
10. 46°

Mixed Exercise 8 – p. 120

1. 75.8 cm^2
2. $\frac{1}{2}$; yes, ∠A can be 30° or 150°
3. (a) 98°   (b) 19.8 cm^2
   (c) 3.96 cm
4. PQ = 8.29 cm, QR = 6 cm,
   RP = 3.46 cm; 9 cm^2
5. (a) 45°   (b) 35°
6. 35°
7. $\dfrac{a\sqrt{3}}{49}$; $\dfrac{2a}{7\sqrt{7}}$
8. $a\sqrt{\dfrac{37}{39}} = 0.974a$

CHAPTER 9

Exercise 9a – p. 126

1.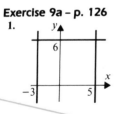

2.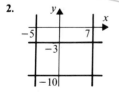

In Questions 3–10, the unshaded region is the one required.

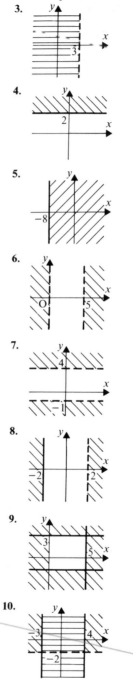

3.

4.

5.

6.

7.

8.

9.

10.

Exercise 9b – p. 129

1. (a) $y = 2x$ (b) $x + y = 0$
   (c) $3y = x$ (d) $4y + x = 0$
   (e) $y = 0$ (f) $x = 0$
2. (a) $2y = x + 2$ (b) $3y + 2x = 0$
   (c) $y = 4x$
3. (a) $2y = x$ (b) $y + 2x - 6 = 0$
   (c) $y = 2x + 5$
4. (a) $2y + x = 0$ (b) $2x - 3y = 0$
   (c) $2x + y = 0$
5. (a) $x - 3y + 1 = 0$
   (b) $2x + y - 5 = 0$
6. (a) $5x - y - 17 = 0$
   (b) $x + 7y + 11 = 0$
7. $3x + 4y - 48 = 0, 5$

Exercise 9c – p. 133

1. (a) $y = 3x - 3$
   (b) $5x + y - 6 = 0$
   (c) $x - 4y - 4 = 0$
   (d) $y = 5$
   (e) $2x + 5y - 21 = 0$
   (f) $15x + 40y + 34 = 0$
2. (a) $3x - 2y + 2 = 0$
   (b) $3x - 2y + 7 = 0$ (c) $x = 3$
3. (a), (c) and (d)
4. $x + y - 7 = 0$
5. (a) $x + 2y - 5 = 0$
   (b) $16x - 6y + 19 = 0$
   (c) $10x - 16y + 23 = 0$
6. $2x + y = 0$
7. $4x + 5y = 0$
8. $5x + 4y = 0$
9. $x + 2y - 11 = 0$
10. $3x - 4y + 19 = 0$

Exercise 9d – p. 135

1. (a) $y = x - 3$
   (b) $y\sqrt{3} = x$
   (c) $x + y - 1 = 0$
   (d) $y = 2\sqrt{3} - 1 - x\sqrt{3}$
   (e) $y\sqrt{3} + x = 3 + \sqrt{3}$
   (f) $3y - x - 14 = 0$
2. (a) $63.4°$ (b) $78.7°$
   (c) $26.6°$ (d) $123.7°$
3. $y = x + 1$

Exercise 9e – p. 138

1. (a) $(1, 5)$ (b) $(6, \frac{19}{2})$
   (c) $(5, 1)$ (d) $(11, -18)$
2. (a) $(\frac{9}{8}, -\frac{13}{8})$ (b) $(\frac{5}{7}, \frac{11}{7})$
   (c) $(\frac{34}{5}, \frac{23}{5})$

Mixed Exercise 9 – p. 140

2. $\frac{1}{4}$ sq units
3. $(-\frac{4}{3}, \frac{11}{3})$, $(-5, 0)$, $(6, 0)$
4. $10x - 26y - 1 = 0$
5. $x + 3y - 11 = 0$, $(\frac{13}{5}, \frac{14}{5})$
6. $[\frac{2}{5}(2 + 2a - b), \frac{1}{5}(4 - 2a + b)]$
7. $y = 2x - 3$
8. $y = x - 4$ (or $y + x = 10$ if line is inclined at $-45°$ to Ox)
9. $8by - 2ax + 8b^2 + 3a^2 = 0$
10. (a) $\sqrt{20}$ (b) $x - 2y + 1 = 0$
11. $(\frac{9}{10}, \frac{17}{10})$ and $(-\frac{18}{10}, \frac{26}{10})$, $(-\frac{27}{10}, -\frac{1}{10})$
   or $(\frac{36}{10}, \frac{8}{10})$, $(\frac{27}{10}, -\frac{19}{10})$
12. (a) $(1, 2), (5, 2), (3, 6)$ (b) 8
13. (a) $y = x$ (b) $(\frac{40}{13}, \frac{40}{13})$
14. $P(11, -2), Q(15, 0)$

CONSOLIDATION B

Multiple Choice Exercise B – p. 144

1. C
2. E
3. A
4. C
5. D
6. B
7. A
8. C
9. A, B
10. A, C
11. B
12. C
13. A, C
14. T
15. T
16. T

Miscellaneous Exercise B – p. 146

1. $y = 3x - 1$; $(\frac{2}{7}, -\frac{1}{7})$
2. (a) $(3, 2)$
   (b) $y + 2x = 8$
3. Area $= \frac{39}{2}$ sq units;
   $4y = 6x - 3$
4. (a) $2y = 3x - 9$
   (b) $x + y = 3$
5. (a) $A(-3, 0)$, $B(0, \frac{3}{2})$
   (b) $2y = x + 3$
   (c) $(\frac{9}{5}, \frac{12}{5})$
6. (a) $90°$ (b) 24.51 km
   (c) 28.92 km   (d) 10.24 km
   (e) $115°$

7. $(4\sqrt{2}, 8\sqrt{2} + 4)$,
$(12 + 4\sqrt{2}, 8\sqrt{2})$, $(0, 4)$

10. 3

11. 2.08 m, 44.3°

12. (a) 43.3°   (b) 53.1°   (c) 111.1°

13. (a) 3 cm   (b) 96°

14. (a) 79.9°   (b) 13.4 cm   (c) 36.7°

15. $2y = x + 10$;
$y = 3x - 5$;
$(5, \frac{5}{2})$, $(2, 6)$

16. $D(-4, 8)$; $2x + 9y = 64$; $(\frac{1}{2}, 7)$

17. $2y + 3x + 11 = 0$
$2y = x + 9$
$2y = 3x - 5$

CHAPTER 10

Exercise 10a – p. 153

1. (a) ∠ADE, ∠ACE   (b) ∠AFE
(c) ∠ADC   (d) ∠ABC
(e) ∠CDE   (f) ∠CAD, ∠CED
(g) ∠BAC, . . .

3. (c) 106.3°   (d) (5, 8), 53.1°

Exercise 10b – p. 157

1. (a) 50°   (b) 40°, 40°   (c) 60°
(d) 54°, 108°

3. (4, 4)

4. 30°

7. $a^2 + b^2 - 9a - b + 14 = 0$

9. $(-\frac{115}{18}, -\frac{29}{6})$, 10.8 units to 3 s.f.

10. 60°, 60°

Exercise 10c – p. 162

1. 61.9°

2. 5 units

3. 67.4°

4. 7.22 cm

5. 60°, 30°, 90°

6. 8.43

7. 16 sq. units

8. 14.9 cm

9. 2.85

10. $\dfrac{2\sqrt{5}}{2}, -\dfrac{2\sqrt{5}}{5}$

12. $(-\frac{3}{4}, -\frac{9}{4})$

13. (a) $2x + y = 0$ (b) $(-\frac{4}{5}, \frac{8}{5})$

14. $x + y - 14 = 0$

CHAPTER 11

Exercise 11a – p. 166

1. $\frac{1}{4}\pi$, $\frac{5}{6}\pi$, $\frac{1}{6}\pi$, $\frac{3}{2}\pi$, $\frac{5}{4}\pi$, $\frac{1}{8}\pi$, $\frac{4}{3}\pi$, $\frac{5}{3}\pi$, $\frac{7}{4}\pi$

2. 30°, 180°, 18°, 45°, 150°, 15°, 22.5°,
240°, 20°, 270°, 80°

3. (Angles in radians, correct to 3 s.f.)
0.611, 0.824, 1.62, 4.07, 0.246, 2.04, 6.46

4. (Angles to the nearest degree.)
97°, 190°, 57°, 120°, 286°, 360°

Exercise 11b – p. 169

1. 2.09 cm, 4.19 cm^2

2. 26.2 cm, 131 cm^2

3. 4.77 cm, 35.8 cm^2

4. 7.96 cm, 79.6 cm^2

5. 18.8 cm, 75.4 cm^2

6. 3.14 cm, 1.57 rad

7. 4.8 cm, 0.96 rad

8. π cm, 6 rad

9. $\frac{5}{8}\pi$, 8 cm

10. 146°

11. 0.283 rad

12. 0.52°

13. (a) 12 cm^2 (b) 23.2 cm^2

14. 14.5 mm^2, 139 mm^2

15. (a) 15.2 cm   (b) 32.5 cm^2

16. 19.1°

17. 85.6 cm

18. 19.6 cm, 108 cm^2

20. $0.979a^2$

21. (a) 135°   (b) 9.41 cm
(c) 35.4 cm   (d) 68.8 cm

CHAPTER 12

Exercise 12a – p. 174

1. (a) yes   (b) yes   (c) yes, $x \neq 1$
(d) no   (e) yes, $x \geqslant 0$ (f) yes
(g) yes   (h) no

2. $-4, -24$

3. 25, 217

4. 1, not defined, 12

5. $1, \dfrac{\sqrt{3}}{2}$

Exercise 12b – p. 177

1. (a) $f(x) \geqslant -3$ (b) $f(x) \geqslant -5$
(c) $f(x) \geqslant 0$ (d) $0 < f(x) \leqslant \frac{1}{2}$

Answers

379

2. (a)

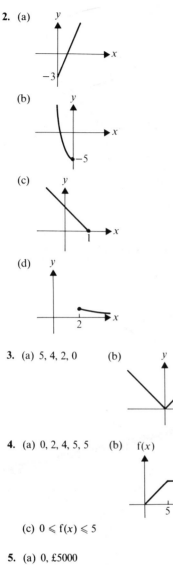

(b)

(c)

(d)

3. (a) 5, 4, 2, 0   (b)

4. (a) 0, 2, 4, 5, 5   (b) f(x)

(c) $0 \leqslant f(x) \leqslant 5$

5. (a) 0, £5000
   (b) $f(x) = 0$ for $0 \leqslant x \leqslant 20\,000$
   $f(x) = \frac{1}{5}(x = 20\,000)$ for $x > 20\,000$

   domain $x \geqslant 0$ (but $x <$ GNP!)
   range $f(x) \geqslant 0$

Exercise 12c – p. 181

1. (a) $\frac{11}{4}$ (b) 3   (c) 4
2. (a) $f(x) \leqslant \frac{29}{4}$ (b) $f(x) \geqslant -2$
   (c) $f(x) \leqslant 1$
3. (a)

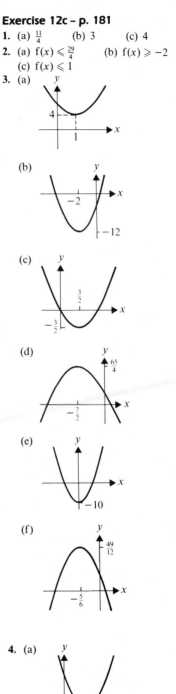

(b)

(c)

(d)

(e)

(f)

4. (a)

(b)

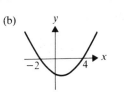

(c)

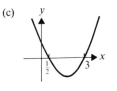

(d)

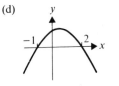

(e)

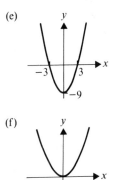

(f)

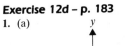

Exercise 12d – p. 183

1. (a)

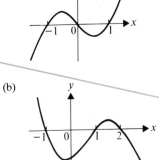

(b)

2.

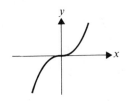

Exercise 12e – p. 183

1. (b) translation $\begin{pmatrix} 0 \\ c \end{pmatrix}$

2. translation $\begin{pmatrix} -c \\ 0 \end{pmatrix}$

3. (a) reflection Ox
   (b) reflection Oy

Exercise 12f – p. 188

1.

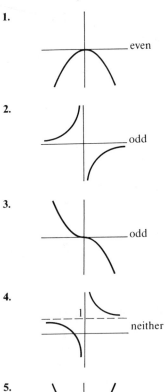

even

2.

odd

3.

odd

4.

neither

5.

even

6.

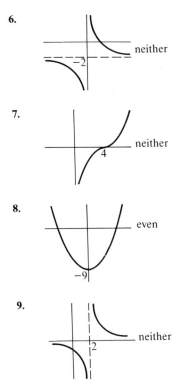

neither

7.

neither

8.

even

9.

neither

11. Reflection in line $y = x$
(there are several alternatives).
12. Reflection in line $y = x$
(there are several alternatives).
13. (5, 2)
14. (b, a)

Exercise 12g – p. 194

1. (a)

f^{-1}(x)

f(x)

(b)

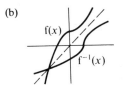

f(x)

f^{-1}(x)

(c)

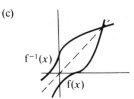

f^{-1}(x)

f(x)

(d)

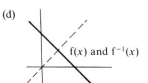

f(x) and f^{-1}(x)

(e)

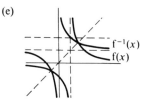

f^{-1}(x)

f(x)

(f)

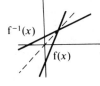

f(x) and f^{-1}(x)

2. (d) and (f)
3. (a) f^{-1} = ($x - 1$)   (b) no
      (c) f^{-1}(x) = $\sqrt[3]{(x - 1)}$
      (d) f^{-1}(x) = $\sqrt{(x + 4)}, x \geqslant -4$
      (e) f^{-1}(x) = $\sqrt[4]{x} - 1, x \geqslant 0$
4. (a) $-\frac{1}{3}$ (b) $\frac{1}{2}$ (c) there isn't one
5. (a) 8   (b) 3   (c) 1

Exercise 12h – p. 195

1. (a) $\dfrac{1}{x^2}$ (b) $(1 - x)^2$ (c) $1 - \dfrac{1}{x}$

      (d) $1 - x^2$ (e) $\dfrac{1}{x^2}$

2. (a) 125   (b) 15   (c) -1
      (d) -1
3. (a) $(1 + x)^2$ (b) $2(1 + x)^2$
      (c) $1 + 4x^2$
4. g(x) = x^2, h(x) = $2 - x$
5. g(x) = x^4, h(x) = ($x + 1$)

6. (a) $f(x) = gh(x)$, $g(x) = \frac{1}{x}$,

$h(x) = 2x + 1$

(b) $f(x) = gh(x)$, $g(x) = x^4$,

$h(x) = 5x - 6$

(c) $f(x) = gh(x)$, $g(x) = \sqrt{x}$,

$h(x) = x^2 - 2$

Mixed Exercise 12 – p. 195

1. (a) $\frac{1}{0}$ is meaningless   (b) $\frac{1}{4}$

(c)

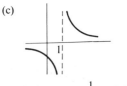

(d) $f^{-1}(x) = 1 - \frac{1}{x}, x \neq 0$

2. (a) $\frac{11}{4}$ when $x = \frac{3}{2}$

(b) $-\frac{41}{8}$ when $x = \frac{7}{4}$

(c) -9 when $x = -2$

3.

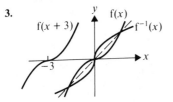

4. (a) g is even, h is odd, f is neither

(b) $9, \frac{1}{9}, 1$

(c) $\frac{1}{2x^2 + 1} \cdot \left(\frac{x + 2}{x}\right)^2$

(d) $f^{-1}(x) = \frac{1}{2}(x - 1)$

g^{-1} does not exist

$h^{-1}(x) = \frac{1}{x}$

(e) $\pm\frac{1}{3}$

(f) No

5. (a)

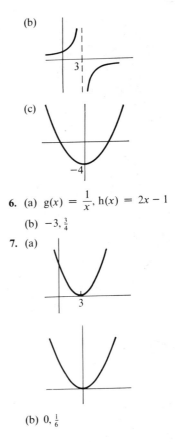

(b)

(c)

6. (a) $g(x) = \frac{1}{x}$, $h(x) = 2x - 1$

(b) $-3, \frac{3}{4}$

7. (a)

(b) $0, \frac{1}{6}$

CHAPTER 13

Exercise 13a – p. 200

1. (a) 3, 2   (b) $-\frac{7}{4}, -\frac{3}{4}$

(c) 4, −4   (d) −1, −8

(e) k, k^2 (f) $(a + 2)/a, -1$

2. (a) $x^2 - 3x + 4 = 0$

(b) $2x^2 + 4x + 1 = 0$

(c) $15x^2 - 5x - 6 = 0$

(d) $4x^2 + x = 0$

(e) $x^2 - ax + a^2 = 0$

(f) $x^2 + (k + 1)x + k^2 - 3 = 0$

(g) $abx^2 - b^2x + ac^2 = 0$

3. (a) $\frac{4}{5}$ (b) $5\frac{1}{2}$ (c) −1

(d) 5   (e) −6   (f) $-\frac{2}{5}$

4. (a) $x^2 - 6x + 11 = 0$

(b) $3x^2 - 2x + 1 = 0$

(c) $x^2 + 2x + 9 = 0$

(d) $3x^2 + 2x + 3 = 0$

(e) $x^2 + 8 = 0$

Exercise 13b – p. 202

1. (a) $Q: x + 1$, $R: -6x + 3$
   (b) $Q: x^3 - x^2 - 4x + 4$, $R: -2$
   (c) $Q: 2x - 4$, $R: 5x - 5$
   (d) $Q: 3x^2 + 6x + 12$, $R: 19$
   (e) $Q: x^2$, $R: -6x^2 + 1$
   (f) $Q: 2x^2 + x + 9$, $R: 29$
   (g) $Q: x - 10$, $R: 32$
   (h) $Q: 5x - 1$, $R: 5x$
   (i) $Q: 3$, $R: -10$
   (j) $Q: 2x^2 + x + 4$; $R: 5$

2. (a) $1 + \dfrac{3}{x + 1}$

   (b) $2 + \dfrac{4}{x - 2}$

   (c) $1 + \dfrac{4}{x^2 - 1}$

   (d) $x + 2 + \dfrac{4}{x - 2}$

   (e) $x + 7 + \dfrac{28}{x - 4}$

   (f) $1 - \dfrac{x + 4}{x(x + 1)}$

Exercise 13c – p. 205

1. (a) 3   (b) 18   (c) 47
   (d) $\frac{35}{16}$ (e) $-\frac{16}{27}$
   (f) $a^3 - 2a^2 + 6$ (g) $c^2 - ac + b$
   (h) $\dfrac{1}{a^4} - \dfrac{2}{a} + 1$

2. (a) yes   (b) no   (c) no
   (d) yes   (e) no   (f) yes

3. (a) $(x - 1)(x + 2)(x + 1)$
   (b) $(x - 2)(x^2 + x + 1)$
   (c) $(x - 1)(x + 1)(x^2 + 1)$
   (d) $(x + 2)(x^2 + x + 1)$
   (e) $(2x - 1)(x^2 + 1)$
   (f) $(3x - 1)(9x^2 + 3x + 1)$
   (g) $(x + a)(x^2 - ax + a^2)$
   (h) $(x - y)(x^2 + xy + y^2)$

4. -7
5. 5
6. -3
7. 1

Exercise 13d – p. 207

1. $(x - 1)(2x - 1)(x + 1)$; $-1, \frac{1}{2}, 1$
2. $f(x) = (x - 2)(x^2 + x + 1) \Rightarrow f(x) = 0$
   only when $x = 2$

3. 8
4. $x = -2.56, 1, 1.56$
5. $(x + 1)(x - 2)(x - 3)$

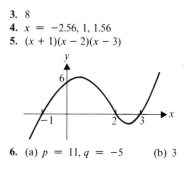

6. (a) $p = 11, q = -5$ (b) 3

CHAPTER 14

Exercise 14a – p. 209

1. $x < \frac{7}{2}$
2. $x > 4$
3. $x > -3$
4. $x > -2$
5. $x < \frac{1}{2}$
6. $x < -3$
7. $x < -\frac{1}{4}$
8. $x > \frac{8}{3}$
9. $x > \frac{3}{8}$

Exercise 14b – p. 210

1. $1 < x < 3$
2. $-4 < x < -\frac{2}{3}$
3. $-1 < x < 2$
4. $-3 < x < 2$
5. $x < -13, x > -2$
6. $2 < x < 5$
7. $-1 < x < 4$
8. $4 < x < 5$

Exercise 14c – p. 212

1. $x > 2$ and $x < 1$
2. $x \geqslant 5$ and $x \leqslant -3$
3. $-4 < x < 2$
4. $x \geqslant \frac{1}{2}$ and $x \leqslant -1$
5. $x > 2 + \sqrt{7}$ and $x < 2 - \sqrt{7}$
6. $-\frac{1}{2} < x < \frac{1}{2}$
7. $-4 \leqslant x \leqslant 2$
8. $x > 1$ and $x < -\frac{2}{5}$
9. $x \geqslant \frac{3}{2}$ and $x \leqslant -5$
10. $x > 4$ and $x < -2$
11. $\frac{1}{2}(-3 - \sqrt{17}) \leqslant x \leqslant \frac{1}{2}(-3 + \sqrt{17})$
12. $x > 7$ and $x < -1$

Exercise 14d – p. 214

1. (a) $-1 < k < 1$
   (b) $k < 0, k > \frac{4}{9}$
2. (a) $p \geqslant 9, p \leqslant 1$
   (b) $p \geqslant 5, p \leqslant 1$
3. $-2 < a < 6$
4. (a) $p < 1, p > 9$
   (b) $p = 1$ and 9
   (c) $1 < p < 9$

Exercise 14e – p. 217

1.
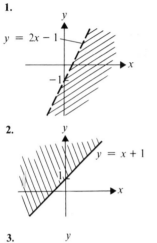
$y = 2x - 1$

2.
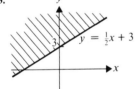
$y = x + 1$

3.
$y = \frac{1}{2}x + 3$

4.
$y = \frac{4}{3} - 2x$

5.
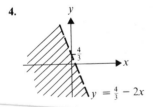
$y = 3 - 2x$

6.
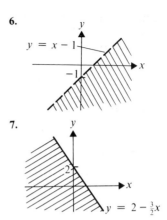
$y = x - 1$

7.
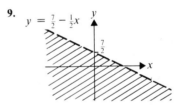
$y = 2 - \frac{3}{2}x$

8.
$y = 5 - 2x$

9.
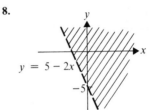
$y = \frac{7}{2} - \frac{1}{2}x$

10.
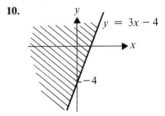
$y = 3x - 4$

Exercise 14f – p. 219

1.
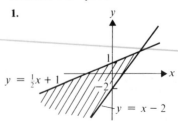
$y = \frac{1}{2}x + 1$
$y = x - 2$

2.

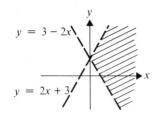

3.

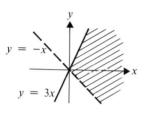

3.

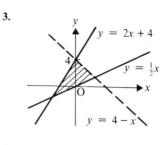

4.

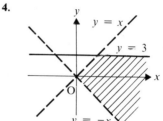

4.

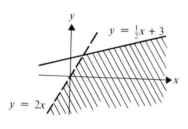

5.

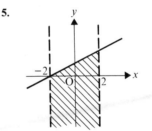

Exercise 14g – p. 221

1.

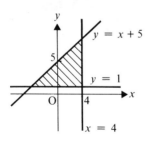

2.

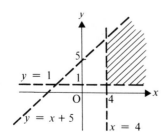

6. (a) 2 sq units
  (b) (1, 1)
7. (a) 9 sq units
  (b) $(-1, -1), (0, -1), (1, -1), (1, 0)$
8. (a) 4 sq units
  (b) $(-1, 1), (0, 1), (1, 1),$
    $(2, 1), (0, 2), (1, 2)$

Mixed Exercise 14 – p. 221
1. $x < 1$
2. $x > \frac{3}{2}$
3. $x > \frac{3}{2}$
4. $x > 3$, or $x < -2$
5. $-\frac{2}{3} < x < \frac{3}{2}$
6. $-\sqrt{13} < x < \sqrt{13}$
7. $x > 3 + \sqrt{2}$, or $x < 3 - \sqrt{2}$
8. $-2 < x < 7$

9. $1 < x \leqslant 4$
10. $-6 < x < -3$
11. $-1 < x < 2$
12. $-8 < x < -3$
14. $k \leqslant 3, k \geqslant 4$
15. (a)

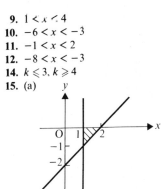

(1, 0), (2, 0), (1, −1)
$\frac{1}{2}$ sq unit

(b)

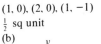
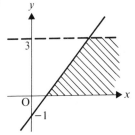

(1, 0), (4, 3)
not finite

(c)

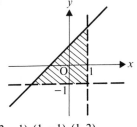

(−2, −1), (1, −1), (1, 2)
$4\frac{1}{2}$ sq units

CONSOLIDATION C

Multiple Choice Exercise C – p. 224

1. C
2. B
3. D
4. D
5. D
6. A
7. E

8. C
9. B
10. E
11. B
12. B, C
13. A, C
14. A
15. B, C
16. F
17. F
18. T
19. F
20. T

Miscellaneous Exercise C – p. 227

1. $-\frac{3}{2} < x < 2$

2.

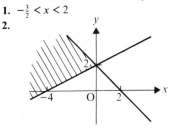

3. $x^2 + 2px + 4q = 0$
4. (a)

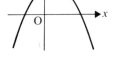

(−1, 0), ($\frac{3}{2}$, 0), (0, 3)
(b) $x \leqslant -1, x \geqslant \frac{3}{2}$

5.

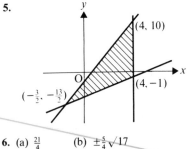

6. (a) $\frac{21}{4}$ (b) $\pm\frac{5}{4}\sqrt{17}$
7. (a) A $= -5$, B $= -1$
       (b) $x = -1, 1\frac{1}{2}, 2$

8. $(0, 1), (0, 2), (0, 3), (0, 4), (1, 1), (1, 2)$
9. (a) $ff:x \rightarrow (10x + 1)/(x + 5), x \neq -5$
  (b) $f^{-1}:x \rightarrow (2x + 1)/(x - 3), x \neq 3$
10. $8x^3 + 24x^2 + 32x + 14$
11. (a) 0   (b) $x < -3, -1 < x < 2$
12. $-\frac{9}{4}$
13. 2
14. (a) $(3, 3)$ (c) $(\frac{14}{3}, \frac{1}{2})$
15. $a = -11, b = 3,$
  $(x - 3)(x - 4)(2x + 3)$
16. (a) $\dfrac{1}{1 - x}$ (b) x (c) $\dfrac{1}{1 - x}$
17. $f \circ g:x \rightarrow (2x + 5)^2 - 3$

$$f(x)$$

$$f \circ g(x)$$

$$6 \leqslant f \circ g(x) \leqslant 166$$
20. $(2, 3), \sqrt{13}$
21. $(12 - 2x)/x$

CHAPTER 15

Exercise 15b – p. 239

1. $5x^4$
2. $11x^{10}$
3. $20x^{19}$
4. $10x^9$
5. $-3x^{-4}$
6. $-7x^{-8}$
7. $-5x^{-6}$
8. $\frac{4}{3}x^{1/3}$
9. 1
10. $-x^{-2}$
11. $\frac{1}{3}x^{-2/3}$
12. $\frac{1}{2}x^{1/2}$ or $\dfrac{1}{2\sqrt{x}}$

13. $\dfrac{-2}{x^3}$
14. $\dfrac{-1}{x^{3/2}}$
15. $\dfrac{-4}{x^5}$
16. $\dfrac{-10}{x^{11}}$
17. $-\frac{1}{4}x^{-5/4}$
18. $\frac{5}{2}\sqrt{x^3}$
19. $\frac{1}{7}x^{-6/7}$
20. px^{p-1}

Exercise 15c – p. 241

1. $15x^2$
2. 7
3. $-\dfrac{8}{x^2}$
4. $\dfrac{5}{2\sqrt{x}}$
5. $\dfrac{3}{2x^4}$
6. $9x^2 - 8x$
7. $4 + \dfrac{4}{x^2}$
8. $-\dfrac{3}{x^2} + \dfrac{1}{3}$
9. $3x^2 - 2x + 5$
10. $6x + \dfrac{4}{x^2}$
11. $3x^2 - 4x - 8$
12. $8x^3 - 8x$
13. $2x + \dfrac{5}{2\sqrt{x}}$
14. $9x^2 - 8x + 9$
15. $\frac{3}{2}x^{1/2} - \frac{1}{2}x^{-1/2} - \frac{1}{2}x^{-3/2}$
16. $\dfrac{1}{2\sqrt{x}} + \dfrac{3\sqrt{x}}{2}$
17. $-\dfrac{2}{x^3} + \dfrac{3}{x^4}$
18. $\dfrac{-1}{2\sqrt{x^3}} + \dfrac{2}{x^2}$
19. $-\frac{1}{2}x^{-3/2} + \frac{9}{2}x^{1/2}$
20. $\frac{1}{4}x^{-3/4} - \frac{1}{5}x^{-4/5}$
21. $-\dfrac{12}{x^4} + \dfrac{3x^2}{4}$

22. $-\dfrac{4}{x^2} - \dfrac{10}{x^3} + \dfrac{18}{x^4}$

23. $\dfrac{3}{2\sqrt{x}} - 3$

24. $1 + 2x^{-2} + 9x^{-4}$

Exercise 15d – p. 242

1. $\dfrac{dy}{dx} = 2x + 2$

2. $\dfrac{dz}{dx} = -4x^{-3} + x^{-2}$

3. $\dfrac{dy}{dx} = 6x + 11$

4. $\dfrac{dy}{dz} = 2z - 8$

5. $\dfrac{ds}{dt} = -\dfrac{3}{2t^4}$

6. $\dfrac{ds}{dt} = \dfrac{1}{2}$

7. $\dfrac{dy}{dx} = 1 - \dfrac{1}{x^2}$

8. $\dfrac{dy}{dz} = \dfrac{5z^2 - 1}{2\sqrt{z}}$

9. $\dfrac{dy}{dx} = 18x^2 - 8$

10. $\dfrac{ds}{dt} = 2t$

11. $\dfrac{ds}{dt} = 1 - \dfrac{7}{t^2}$

12. $\dfrac{dy}{dx} = -\dfrac{3\sqrt{x} + 28}{2x^3}$

Exercise 15e – p. 244

1. $2; -\frac{1}{2}$

2. $-\frac{1}{3}; 3$

3. $\frac{1}{4}; -4$

4. $6; -\frac{1}{6}$

5. $1; -1$

6. $5; -\frac{1}{5}$

7. $11; -\frac{1}{11}$

8. $-11; \frac{1}{11}$

9. $4; -\frac{1}{4}$

10. $\frac{4}{27}; -\frac{27}{4}$

11. $\frac{5}{4}; -\frac{4}{5}$

12. $2; -\frac{1}{2}$

13. $(2, 2)$ and $(-2, 4)$

14. $(1, 0)$ and $(-\frac{1}{3}, \frac{4}{27})$

15. $(3, 0)$ and $(-3, 18)$

16. $(-1, -2)$ and $(1, 2)$

17. $(1, -16)$

18. $(-1, \frac{1}{4})$

19. $(0, -5)$

20. $(1, -2)$ and $(-1, 2)$

Exercise 15f – p. 246

1. $6(3x + 1)$

2. $4(x - 3)^3$

3. $20(4x + 5)^4$

4. $21(2 + 3x)^6$

5. $18(6x - 2)^2$

6. $-10(4 - 2x)^4$

7. $-10(1 - 5x)$

8. $-6(3 - 2x)^2$

9. $-12(4 - 3x)^3$

10. $-3(3x + 1)^{-2}$

11. $-8(2x - 5)^{-5}$

12. $10(1 - 2x)^{-6}$

13. $(2x + 3)^{-1/2}$

14. $-(8 - 3x)^{-2/3}$

15. $\dfrac{-2}{\sqrt{(1 - 4x)}}$

16. $\dfrac{7}{(2 - 7x)^2}$

17. $\dfrac{1}{2(3 - x)^{3/2}}$

18. $\dfrac{10}{(1 - 5x)^3}$

Mixed Exercise 15 – p. 246

1. $6x + 1$

2. (a) $-3x^{-4} - 3x^2$

    (b) $\frac{1}{2}x^{-1/2} + \frac{1}{2}x^{-3/2}$

    (c) $-\dfrac{2}{x^3} - \dfrac{6}{x^4}$

3. (a) $\frac{3}{2}x^{1/2} - \frac{2}{3}x^{-1/3} - \frac{1}{3}x^{-4/3}$

    (b) $\dfrac{1}{2\sqrt{x}} + \dfrac{1}{x^2} - \dfrac{3}{x^4}$

    (c) $-\frac{8}{3}(1 - 8x)^{-2/3}$

    (d) $\dfrac{-12}{(3x - 2)^5}$

4. (a) 5 (b) 5 (c) 17

5. (a) 5 (b) 6

    (c) 12 (d) $-\frac{3}{4}$

6. (a) 1 (b) 7 and -7

7. (a) -3 (b) $(-\frac{1}{2}, 0)$ and $(2, 0)$

    (c) -5 and 5

8. (a) $(1, 10)$ and $(-1, 6)$

 (b) $(\frac{1}{3}, 7\frac{7}{9})$ and $(-\frac{1}{3}, 8\frac{2}{9})$

9. (a) $4x^3 - 2x$ (b) $6x + 8$

 (c) $\dfrac{x + 3}{2x\sqrt{x}}$

10. $-\frac{1}{2}$

11. $(-2, 4)$

12. $\frac{1}{5}$ and $-\frac{1}{5}$

13. $(1, 9)$ and $(3, 11)$

14. 1

15. (b), (d)

6. $x = \pm 1$

7. $x = 4$

8. $x = \pm 3$

9. $x = 1, x = -\frac{4}{3}$

10. $x = \pm\frac{5}{9}\sqrt{3}$

11. $x = 2$

12. $x = \pm\frac{2}{3}\sqrt{3}$

13. $(3, 3), (-3, -3)$

14. $(1, -7), (\frac{1}{3}, -\frac{185}{27})$

15. $(\frac{1}{2}, -\frac{25}{4})$

16. $(4, 0)$

17. $(1, 2)$

18. $(4, 10), (-4, 6)$

CHAPTER 16

Exercise 16a – p. 250

1. (a) $y = 2x - 5$

 (b) $2y + x + 5 = 0$

2. (a) $y = 4x - 2$

 (b) $4y + x + 8 = 0$

3. (a) $y + x + 2 = 0$ (b) $y = x$

4. (a) $y = 5$ (b) $x = 0$

5. (a) $y + x = 3$ (b) $y = x - 1$

6. (a) $y = 19x + 26$

 (b) $19y + x + 230 = 0$

7. (a) $y = 8x - 7$

 (b) $x + 8y = 9$

8. $4y + x + 12 = 0$

9. $y = 7x - 29$

10. $y + x = 1, 2y = 2x - 3; (\frac{5}{4}, -\frac{1}{4})$

11. $4y - x + 1 = 0, 4y + x - 5 = 0$

12. $y = 5x - 1, 3y + 9x + 19 = 0$

13. $y = 7x - 4, y + 5x + 28 = 0$

14. $(2, 8), y = 8x - 8$

15. $(\frac{1}{2}, -\frac{1}{4})$

16. $y + x + 1 = 0$

17. $2y = x + 2$

18. $k = -\frac{7}{2}$

19. $8y + 121 = 0$

20. $(1, -1)$

21. $p = 12, q = 8; (-2, 24)$

Exercise 16b – p. 253

1. $x = 0$

2. $x = \frac{3}{4}$

3. $x = 0, x = \frac{8}{3}$

4. $x = \frac{3}{2}$

5. $x = 0, x = \frac{4}{3}$

Exercise 16c – p. 257

1. $(1, 1)$

2. $(-1, -2)$ min; $(1, 2)$ max

3. $(3, 6)$ min; $(-3, -6)$ max

4. $(0, 0)$ max; $(\frac{10}{3}, -\frac{500}{27})$ min

5. $(0, 0)$ min

6. $(\frac{5}{2}, 0)$ min

7. $(0, 0)$ inflex

8. $(0, 0)$ min

9. $(\frac{5}{4}, -\frac{49}{8})$ min

10. $(-1, 4)$ max; $(1, -4)$ min

11. -2 max; 2 min

12. $2\frac{3}{4}$ min

13. 54 max; -54 min

14. $-\frac{5}{16}$ min; 0 max; -2 min

Exercise 16d – p. 261

1. $800\,\text{m}^2$; $20\,\text{m} \times 40\,\text{m}$

2. $20\,\text{cm} \times 20\,\text{cm} \times 10\,\text{cm}$

3. $r = \sqrt{(9 - h^2)}$; $12\pi\sqrt{3}\ \text{cm}^2$

4. 5 cm square

5. $\sqrt{35}$ cm square

6. $a = 1, b = -2, c = 3$

7. $p = q = 1, r = 2; (-1, 0)$

8. $5y = x^2 + 4x + 9$

Mixed Exercise 16 – p. 262

1. $11; y = 11x - 6; (\frac{2}{3}, \frac{4}{3})$

2. $2y = x - 1; (-\frac{3}{2}, -\frac{5}{4})$

3. $(2, 14), (-2, -14)$

4. $y = 3x + 6, y = 3x + 2$

5. $x + 2y + 10 = 0$

6. ($\cdot$ 2, 16) max, (2, −16) min;

7. min 2, max −2;

9. $h = \frac{1}{2}(7 - 2r - \pi r)$
10. 4 m × 4 m × 2 m
11. 12.5 cm^2

CHAPTER 17

Exercise 17a – p. 269

1. $\frac{1}{2}\sqrt{3}$
2. 0
3. $-\frac{3}{2}\sqrt{3}$
4. $\frac{1}{2}$
5. $-\frac{1}{2}\sqrt{3}$
6. $\frac{1}{2}$
7. $-1/\sqrt{2}$
8. $-\frac{1}{2}\sqrt{3}$
9. $\frac{1}{2}\pi, \frac{5}{2}\pi, \frac{9}{2}\pi$
10. $-\frac{1}{2}\pi, -\frac{5}{2}\pi$
11. sin 55°
12. −sin 70°
13. −sin 60°
14. −sin $\frac{1}{6}\pi$

15.

16.

17.

18.

19.

20.

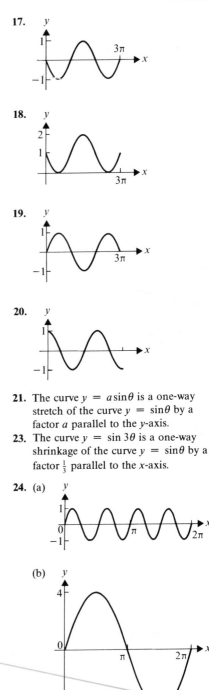

21. The curve $y = a\sin\theta$ is a one-way stretch of the curve $y = \sin\theta$ by a factor a parallel to the y-axis.
23. The curve $y = \sin 3\theta$ is a one-way shrinkage of the curve $y = \sin\theta$ by a factor $\frac{1}{3}$ parallel to the x-axis.

24. (a)

(b)

Exercise 17b – p. 273

1. (a) $-\cos 57°$ (b) $-\cos 70°$
(c) $\cos 20°$ (d) $-\cos 26°$

2. (a) $-\dfrac{\sqrt{3}}{2}$ (b) 0 (c) $-\dfrac{1}{\sqrt{2}}$

(d) 1 (e) $-\frac{1}{2}$ (f) $\frac{1}{2}$

(h) $\dfrac{-3}{\sqrt{2}}$ (h) $\dfrac{-1}{\sqrt{2}}$

3. (a)

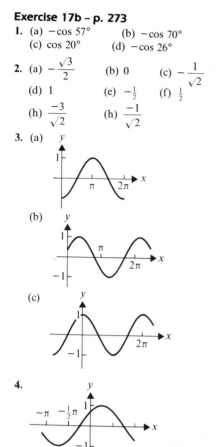

(b)

(c)

4.

(a) $\theta = \frac{1}{4}\pi$ (b) $\theta = -\frac{3}{4}\pi$
(c) $\theta = -\frac{1}{4}\pi$ and $\frac{3}{4}\pi$

5.

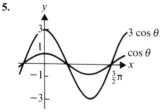

$3\cos\theta$

$\cos\theta$

6.

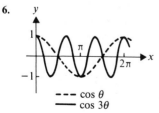

$---\ \cos\theta$
$---\ \cos 3\theta$

7.

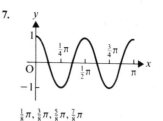

$\frac{1}{8}\pi, \frac{3}{8}\pi, \frac{5}{8}\pi, \frac{7}{8}\pi$

8.

$\sin\theta = \cos(\theta - \frac{1}{2}\pi)$
$\cos\theta = -\sin(\theta - \frac{1}{2}\pi)$

9. (a) $\frac{1}{4}\pi, \frac{3}{4}\pi, \frac{5}{4}\pi, \frac{7}{4}\pi$

(b) $\frac{1}{2}\pi, \frac{3}{2}\pi$

10. $\theta \approx \frac{5}{8}\pi \ (\pi = 1.95 \text{ rad})$

Exercise 17c – p. 275

1. (a) 1 (b) $-\sqrt{3}$ (c) $\sqrt{3}$
(d) -1

2. (a) $\tan 40°$ (b) $-\tan\frac{2}{7}\pi$
(c) $-\tan 50°$ (d) $\tan\frac{2}{5}\pi$

3. (a) $\frac{1}{4}\pi, \frac{5}{4}\pi$
(b) $\frac{3}{4}\pi, \frac{7}{4}\pi$
(c) $0, \pi, 2\pi$
(d) $\frac{1}{2}\pi, \frac{3}{2}\pi$

Exercise 17d – p. 278

1. (a) 0.412 rad, 2.73 rad, -5.87 rad, -3.55 rad
(b) $-\frac{4}{3}\pi, -\frac{2}{3}\pi, \frac{2}{3}\pi, \frac{4}{3}\pi$
(c) 0.876 rad, 4.02 rad, -2.27 rad, -5.41 rad

2. (a) $141.3°, 321.3°, 501.3°, 681.3°$
(b) $191.5°, 348.5°, 551.5°, 708.5°$
(c) $84.3°, 275.7°, 444.3°, 635.7°$

3. (a) $36.9°$ (b) $323.1°$ (c) 0.464 rad

4. $0, \pi, 2\pi$

5. $11.8°, 78.2°, 191.8°, 258.2°$

6. $\frac{1}{3}\pi, \pi, \frac{5}{3}\pi$

Exercise 17e – p. 281

1. (a) 60°, 300°   (b) 59.0°, 239.0°
   (c) 41.8°, 138.2°
2. (a) −140.2°, 39.8°
   (b) −131.8°, 131.8°
   (c) −150°, −30°
3. $-\frac{1}{2}\pi, \frac{1}{2}\pi$
4. (a) 1   (b) $-\sqrt{2}$ (c) −2
5.

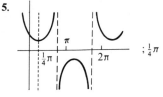

$; \frac{1}{4}\pi$

6.

$; -\frac{1}{12}\pi, \frac{11}{12}\pi$

8. $\cot\theta = \tan\left(\frac{1}{2}\pi - \theta\right)$

Exercise 17f – p. 283

1.

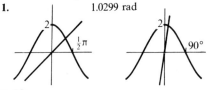

1.0299 rad

2. No
3. (a) −1.895 49, 0, 1.895 49
   (b) 0, 0.8767 rad   (c) 1.2834 rad

CHAPTER 18

Exercise 18a – p. 287

	$\sin\theta$	$\cos\theta$	$\tan\theta$
1. (a)	$-\frac{12}{13}$	$-\frac{5}{13}$	$\frac{12}{5}$
(b)	$\frac{3}{5}$	$-\frac{4}{5}$	$-\frac{3}{4}$
(c)	$\frac{7}{25}$	$\frac{24}{25}$	$\frac{7}{24}$
(d)	0	±1	0

2. $\tan^4 A$
3. 1
4. $\sec\theta \csc\theta$
5. $\sec^2\theta$
6. $\tan\theta$

7. $\sin^3\theta$
8. $x^2 - y^2 = 16$
9. $b^2x^2 - a^2y^2 = a^2b^2$
10. $y^2(4 + x^2) = 36$
11. $(1 - x)^2 + (y - 1)^2 = 1$
12. $y^2(x^2 - 4x + 5) = 2$
13. $x^2(b^2 - y^2) = a^2b^2$

Exercise 18b – p. 290

1. 57.7°, 122.3°, 237.7°, 302.3°
2. 190.1°, 349.9°
3. 38.2°, 141.8°
4. 30°, 150°
5. 30°, 150°
6. 0°, 131.8°, 228.2°, 360°
7. ±0.723 rad
8. −0.314 rad, −2.83 rad
9. $-\frac{3}{4}\pi$, −0.245 rad, $\frac{1}{4}\pi$, 2.90 rad
10. $-\pi, -\frac{1}{3}\pi, \frac{1}{3}\pi, \pi$
11. $-\pi, -\frac{2}{3}\pi, 0, \frac{2}{3}\pi, \pi$
12. $-\frac{1}{2}\pi, \frac{1}{6}\pi, \frac{1}{2}\pi, \frac{5}{6}\pi$
13. $-\pi, -\frac{1}{6}\pi, 0, \frac{1}{6}\pi, \pi$
14. $-\frac{1}{2}\pi, \frac{1}{2}\pi$

Exercise 18c – p. 292

1. 22.5°, 112.5°, 202.5°, 292.5°
2. 40°, 80°, 160°, 200°, 280°, 320°
3. no values in the given range
4. 12°, 60°, 84°, 132°, 156°, 204°, 228°, 276°, 300°, 348°
5. no values in the given range
6. 25.5°, 154.5°, 205.5°, 334.5°
7. 135°, 315°
8. 240°
9. 60°, 240°
10. 0, 120°, 360°
11. $-\frac{7}{12}\pi, \frac{1}{12}\pi$
12. $-\frac{11}{12}\pi, \frac{1}{12}\pi$
13. $0, \frac{2}{3}\pi$
14. $-\frac{1}{3}\pi, \pi$
15. 149.5°, 59.5°, 30.5°, 120.5°
16. −105.2°, 14.8°, 134.8°, −74.8°, 45.2°, 165.2°
17. ±63.6°
18. $\frac{1}{6}\pi, \frac{5}{12}\pi, \frac{2}{3}\pi, \frac{11}{12}\pi, \frac{7}{6}\pi, \frac{17}{12}\pi, \frac{5}{3}\pi, \frac{23}{24}\pi$
19. $\frac{1}{15}\pi, \frac{1}{3}\pi, \frac{7}{15}\pi, \frac{11}{15}\pi, \frac{13}{15}\pi, \frac{17}{15}\pi, \frac{19}{15}\pi, \frac{23}{15}\pi,$ $\frac{5}{3}\pi, \frac{29}{15}\pi$
20. $\frac{3}{2}\pi$

Mixed Exercise 18 – p. 293

1. $x^2 + \dfrac{1}{y^2} = 1$

2. $\sin \beta = \pm \dfrac{\sqrt{3}}{2}$, $\tan \beta = \pm \sqrt{3}$

3. $\dfrac{2}{\sin^2 \theta}$; $\frac{1}{4}\pi, \frac{3}{4}\pi, \frac{5}{4}\pi, \frac{7}{4}\pi$

4. $60°, 109.5°$

6. $-\frac{29}{36}\pi, -\frac{17}{36}\pi, -\frac{5}{36}\pi, \frac{7}{36}\pi, \frac{19}{36}\pi, \frac{31}{36}\pi$

7. (a) $(x-2)^2 + (y+1)^2 = 1$
   (b) $(x+3)^2 = 1 + (2-y)^2$

8. $\frac{1}{8}\pi, \frac{3}{8}\pi, \frac{5}{8}\pi, \frac{7}{8}\pi$

10. $\sin^2 A$

11. $0, 180°, 360°, 70.5°, 289.5°$

12. $\sec^2 \theta \tan^2 \theta$

CONSOLIDATION D

Multiple Choice Exercise D – p. 297

1. C
2. E
3. B
4. D
5. A
6. C
7. B
8. B
9. A, B
10. B, C
11. B, C
12. B
13. A, C
14. B
15. B
16. B, C
17. F
18. T
19. T
20. F

Miscellaneous Exercise D – p. 300

1. $3\frac{19}{27}$ at $x = -\frac{1}{3}$; -9 at $x = 2$;
   crosses at $(-1, 0)$, $(\frac{1}{2}, 0)$, $(3, 0)$
   and $(0, 3)$

2. Least value is -2, greatest value is $\frac{1}{4}$

3. $\pm\frac{2}{3}\pi$

4. $x - 9y + 73 = 0$

5. (a) $12(4x+1)^2$

   (b) $\dfrac{3}{2(2-3x)^{3/2}}$

   (c) $\dfrac{2}{3x^{1/3}} - \dfrac{2}{3x^{5/3}}$

6. (a) $y = 4x + 4$
   (b) $x = -2$

7. $x = \dfrac{10}{\sqrt{3}}$; Min $C = 20\sqrt{3}$

   (i.e. £35 000 to nearest £1000)

8. $1 < t < 4$

9. 60 km/hr; £50

11. (a) $x = \frac{1}{3}$; $V_{max} = \frac{16}{27}$

   (b)

12. 20 m s^{-1}

13. (a) $V = \pi r(27 - r^2)$
    (b) $r = 3$

14. $26°, 154°$

15. (a) $53.1°$
    (b) $53.6°$

16. $(x-1)$ and $(3x+2)$
    $90°, 221.8°, 318.2°$

17. (a) $\pm\frac{1}{3}\pi, \pm\frac{1}{2}\pi$
    (b) $0, \frac{1}{4}\pi$

CHAPTER 19

All general integrals in this chapter require the addition of **K**, the constant of integration.

Exercise 19a – p. 304

1. $\frac{1}{5}x^5$

2. $\frac{1}{8}x^8$

3. $\frac{1}{4}x^4$

4. $\frac{1}{12}x^{12}$

5. $-\dfrac{1}{x}$

6. $\dfrac{-1}{4x^4}$

7. $\frac{2}{3}x^{3/2}$

8. $\frac{4}{7}x^{7/4}$

9. $\frac{3}{4}x^{4/3}$

10. $-\frac{1}{2}x^{-2}$

11. $2x^{1/2}$

12. $\frac{1}{2}x^2$

Exercise 19b – p. 305

1. $\frac{1}{6}x^6 + \frac{2}{3}x^{3/2}$

2. $-\dfrac{1}{4x^4} - \frac{1}{3}x^3$

3. $\frac{4}{5}x^{5/4} + \frac{1}{5}x^5$

4. $-\frac{1}{2}x^{-2} - \frac{1}{4}x^4$

5. $\dfrac{-2}{3x^{3/2}} + \frac{5}{7}x^{7/5}$

6. $2x^{1/2} + \frac{2}{3}x^{3/2}$

7. $\frac{1}{2}x^2 + \dfrac{1}{x}$

8. $\frac{3}{2}x^{2/3} + \frac{2}{5}x^{5/2}$

Exercise 19c – p. 306

1. $\frac{1}{16}(4x + 1)^4$

2. $\frac{1}{24}(2 + 3x)^8$

3. $\frac{1}{30}(5x - 4)^6$

4. $\frac{1}{24}(6x - 2)^4$

5. $-\frac{1}{3}(4 - x)^3$

6. $-\frac{1}{8}(3 - 2x)^4$

7. $-\frac{1}{3}(3x + 1)^{-1}$

8. $-\frac{1}{6}(2x - 5)^{-3}$

9. $\frac{1}{3}(4 - 3x)^{-1}$

10. $\frac{1}{3}(2x + 3)^{3/2}$

11. $\frac{1}{4}(3x - 1)^{4/3}$

12. $(2x - 1)^{1/2}$

13. $-\frac{2}{3}(8 - 3x)^{1/2}$

14. $-\frac{1}{6}(1 - 4x)^{3/2}$

15. $-\frac{1}{5}(3 - 2x)^{5/2}$

16. $\dfrac{1}{7(2 - 7x)}$

17. $-2\sqrt{(3 - x)}$

18. $-\frac{3}{10}(1 - 5x)^{2/3}$

Exercise 19d – p. 311

1. $\frac{1}{5}$

2. $\frac{1}{2}$

3. $\frac{14}{3}$

4. $\frac{15}{64}$

5. 2

6. $16\frac{1}{3}$

7. 4

8. $\frac{2}{7}(8\sqrt{2}, -1)$

9. $26\frac{2}{3}$

10. $12\frac{2}{3}$

11. 15

12. -2

13. $\frac{1}{2}$

14. 6

15. $6\frac{1}{7}$

16. 15

Exercise 19e – p. 312
Answers in square units.

1. $5\frac{1}{3}$

2. $12\frac{2}{3}$

3. $2\frac{2}{3}$

4. $13\frac{1}{2}$

5. $5\frac{1}{3}$

6. 60

7. $5\frac{1}{3}$

8. $4\frac{7}{8}$

9. 24

Exercise 19f – p. 315

1. $2\frac{1}{3}$

2. $2\frac{1}{4}$

3. $28\frac{1}{2}$

4. $15\frac{1}{4}$

5. $2\frac{1}{6}$

6. $\frac{4}{3}$

7. $\frac{1}{3}$

8.

(a) $\frac{1}{4}$ (b) $\frac{1}{4}$ (c) $\frac{1}{2}$

9.

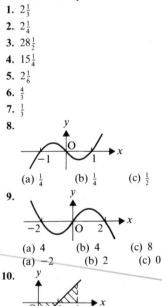

(a) 4   (b) 4   (c) 8

(a) -2 (b) 2   (c) 0

10.

Exercise 19g – p. 316

1. (a) $49\frac{1}{2}$ (b) 2   (c) 3
2. $2\frac{1}{3}$
3. $5\frac{1}{3}$
4. 36
5. $4\frac{1}{2}$
6. (a) $42\frac{2}{3}$ (b) $42\frac{2}{3}$
7. $A = \int_0^1 x \, dy$ $B = \int_0^1 y \, dx$
   $A + B = 1$

Mixed Exercise 19 – p. 317

1. $\frac{1}{3}x^3 + \dfrac{1}{x} + K$
2. $\frac{2}{9}(3x + 7)^{3/2} + K$
3. $\frac{2}{3}x^{3/2} + 2x^{1/2} + K = \frac{2}{3}\sqrt{x}(x + 3) + K$
4. $\frac{1}{4}\ln(4x - 3) + K$
5. $\frac{1}{4}x^4 - \dfrac{1}{2(1 - x)^2} + K$
6. $\frac{2}{5}\sqrt{x}(x^2 - 5) + K$
7. $\frac{1}{5}$
8. 18
9. $33\frac{3}{4}$
10. $20\frac{5}{6}$
11. 27
12. $\frac{3}{4}$
13. 16
14. $10\frac{2}{3}$
15. $\frac{1}{2}$
16. (b) $-57\frac{1}{6}$ (c) $57\frac{1}{6}$

Exercise 20a – p. 318

1. (a) $\dfrac{dq}{dn}$ (b) $\dfrac{dg}{da}$ (c) $\dfrac{dA}{dt}$

Exercise 20b – p. 322

1. (a) $v = 12,\ a = 6$
   (b) 4 units
2. (a) 15 m   (b) $-2\ \text{m s}^{-1}$
   (c) $2\ \text{m s}^{-1}$
3. (a) $v = 6\ \text{m s}^{-1};\ v = 0;$
   $v = -6\ \text{m s}^{-1}$
   When $t = 6$, P is moving in the direction Ox, away from O.

P comes to rest when $t = 7$ and then reverses direction and moves back towards O.
    (b) 147 m
4. (a) $\frac{1}{2}\ \text{m s}^{-2}$ (b) 36 m
5. 129 m
6. $v = \frac{1}{2}t^2 - 6t;\ s = \frac{1}{6}t^3 - 3t^2;$
   $v_6 = -18\ \text{m s}^{-1};\ s_6 = -72\ \text{m}$
7. $6t\ \text{m s}^{-1};\ 504\ \text{m}\ (t = 2\ \text{at A})$
8. $v = 10(t - 1)^{3/2} - 8;$
   $s = 4(t - 1)^{5/2} - 8t + 12$
9. $v = \frac{1}{3}t + 4;$
   $s = \frac{1}{6}t^2 + 4t;\ 270\ \text{m}$
10.
$$v = \frac{1}{2}\left(1 - \frac{1}{t^2}\right);\ \frac{3}{8}\ \text{m s}^{-1}$$
$$s = \frac{1}{2}\left(t + \frac{1}{t}\right) - 1;\ \frac{1}{4}\ \text{m}$$
11. $8\frac{1}{3}\ \text{m}$
12. (a) $s = 2t - \dfrac{1}{t + 1} + 1$
   (b) $v \to 2$ as $t \to \infty$

CHAPTER 21

Exercise 21a – p. 327

1. (a) $\displaystyle\sum_{r=1}^{5} r^3$ (b) $\displaystyle\sum_{r=1}^{10} 2r$
   (c) $\displaystyle\sum_{r=2}^{50} \frac{1}{r}$ (d) $\displaystyle\sum_{r=0}^{\infty} \frac{1}{3^r}$
   (e) $\displaystyle\sum_{r=0}^{7} (-4 + 3r)$
   (f) $\displaystyle\sum_{r=0}^{\infty} \left(\frac{8}{2^r}\right) = \sum_{r=0}^{\infty} \frac{1}{2^{r-3}}$
2. (a) $1 + \frac{1}{2} + \frac{1}{3} + \ldots$
   (b) $0 + 2 + 6 + 12 + \ldots + 30$
   (c) $2 + \frac{1}{2} + \frac{4}{15} + \ldots + \frac{22}{861}$
   (d) $1 + \frac{1}{2} + \frac{1}{5} + \ldots$
   (e) $0 + 0 + 6 + 24 + 60 + \ldots + 720$
   (f) $-1 + a - a^2 + \ldots$
3. (a) 8; 9   (b) 9; 10   (c) -1; 6
   (d) $\frac{1}{420}$; ∞ (e) $(\frac{1}{2})^n$; ∞
   (f) -48; 23   (g) 4; 10

Exercise 21b – p. 332

1. (a) $9, 2n - 1$ (b) $16, 4(n - 1)$
   (c) $15, 3n$ (d) $17, 3n + 2$
   (e) $-2, 8 - 2n$
   (f) $p + 4q, p + (n - 1)q$

(g) $18, 8 + 2n$ (h) $17, 4n - 3$

(i) $0, \frac{1}{2}(5 - n)$ (j) $8, 3n \quad 7$

2. (a) 100 (b) 180 (c) 165

(d) 185 (e) -30

(f) $5(2p + 9q)$ (g) 190

(h) 190 (i) $-\frac{5}{2}$ (j) 95

3. $a = 27.2, d = -2.4$

4. $d = 3; 30$

5. $1, \frac{1}{2}, 0; -8\frac{1}{2}$

6. (a) $28\frac{1}{2}$ (b) 80 (c) 400

(d) 80 (e) 108 (f) $3n(1 - 6n)$

(g) 40 (h) $2m(m + 3)$

7. $4, 2n - 4$

9. $2, 364$

10. 39

11. 64

12. (a) $1, 5$ (b) 270

13. (a) $a = 21, d = -3$

(b) less than 4 or more than 11

Exercise 21c – p. 337

1. (a) $32, 2^n$ (b) $\frac{1}{8}, \dfrac{1}{2^{n-2}}$

(c) $48, 3(-2)^{n-1}$

(d) $\frac{1}{2}, (-1)^{n-1}(\frac{1}{2})^{n-4}$ (e) $\frac{1}{27}, (\frac{1}{3})^{n-2}$

2. (a) 189 (b) -255

(c) $2 - (\frac{1}{2})^{19}$ (d) $781/125$

(e) $341/1024$ (f) 1

3. $\frac{1}{2}, 2$

4. $-\frac{1}{2}$

5. $-\frac{1}{2}, 1/1024$

6. 13.21 to 4 s.f.

7. (a) $(x - x^{n+1})/(1 - x)$

(b) $(x^n - 1)/x^{n-2}(x - 1)$

(c) $(1 + (-1)^{n+1}y^n)/(1 + y)$

(d) $x(2^n - x^n)/2^{n-1}(2 - x)$

(e) $[1 - (-2)^n x^n]/(1 + 2x)$

8. $\frac{8}{3}(1 - (\frac{1}{4})^n), 4$

9. 62 or 122

10. 8.493 to 4 s.f.

11. 8 (last repayment is less than £2000)

12. £23.31

Exercise 21d – p. 342

1. (a) yes   (b) no   (c) yes

(d) yes   (e) no   (f) yes

2. (a) 6 (c) $13\frac{1}{3}$ (d) $\frac{5}{9}$ (f) $\frac{9}{4}$

3. (a) $161/990$ (b) $34/99$ (c) $7/330$

4. $\frac{1}{2}$

5. $8, 4, 2, 1$

Mixed Exercise 21 – p. 342

1. $\frac{2}{3}$

2. $\frac{1}{2}(1 + 3^{11}) = 88\,574$

3. $\dfrac{ab^4(1 - b^{2(n-1)})}{1 - b^2}$

4. $2(n + 5)(n - 4)$

5. $\dfrac{e(1 - e^n)}{1 - e}$

6. $\frac{1}{2}n(7 + 3n)$

7. $-3n^2$

8. 1

9. $n(n + 1)$

10. $\frac{17}{330}$

11. $\frac{10}{99}$

12. $1, 7, 19, 37; 3n^2 - 3n + 1$

13. $16, -8$

14. 2

15. 1

16. $a = 2 \pm \sqrt{2}, r = \frac{1}{4}(2 \pm \sqrt{2})$

17. £805 056.27

18. £6391.62

CHAPTER 22

Exercise 22a – p. 348

1. (a) $1 + 36x + 594x^2 + 5940x^3$

(b) $1 - 18x + 144x^2 - 672x^3$

(c) $1 + 35x + 525x^2 + 4375x^3$

(d) $1 - \frac{20}{3}x + \frac{190}{9}x^2 - \frac{380}{9}x^3$

(e) $1 - 4x + \frac{20}{3}x^2 - \frac{160}{27}x^3$

(f) $1 + 12x + \frac{342}{5}x^2 + \frac{6156}{25}x^3$

2. (a) $1024 + 5120x + 11\,520x^2$

(b) $128 - 672x + 1512x^2$

(c) $(\frac{3}{2})^9 + \dfrac{3^{10}}{2^7}x + \dfrac{3^9}{2^3}x^2$

3. (a) $336x^2$ (b) $-10x$

(c) $1760x^3$ (d) $9x^2$

(e) $-455(2)^{12}x^3$ (f) $7920x^4$

(g) $63x^5$ (h) $3360p^6q^4$

(i) $56a^3b^5$ (j) $16a^7b$

4. (a) $1 - 8x + 27x^2$

(b) $1 + 19x + 160x^2$

(c) $2 - 19x + 85x^2$

(d) $1 - 68x + 2136x^2$

5. (a) $64x^6 - 192x^5 + 240x^4 - 160x^3$
$+ 60x^2 - 12x + 1$

(b) $243x^5 - 810x^4 + 1080x^3 - 720x^2$
$+ 240x - 32$

(c) $x^9 - 36x^8 + 576x^7 - 5376x^6$
   $+ 32256x^5 - 129024x^4$
   $+ 344064x^3 - 589824x^2$
   $+ 589824x - 262144$
6. $1 + 4x - 11x^2 - 70x^3$
7. $5^5(25 - 90x + 301x^2)$
8. $X = x(2 - x), 1 + 40x + 740x^2$

Exercise 22b – p. 353
1. 0.8171
2. 1.062
6. (a) $1 - 50x$
   (b) $256 - 1024x$
   (c) $1 - 19x$
7. $a = 114, b = 16, c = 1$
8. $y = 1 - 4x$
9. (a) $a = 115, b = -28, c = 3$
   (b) 0.0314 (4 s.f.)
   (c) $|x| \leqslant 0.068$

CONSOLIDATION E

Multiple Choice Exercise E – p. 358
1. D
2. B
3. C
4. D
5. A
6. E
7. B
8. C
9. E
10. C
11. B
12. A
13. B, C
14. A
15. A, B
16. F
17. T
18. F
19. T
20. T

Miscellaneous Exercise E – p. 361
1. (a) $2x + 2$ (b) $\frac{1}{3}x^3 + x^2 + C$
2. 19.4 m s^{-1}
3. $a = 32, b = 80, c = 40$
4. (a) A(2, 0), B(−2, 32)
   (b) (−4, 0)
   (c) 108 sq units

5. (a) $b = \frac{3}{5}a, c = \frac{1}{5}a$
   (b) $f = e^{3/5}, g = e^{1/5}$
6. (a) $-\frac{1}{4}$ m s^{-1} (b) $\dfrac{-1}{8\sqrt{2}}$
7. $a = 2, b = \frac{80}{3}$
8. $(2 + x)^4 = 16 + 32x + 24x^2 + 8x^3 + x^4$,
   $(1 - 3x)^4 = 1 - 12x + 54x^2 - 108x^3$
   $+ 81x^4; 504$
9. (a) $13\frac{1}{2}$
   (b)

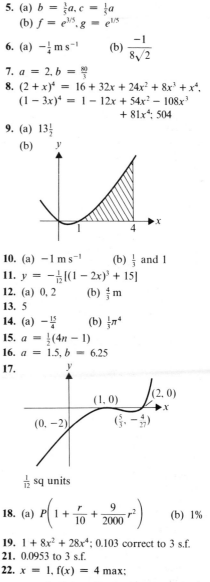

10. (a) -1 m s^{-1} (b) $\frac{1}{3}$ and 1
11. $y = -\frac{1}{12}[(1 - 2x)^3 + 15]$
12. (a) 0, 2   (b) $\frac{4}{3}$ m
13. 5
14. (a) $-\frac{15}{4}$ (b) $\frac{1}{3}\pi^4$
15. $a = \frac{1}{2}(4n - 1)$
16. $a = 1.5, b = 6.25$
17.

$\frac{1}{12}$ sq units

18. (a) $P\left(1 + \dfrac{r}{10} + \dfrac{9}{2000}r^2\right)$ (b) 1%
19. $1 + 8x^2 + 28x^4; 0.103$ correct to 3 s.f.
21. 0.0953 to 3 s.f.
22. $x = 1, f(x) = 4$ max;
   $x = 3; f(x) = 0$ min; $6\frac{3}{4}$ sq units

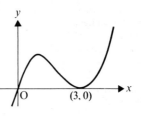

INDEX